Molecular Breeding and Germplasm Improvement of Rice

Molecular Breeding and Germplasm Improvement of Rice

Guest Editors

Xiangjin Wei
Yingxin Zhang
Weixun Wu
Guiai Jiao

MDPI

Basel • Beijing • Wuhan • Barcelona • Belgrade • Novi Sad • Cluj • Manchester

Guest Editors

Xiangjin Wei
State Key Laboratory of Rice
Biology and Breeding
China National Rice
Research Institute
Hangzhou
China

Yingxin Zhang
State Key Laboratory of Rice
Biology and Breeding
China National Rice
Research Institute
Hangzhou
China

Weixun Wu
State Key Laboratory of Rice
Biology and Breeding
China National Rice
Research Institute
Hangzhou
China

Guiai Jiao
State Key Laboratory of Rice
Biology and Breeding
China National Rice
Research Institute
Hangzhou
China

Editorial Office

MDPI AG
Grosspeteranlage 5
4052 Basel, Switzerland

This is a reprint of the Special Issue, published open access by the journal *Plants* (ISSN 2223-7747), freely accessible at: www.mdpi.com/journal/plants/special_issues/Rice_Germplasm_Improvement.

For citation purposes, cite each article independently as indicated on the article page online and using the guide below:

Lastname, A.A.; Lastname, B.B. Article Title. *Journal Name* **Year**, *Volume Number*, Page Range.

ISBN 978-3-7258-3010-7 (Hbk)
ISBN 978-3-7258-3009-1 (PDF)
https://doi.org/10.3390/books978-3-7258-3009-1

Contents

About the Editors

Xiangjin Wei

Dr. Xiangjin Wei has received his doctor's degree in genetics from Nanjing Agricultural University and is currently working at the China National Rice Research Institute. He is mainly engaged in the mechanism analysis of rice quality, abiotic stress resistance and other important agronomic traits, and the cultivation of new rice varieties with high yield, high quality, and resistance to abiotic stress. He has published over 60 research papers in *Molecular Plant*, *National Science Review*, *Plant Biotechnol J*, *New Phytologist*, *Plant Communications* and other important journals, and he holds 9 national invention patents of China. He contributed to over 10 new rice varieties and received 5 awards, including the National Science and Technology Progress Award of China.

Yingxin Zhang

Dr. Yingxin Zhang has received his doctor's degree genetics from Nanjing Agricultural University and is currently working at the China National Rice Research Institute. He is mainly engaged in developing disease-resistant, high-yield hybrid rice varieties and creating specific germplasm materials. He has published 33 papers as the first or corresponding author in journals like *Plant Cell*, *Plant Physiology*, and *Theoretical and Applied Genetics*. He holds four invention patents and one utility model patent as the first author. He is involved in a major agricultural research project and a National Natural Science Foundation of China project and has contributed to the cultivation of one nationally approved and two provincially approved rice varieties.

Weixun Wu

Dr. Weixun Wu has received his doctor's degree in genetics from Nanjing Agricultural University and is currently working at the China National Rice Research Institute. He is mainly engaged in the cloning and functional research of genes related to rice heading date and salt-alkali tolerance. He presided over a number of NSFC projects. Since 2013, he has published more than 50 papers in academic journals such as *PNAS*, *Journal of advanced research*, *Theor Appl Genet*, *Plant Science*, and *The Crop Journal*, including 14 first/co-first/corresponding authors papers, and 4 invention patents have been granted. In October 2017, he was selected into the "Young Talents Program" of the Chinese Academy of Agricultural Sciences.

Guiai Jiao

Dr. Guiai Jiao has received her doctor's degree in crop genetics and breeding from Hunan Agricultural University and is currently working at the China National Rice Research Institute. She is focused on enhancing rice quality, developing rapid evaluation technology, studying the link between rice starch structure and cooking quality, and understanding key rice quality traits. She has received over 10 awards, including the National Science and Technology Progress Second Prize of China and Zhejiang Province Science and Technology Progress First Prize. She has led 4 projects, published over 20 papers, and held 10 authorized invention patent.

plants MDPI

Editorial

Progress in Understanding and Enhancing Rice Tolerance to Biotic and Abiotic Stresses

Weixun Wu, **Yingxin Zhang**, **Guiai Jiao** and **Xiangjin Wei** *

State Key Laboratory of Rice Biology and Breeding, China National Center for Rice Improvement, China National Rice Research Institute, Hangzhou 311400, China; wuweixun@caas.cn (W.W.); zhangyingxin@caas.cn (Y.Z.); jiaoguiai@caas.cn (G.J.)
* Correspondence: weixiangjin@caas.cn

Rice growth and development occur in several distinct stages: a seedling stage, a vegetative stage, a reproductive stage, and maturity. Throughout these stages, rice is frequently subjected to various biotic stresses, such as diseases, pests, and weeds, as well as abiotic stresses, including extreme temperatures, drought, waterlogging, salinity, heavy metals, and nutrient deficiencies. These stressors negatively impact rice yield and quality to varying degrees, resulting in significant annual economic losses for the global rice industry. Consequently, developing rice varieties with enhanced adaptability remains necessary.

Among biotic stresses, disease is a primary threat in rice cultivation, with rice blast, sheath blight, and rice false smut numbering among the most detrimental. The main strategies for disease control include using disease-resistant cultivars and chemical interventions. In resistance breeding, combining multiple resistance genes can widen disease resistance and improve overall crop resilience. Peng et al. [1] successfully introduced three resistance genes, *Pigm*, *Pi48*, and *Pi49*, into the photothermosensitive male-sterile line Chuang5S using marker-assisted selection (MAS) and the RICE10K SNP chip. This approach allowed the rapid breeding of an enhanced line that retained a genetic profile similar to that of Chuang5S. The resulting rice varieties demonstrated improved blast resistance without compromised yields or agronomic traits, indicating the promising potential of using gene stacking to provide broad-spectrum disease resistance in rice breeding.

Lesion-mimic mutants (LMMs) in rice, often associated with hypersensitive response (HR), are significantly influenced by environmental factors such as temperature, light, and humidity. Certain chloroplast maculoid mutants activate defense responses, augmenting disease resistance. Liu et al. characterized the rice spot-like mutant *spl42*, which is associated with *SPOTTED LEAF42* (*SPL42*). This gene encodes porphobilinogen deaminase (PBGD), an enzyme crucial in the porphyrin biosynthesis pathway during chlorophyll synthesis. Mutations in *SPL42* reduced PBGD activity, altering the expression of several genes involved in chlorophyll biosynthesis and defense responses. The *spl42* mutant displayed red-brown spotted leaves, reduced chlorophyll and carotenoid contents, and an abnormal leaf ultrastructure [2]. These findings explain *SPL42*'s role in chlorophyll synthesis and leaf development in rice, providing insights helpful for enhancing rice yield and quality.

Weeds in paddy fields constitute a major biological stressor for rice cultivation, and waterlogging treatment is a common strategy for weed management. Developing flood-resistant rice varieties is therefore essential for effective weed management and can decrease reliance on pesticides. Kostylev et al. [3] successfully incorporated the *Sub1A* gene for long-term waterlogging tolerance into economically viable rice varieties, using anther culture to produce dihaploids resilient to extended waterlogging. Evaluation of these samples revealed that certain samples exhibited rapid expansion of the first leaf and enhanced nutrient accumulation under flooded conditions, indicating enhanced flood resistance. These results contribute valuable genetic resources for rice breeding, helping to produce varieties with improved flood tolerance.

Citation: Wu, W.; Zhang, Y.; Jiao, G.; Wei, X. Progress in Understanding and Enhancing Rice Tolerance to Biotic and Abiotic Stresses. *Plants* **2024**, *13*, 3206. https://doi.org/10.3390/plants13223206

Received: 5 November 2024
Accepted: 7 November 2024
Published: 15 November 2024

Rice is also significantly impacted by abiotic stressors, particularly extreme temperatures, waterlogging, and high salinity. The optimal temperatures for rice vary across growth stages, and deviations from these optimal temperatures can severely affect rice development. The maximum temperature during the rice-filling period should not exceed 35 °C. Elevated temperatures during this stage accelerate the accumulation of stored substances in seeds, resulting in grain-filling defects and reductions in yield and quality. Li et al. [4] studied the effects of high temperatures on seed development and successfully cloned *OsLEA1b*, a gene that responds to high temperatures. Expressed in rice endosperm, *OsLEA1b* encodes a protein involved in the heat stress response during grain filling. The *oslea1b* mutant exhibited an abnormal starch grain structure, a white endosperm phenotype, and significant reductions in grain weight and the number of grains per panicle. The starch composition also changed in the mutant, with fewer short-chain starches and more long-chain starches, especially under high temperatures. This study demonstrates *OsLEA1b*'s regulatory role in starch synthesis under high-temperature conditions, impacting rice quality and yield.

Mesocotyl elongation, crucial for waterlogging resilience during the seedling stage, helps establish deep sowing tolerance and enhances drought resilience in directly seeded rice. Varieties with longer mesocotyls can better break the soil, reducing anaerobic stress and increasing emergence and survival rates. Feng et al. [5] studied mesocotyl length in the Rice Diversity Panel 1 (RDP1) and Hanyou 73 (HY73) recombinant inbred lines (RILs) under dark germination conditions. They then employed genome-wide association studies (GWAS) of RDP1 and linkage mapping of HY73 RIL to identify genetic loci associated with mesocotyl elongation. Dynamic RNA sequencing was used to further validate the candidate genes, integrating phenotypic and genomic data to inform future rice breeding. This research provides insights into genetic factors influencing mesocotyl elongation, offering guidance for developing rice varieties with enhanced adaptability.

Soil salinization poses a significant challenge regarding rice yield and quality, necessitating insights into molecular mechanisms that enhance salt tolerance for breeding purposes. The *CIPK* gene family has been shown to play diverse roles in modulating salt tolerance, with *OsCIPK24* being linked specifically to this trait. Recent findings made by Zhou et al. [6] demonstrated that *OsCIPK9* knockout mutants exhibited improved salt tolerance, while overexpressing *OsCIPK9* resulted in heightened salt sensitivity. This study also found that *OsCIPK9* interacts with *OsSOS3*, suggesting the former's regulator role in salt-related transporters. RNA-seq data showed a pronounced *OsCIPK9* response to salt stress, likely through downstream salt tolerance genes. These findings demonstrate that *OsCIPK9* functions as a negative regulator of salt tolerance, offering insights for further research on salt tolerance mechanisms in rice. CRISPR technology could expedite assessments of other *OsCIPK* family genes, contributing to the development of a clearer picture of the regulatory pathway for rice salt tolerance.

YT521-B homology (YTH) domain proteins are essential for rice growth and environmental stressor responses, particularly in RNA modifications like N6-methyladenosine (m6A), which are crucial for gene regulation. In a recent study, Ma et al. [7] used CRISPR-Cas9 to generate YTH-deficient rice mutants, which displayed reduced plant heights, fewer spikelets per panicle, and lower setting rates and grain weights, indicating YTH protein's importance in rice growth and development. Additionally, certain YTH protein mutants exhibited variable tolerance to saline–alkali stress, demonstrating the YTH protein's role in stress response.

The cyclic nucleotide-gated channel (CNGC) family significantly modulates plant immunity and abiotic stress responses. Wang et al. [8] identified 16 *CNGC* genes in rice, examining their chromosomal locations, physicochemical properties, and subcellular localization. By analyzing haplotype (gcHap) diversity across 3010 genomes, the researchers explored genetic variation and selection pressures acting on *CNGC* genes in rice populations. Their study reveals the advantageous alleles of *OsCNCGs* that may enhance rice yield and abiotic stress tolerance, marking these alleles as key resources for future rice breeding.

Heading date is an important agronomic trait that affects rice maturity, adaptability, and yield. Identifying genes associated with heading date can improve rice's regional and seasonal adaptability. Despite numerous quantitative trait loci (QTLs) linked to heading date, few have been successfully cloned. Sohail et al. [9] used 76 chromosome segment substitution lines (CSSLs) derived from Zhonghui9308 (ZH9308) and XieqingzaoB (XQZB) to identify 14 QTLs associated with heading date, including a fine-mapped *qHD7b* on chromosome 7. Their results showed that the $qHD7b^{XQZB}$ allele was non-functional at the *Ghd7* locus, while the $qHD7b^{ZH9308}$ allele was functional. The functional *Ghd7* allele can extend the heading date in early-maturing varieties, potentially enhancing yield. Conversely, the non-functional *Ghd7* allele can accelerate maturation in late-maturing varieties. Additionally, several rice heading date genes also regulate abiotic stress responses, such as salinity and alkalinity tolerance.

The substantial advancements made over the past decade have broadened our understanding of rice responses to biotic and abiotic stresses. Breeding resistant varieties remains an effective solution to these challenges. Using well-characterized germplasm resources and marker-assisted selection (MAS) to introduce resistance genes has been effective in breeding programs. Consequently, the investigation of genes and haplotypes that have adapted to diverse environments provides valuable genetic resources for breeding additional resistant varieties. Recently, gene-editing and high-throughput sequencing technologies—such as Kompetitive Allele Specific PCR (KASP), whole-genome sequencing, targeted sequencing, and single-nucleotide polymorphism (SNP) breeding chips—have been widely employed, accelerating the development of resistant varieties and reducing breeding timelines.

This Special Issue aims to provide a comprehensive overview of advancements in the resistance of rice to biotic and abiotic stresses, covering recent research on molecular breeding and germplasm improvement. We believe that the articles and insights discussed here will provide readers with relevant knowledge of resistance resources with which to sustainably increase crop yields—a necessity for ensuring future food security. Combining modern biotechnology with traditional breeding methods, these studies offer actionable strategies for addressing stress-related challenges in rice production and supporting sustainable agriculture.

Author Contributions: X.W. Conceived and designed the editorial; W.W. drafted the manuscript; X.W., W.W., Y.Z. and G.J. jointly revised the manuscript and approved the final version. They take responsibility for their contributions and will address any questions about the work's accuracy or integrity. All authors have read and agreed to the published version of the manuscript.

Funding: This work was supported by STI 2030—Major Projects (2023ZD04066).

Conflicts of Interest: The authors declare there are no conflicts of interest.

References

1. Peng, P.; Jiang, H.; Luo, L.; Ye, C.; Xiao, Y. Pyramiding of Multiple Genes to Improve Rice Blast Resistance of Photo-Thermo Sensitive Male Sterile Line, without Yield Penalty in Hybrid Rice Production. *Plants* **2023**, *12*, 1389. [CrossRef] [PubMed]
2. Liu, L.; Wang, Y.; Tian, Y.; Song, S.; Wu, Z.; Ding, X.; Zheng, H.; Huang, Y.; Liu, S.; Dong, X.; et al. Isolation and Characterization of *SPOTTED LEAF42* Encoding a Porphobilinogen Deaminase in Rice. *Plants* **2023**, *12*, 403. [CrossRef] [PubMed]
3. Kostylev, P.; Kalinina, N.; Vozhzhova, N.; Golubova, V.; Chertkova, N. Creation of Rice Doubled Haploids Resistant to Prolonged Flooding Using Anther Culture. *Plants* **2023**, *12*, 3681. [CrossRef]
4. Li, G.; Cao, R.; Ma, L.; Jiao, G.; Chen, P.; Dong, N.; Li, X.; Duan, Y.; Li, X.; Zhu, M.; et al. *OsLEA1b* Modulates Starch Biosynthesis at High Temperatures in Rice. *Plants* **2023**, *12*, 4070. [CrossRef] [PubMed]
5. Feng, F.; Ma, X.; Yan, M.; Zhang, H.; Mei, D.; Fan, P.; Xu, X.; Wei, C.; Lou, Q.; Li, T.; et al. Identification of Genetic Loci for Rice Seedling Mesocotyl Elongation in Both Natural and Artificial Segregating Populations. *Plants* **2023**, *12*, 2743. [CrossRef] [PubMed]
6. Zhou, Z.; Tang, W.; Sun, Z.; Li, J.; Yang, B.; Liu, Y.; Wang, B.; Xu, D.; Yang, J.; Zhang, Y. OsCIPK9 Interacts with OsSOS3 and Affects Salt-Related Transport to Improve Salt Tolerance. *Plants* **2023**, *12*, 3723. [CrossRef] [PubMed]
7. Ma, W.; Cui, S.; Lu, Z.; Yan, X.; Cai, L.; Lu, Y.; Cai, K.; Zhou, H.; Ma, R.; Zhou, S.; et al. YTH Domain Proteins Play an Essential Role in Rice Growth and Stress Response. *Plants* **2022**, *11*, 2206. [CrossRef] [PubMed]

8. Wang, X.; Wu, F.; Zhang, J.; Bao, Y.; Wang, N.; Dou, G.; Meng, D.; Wang, X.; Li, J.; Shi, Y. Identification of the CNGC Gene Family in Rice and Mining of Alleles for Application in Rice Improvement. *Plants* **2023**, *12*, 4089. [CrossRef] [PubMed]
9. Sohail, A.; Shah, L.; Liu, L.; Islam, A.; Yang, Z.; Yang, Q.; Anis, G.B.; Xu, P.; Khan, R.M.; Li, J.; et al. Mapping and Validation of *qHD7b*: Major Heading-Date QTL Functions Mainly under Long-Day Conditions. *Plants* **2022**, *11*, 2288. [CrossRef] [PubMed]

plants

MDPI

Article

G1 Interacts with OsMADS1 to Regulate the Development of the Sterile Lemma in Rice

Huimin Fang [1,†], Hualan Chen [2,3,†], Jianing Wang [1], Ning Li [2,3], Long Zhang [2,3] and Cunxu Wei [2,3,*]

1 Guangling College, Yangzhou University, Yangzhou 225000, China; hmfang@yzu.edu.cn (H.F.); w2669941371@163.com (J.W.)
2 Key Laboratory of Crop Genetics and Physiology of Jiangsu Province, Key Laboratory of Plant Functional Genomics of the Ministry of Education, Yangzhou University, Yangzhou 225009, China; chenhualan0210@163.com (H.C.); lining20222024@126.com (N.L.); zhanglong@yzu.edu.cn (L.Z.)
3 Co-Innovation Center for Modern Production Technology of Grain Crops of Jiangsu Province, Joint International Research Laboratory of Agriculture & Agri-Product Safety of the Ministry of Education, Yangzhou University, Yangzhou 225009, China
* Correspondence: cxwei@yzu.edu.cn; Tel.: +86-514-8799-7217; Fax: +86-514-8797-1747
† These authors contributed equally to this work.

Abstract: Flower development, as the basis for plant seed development, is principally conserved in angiosperms. At present, a number of genes regulating flower organ differentiation have been identified, and an ABCDE model has also been proposed. In contrast, the mechanism that regulates the development of the sterile lemma remains unclear. In this study, we identified and characterized a rice floral organ mutant, M15, in which the sterile lemma transformed into a lemma-like organ. Positional cloning combined with a complementary experiment demonstrated that the mutant phenotype was restored by *LONG STERILE LEMMA1*/(*G1*). *G1* was expressed constitutively in various tissues, with the highest expression levels detected in the sterile lemma and young panicle. G1 is a nucleus-localized protein and functions as a homomer. Biochemical assays showed that G1 physically interacted with OsMADS1 both in vitro and in vivo. Interestingly, the expression of *G1* in M15 decreased, while the expression level of *OsMADS1* increased compared with the wild type. We demonstrate that G1 plays a key role in sterile lemma development through cooperating with OsMADS1. The above results have implications for further research on the molecular mechanisms underlying flower development and may have potential applications in crop improvement strategies.

Keywords: ALOG; DUF640; empty glume; floral organ; protein interaction

Citation: Fang, H.; Chen, H.; Wang, J.; Li, N.; Zhang, L.; Wei, C. G1 Interacts with OsMADS1 to Regulate the Development of the Sterile Lemma in Rice. *Plants* **2024**, *13*, 505. https://doi.org/10.3390/plants13040505

Academic Editors: Xiangjin Wei, Yingxin Zhang, Weixun Wu and Guiai Jiao

Received: 26 October 2023
Revised: 4 February 2024
Accepted: 8 February 2024
Published: 11 February 2024

1. Introduction

Flower development is one of the most significant events in plants' transition from vegetative to reproductive growth. Based on analyses of flower mutants of *Antirrhinum majus* and *Arabidopsis thaliana*, the ABC model was proposed to explain how genes determine the identity of the floral meristem and control floral organ differentiation [1]. In recent years, this model has been further supplemented and developed into the ABCDE model and successfully applied to a variety of plants including rice [2–5].

Rice (*Oryza sativa*), as a staple food for more than half of the population, is one of the most important crops in the world. The development of rice floral organs is closely related to the number of grains per panicle and affects the rice yield. The inflorescence of rice is named a panicle, with a spikelet as its structural unit. A typical rice spikelet comprises one fertile floret, two empty glumes (also called sterile lemmas) and two rudimentary glumes, and the floret consists of one lemma, one palea, two lodicules, six stamens and one pistil (Figure 1A,C) [6]. Recently, a number of genes belonging to the ABCDE model that control the development of rice floral organs were identified and characterized in rice, most of which belong to the MADS box gene family [7].

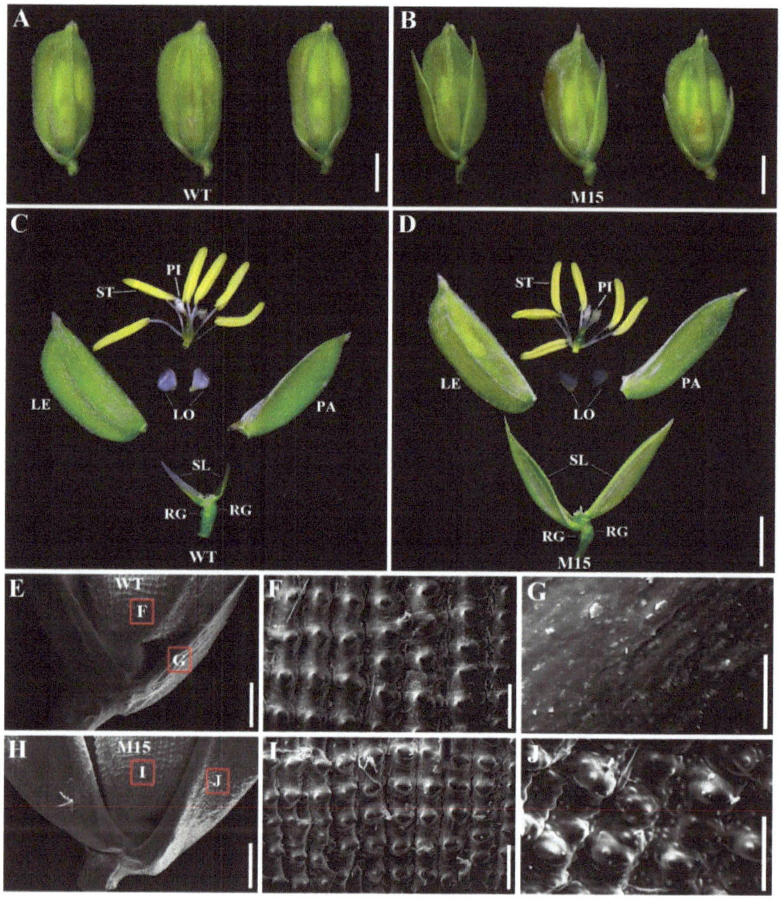

Figure 1. The spikelet structures of wild-type (WT) and mutant (M15) rice. (**A–D**) Phenotypes of WT (**A**,**C**) and M15 (**B**,**D**) spikelets at the heading stage. (**E–J**) Scanning electron microscope analysis of WT (**E–G**) and M15 (**H–J**) lemmas and sterile lemmas. RG: rudimentary glume; LE: lemma; LO: lodicule; PA: palea; PI: pistil; SL: sterile lemma; ST: stamen. Scale bars = 2 mm (**A–D**), 500 μm (**E**,**H**), 100 μm (**F**,**I**), and 50 μm (**G**,**J**).

Plant MADS box transcription factors are MIKC-type proteins, including highly conserved MADS (M) box DNA-binding domain, intervening (I) domain, keratin-like (K) domain and C-terminal (C) domain [8]. The rice A-function gene, *Degenerative palea/OsMADS15*, determines the characteristics of the lemma and palea [9]. *OsMADS2*, *OsMADS4* and *SUPERWOMAN1(SPW1)/OsMADS16* belong to B-class genes. Both OsMADS2 and OsMADS4 interact with OsMADS16 to regulate the development of the lodicule and stamen [10,11]. Two C-class genes, *OsMADS3* and *OsMADS58*, belong to the *AGAMOUS* (*AG*) subfamily. *OsMADS3* plays a key role in regulating stamen development, and *OsMADS58* mainly functions in the development of the carpel [12]. Rice *OsMADS13* and *OsMADS21*, as the homologs of Arabidopsis *SEEDSTICK* (*STK*) and *FLORAL BINDING PROTEIN* (*FBP*), are two D-class genes. *OsMADS13* regulates the ovule identity specification and also has a function in floral meristem determinacy [13,14]. *OsMADS21* has lost the ability to determine ovule identity during evolution [13]. To date, at least five E-class genes have been identified in rice, including *OsMADS1/LEAFY HULL STERILE1* (*LHS1*), *OsMADS5*, *OsMADS7*, *OsMADS8* and *OsMADS34/(PANICLE PHYTOMER2)PAP2* [15–18]. OsMADS1/LHS1 mainly specifies the identity of the lemma and palea and the development of the inner floral organ [19,20]. OsMADS7 and OsMADS8, two homologs of Arabidopsis SEPALLATA3 (SEP3), are shown to be involved in floral development redundantly [16]. OsMADS34 has versatile functions in lemma/palea, lodicule, stamen, carpel and spikelet development [21].

The rice floret is flanked by a pair of glume-like organs which are regarded as vestigial organs of two lower florets during the evolution of *Oryza*. The glume-like organs, which are called empty glumes or sterile lemmas in different publications in the rice literature, are not usually observed in the spikelets of maize and wheat [22]. The ectopic expression of rice *OsMADS1* causes spikelet alteration with elongated sterile lemmas [23]. The loss of function of *OsMADS34* leads to large sterile lemmas [24]. In addition to two E-class genes, *OsMADS1 and OsMADS34*, the *LONG STERILE LEMMA1(G1)/Elongated empty glume* (*ELE*) gene is also involved in the development of sterile lemmas [22,25]. *G1/ELE* encodes a protein with a domain of unknown function 640 (DUF640), and its mutation results in a long sterile lemma [22,25]. However, the mechanism of G1 controlling sterile lemma development remains poorly understood, and whether G1, OsMADS1 and OsMADS34 jointly regulate sterile lemmas remains to be identified.

In this study, as part of our continuous effort to understand the molecular machinery responsible for sterile lemma development in rice, we isolated a novel allele of *G1* which confers grain with long sterile lemma. Meanwhile, the functional characteristics of G1 including expression pattern, subcellular localization, and interacting proteins were characterized and identified. Our objective was to provide valuable information for understanding the functions of G1 and OsMADS1 in flower development and identities of sterile lemma.

2. Results

2.1. The M15 Mutant Shows Elongated Sterile Lemmas

In WT spikelets, the length of the sterile lemma was only one-quarter of the lemma/palea (Figure 1A). We obtained a ^{60}Co-induced rice mutant, M15, which showed long and large sterile lemmas (Figure 1B). In most spikelets, the two sterile lemmas nearly reached the length of the lemma/palea. Except for the change in the sterile lemma, the lemma, palea, stamen, pistil and lodicule of the M15 mutant were not significantly different from those of the WT (Figure 1C,D), suggesting that the M15 mutation specifically affects the development of the sterile lemma. Furthermore, scanning electron microscopy was applied to compare the surface structures of the spikelet organs of WT and M15 rice (Figure 1E–J). The epidermal cells of the WT lemma were characterized with rows of round bulges tipped with sharp projections (Figure 1F), while the surface of the sterile lemma was smooth (Figure 1G). The surface morphology of the M15 lemma was largely comparable with the WT (Figure 1I), while the elongated sterile lemma of M15 was transformed into the lemma or palea, displaying a bulge similar to those in the lemma and palea (Figure 1J). These results collectively suggested that the sterile lemma in M15 is not only larger but also mimics the lemma. In addition, compared with the wild type, the grain length of M15 is significantly reduced, while there is no significant change in grain width or thickness (Figure S1).

2.2. Positional Cloning and Complementation of the Mutated Gene in M15

To identify the mutant locus controlling the long sterile lemma phenotype, M15 was crossed with the indica variety Dular to produce an F_2 mapping population. Twenty long sterile lemma plants and twenty normal individual plants were selected in the F_2 population to generate two DNA pools. Then, 170 polymorphic insertion–deletion (InDel) markers on rice 12 chromosomes were applied to genotype the two pools for preliminary mapping. The gene was mapped to chromosome 7 between InDel markers 7-1 and 7-2, with genetic distances of 2.2 cM and 24.2 cM, respectively (Figure 2A). It has been reported that the *G1* gene regulates the development of glumes and is located at 8.8 cM of chromosome 7 [22], which is included in the markers of 7-1 and 7-2. Therefore, *G1* (*Os07g0139300*) is preferred as a candidate gene. Gene sequencing showed that the *G1* genomic sequences of M15 carried a single nucleotide substitution of guanine (G) to adenine (A) compared with the WT, leading to a non-synonymous mutation from glycine (G) to aspartic acid (D) at the 107th amino acid (Figure 2B,C). The *G1* gene contains only one exon and encodes a protein composed of 276 amino acid residues that harbors a DUF640 domain in the middle terminus (Figure 2C).

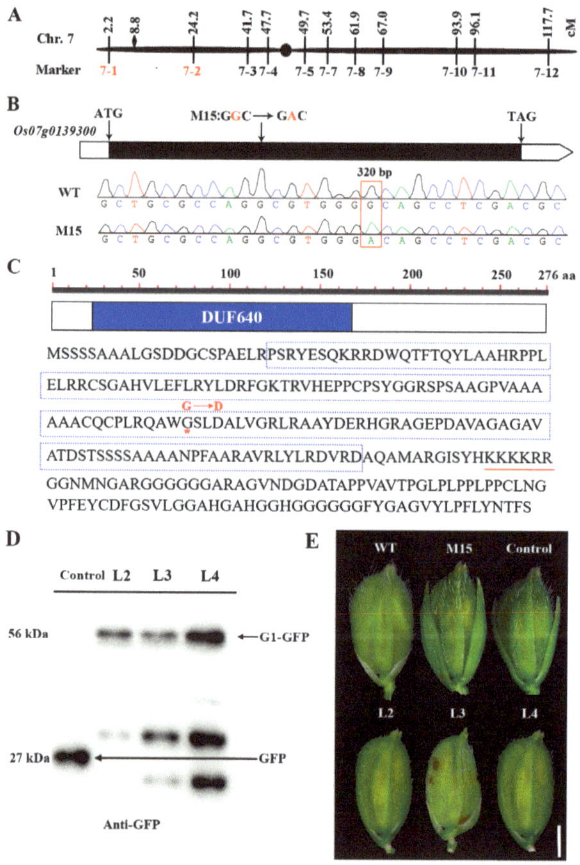

Figure 2. Identification of mutated gene and complementation of the M15 mutant. (**A**) The mutated gene was mapped to the short arm of chromosome 7 between markers 7-1 and 7-2. cM: centimorgan. (**B**) Mutation site and sequence chromatograms in *Os07g0139300*. (**C**) Amino acid sequence of G1. The DUF640 domain is marked with red font and the amino acids indicate the putative nucleus localization signal, which is underlined in red. The red asterisk represents that a single nucleotide substitution of guanine (G) to adenine (A) leads to a non-synonymous mutation from glycine (G) to aspartate (D). (**D**) Western blot analysis of G1-GFP fusion proteins from transgenic fragment-positive lines. The transgenic plants of free GFP were used as controls. L: line. (**E**) The functional complementation of the *G1* gene completely rescues the sterile lemma appearance. The phenotype of the sterile lemma in three positive transgenic lines was restored, while the control line still showed the phenotype of the M15 mutant. Scale bar = 2 mm.

To test whether *Os07g0139300* was responsible for the M15 mutation, a p35S:G1-GFP recombinant plasmid was transferred into the M15 *calli*. The G1-GFP fusion protein was detected in positive transgenic lines using anti-GFP antibodies (Figure 2D). The phenotype of a sterile lemma in three positive transgenic lines was restored to the WT (Figure 2E), suggesting that G1-GFP was functional. In summary, the mutation of *G1* leads to the long sterile lemma of M15.

2.3. Expression Patterns of G1 and Subcellular Localization of G1

G1 was highly expressed in the early stage of spikelet development but was relatively low in seeds, leaves and roots, according to the Rice eFP Browser (http://bar.utoronto.ca/efprice/, accessed on 4 February 2024, Figure 3A). We then tested the *G1* expression pattern in various organs from the WT using a qRT-PCR. *G1* was expressed constitutively in all tested organs, including the stem, leaf, leaf sheath, lemma, palea, sterile lemma, stamen, pistil, endosperm and panicle, with the highest expression level detected in the sterile lemma and panicle (Figure 3B). Further, *G1* had the highest expression level during early panicle development, but this dropped dramatically as the spikelets developed from the P1 stage to the P4 stage (Figure 3B).

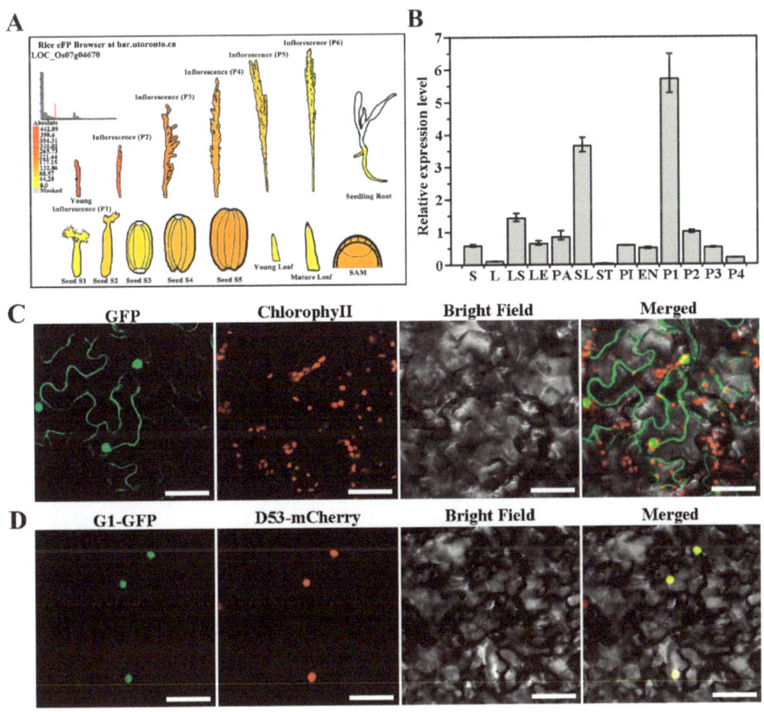

Figure 3. Spatiotemporal expression of G1 and subcellular localization of G1. (**A**) Expression patterns of G1 based on the rice eFP browser. (**B**) qRT-PCR analysis of G1 expression in various tissues. (**C**) Transient expression of free GFP in *N. benthamiana* leaves. (**D**) Transient expression of G1-GFP in *N. benthamiana* leaves. G1-GFP protein is located in the nuclei, overlapping with the nucleus marker D53-mCherry. GFP (green), chlorophyll autofluorescence (red) (C), mCherry (red) (D), bright-field images, and an overlay of green and red signals are shown. S: stem; L: leaf; LS: leaf sheath; LE: lemma, PA: palea, SL: sterile lemma, ST: stamen, PI: pistil. These tissues are taken from wild type plants just undergoing heading. EN: endosperm at 9 days after flowering, P1: 0.5~2 cm panicles, P2: 3~5 cm panicles, P3: 6 cm panicles, P4: panicles before heading. Data represent means and standard errors of three replicates. Scale bars = 50 μm (**C**,**D**).

A WoLF PSORT [26] software analysis revealed that the G1 protein contains a nucleus localization signal (KKKKRR, Figure 2C). To determine the subcellular localization of G1, p35S:G1-GFP was expressed in *N. benthamiana* leaves. Free GFP was distributed evenly in the cytoplasm and nuclei, whereas G1-GFP fluorescence was detected only in the nuclei (Figure 3C) and co-localized with red signals from the nucleus marker D53-mCherry (Figure 3D). Therefore, G1 is a nucleus-localized protein.

2.4. G1 Protein Forms Homodimers

The G1 protein has transcription factor activity, according to the RiceData Browser (http://www.ricedata.cn, accessed on 4 February 2024). A transactivation assay was then performed using the full-length sequence and different truncations of G1 fused to the GAL4 DNA-binding domain in the Y2HGold yeast strain. Yeast transformants containing BD-G1, BD-G1$^{1–169}$, BD-G1$^{22–276}$, BD-G1$^{1–21}$, BD-G1$^{22–169}$ and BD-G1$^{170–276}$ constructs grew well on an SD-Trp medium, whereas the growth of yeast transformants was completely inhibited on SD-Trp-His-Ade media (Figure S2). In summary, the G1 protein has no transactivation activity in yeast. Previous studies demonstrated that transcription factors can form homo- or heterodimers to function. A yeast two-hybrid assay showed that G1 can interact with itself (Figure 4A). Furthermore, the G1$^{170–276}$ region (170–276 amino acid residues) in the C-terminus, rather than the N-terminus and DUF640, was required for the self-interaction of G1 (Figure 4A). A GST pull-down assay also confirmed self-interaction in vitro (Figure 4B). In addition, a BiFC analysis also showed that G1 can physically interact with itself in the nuclei of leaves from *N. benthamiana* (Figure 4C). In summary, we concluded that G1 is capable of forming homodimers.

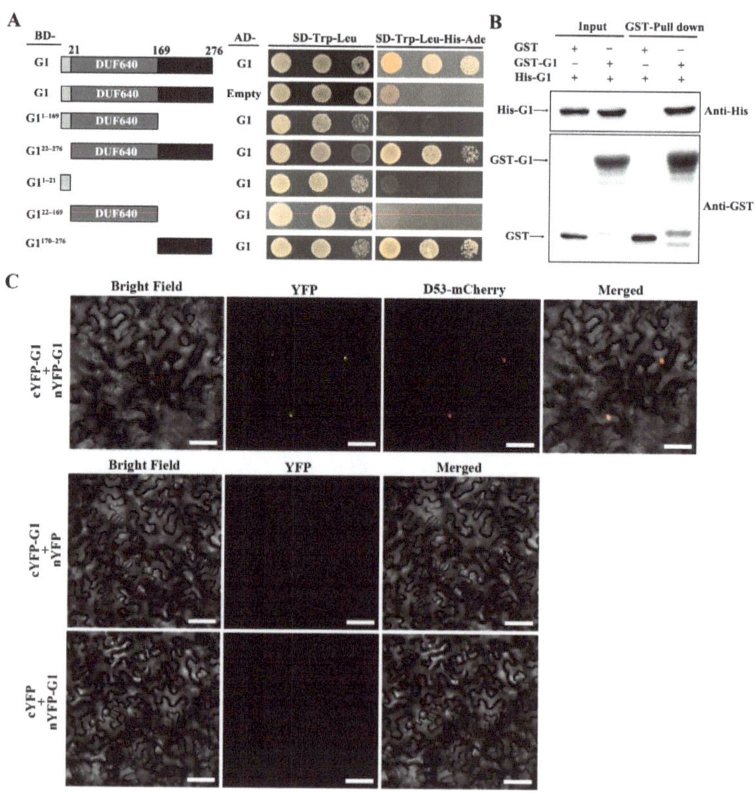

Figure 4. G1 forms a homocomplex. (**A**) Yeast two-hybrid assay showing the interaction between G1 and itself. (**B**) Pull-down assay showing the direct interaction between GST-G1 and His-G1 in vitro. (**C**) BiFC assay showing that G1 interacts with G1 in *N. benthamiana* leaf cells. Scale bars = 50 μm.

2.5. G1 Interacts with OsMADS1

Previous studies demonstrated that OsMADS1 and OsMADS34 also participate in the development of sterile lemmas in rice [23,24]. In addition, earlier studies showed that OsMADS1 and OsMADS34 are localized to the nuclei [24,27]. Therefore, we speculated that G1, OsMADS1 and OsMADS34 interact with each other. To evaluate this possibility, we used the yeast two-hybrid assay and found that G1 and G1$^{170–276}$ interact with OsMADS1 but not with OsMADS34 (Figure 5A). A pull-down assay was performed to confirm the physical interaction between G1 and OsMADS1 in vitro (Figure 5B). A BiFC experiment was performed to further confirm the interactions in vivo. In the BiFC assay, YFP fluorescence was produced in *N. benthamiana* leaves co-infected with *Agrobacterium* containing cYFP-G1 and nYFP-OsMADS1 or cYFP-OsMADS1 and nYFP-G1 (Figure 5C). These results indicated that G1 interacts with OsMADS1.

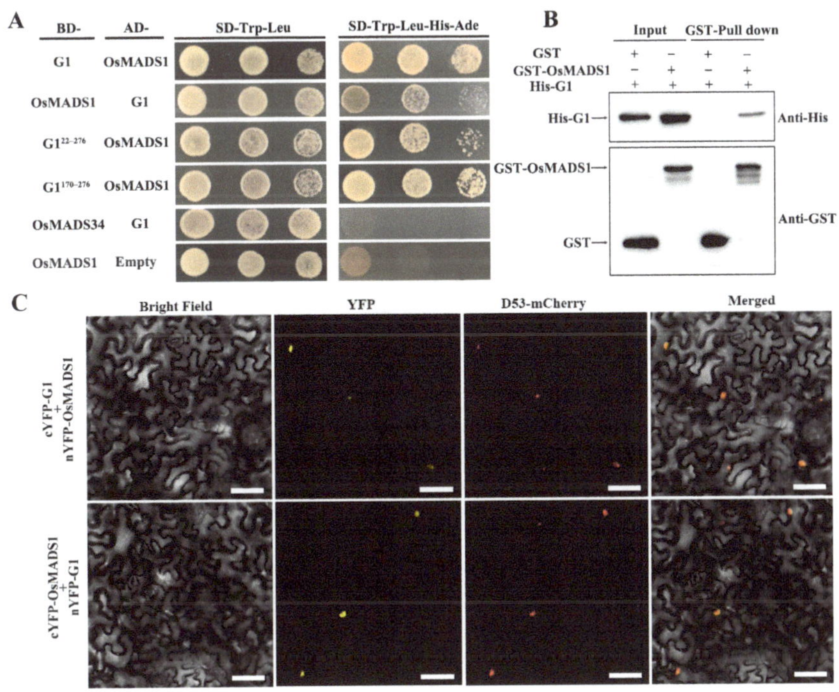

Figure 5. G1 interacts with OsMADS1. (**A**) Yeast two-hybrid assay showing the interaction between G1 and OsMADS1. (**B**) Pull-down assay showing the interaction between GST-OsMADS1 and His-G1 in vitro. (**C**) BiFC assay showing that G1 interacts with OsMADS1 in *N. benthamiana* leaf cells. Scale bars = 50 μm.

2.6. Expression Patterns of OsMADS1 and OsMADS34 in Rice

To investigate the correlation between gene expression pattern and biological function, we examined the temporal and spatial expression patterns of *OsMADS1* and *OsMADS34*. As shown in Figure 6A, *OsMADS1* was highly expressed in the lemma, palea, sterile lemma, pistil and panicle, with low levels of expression in the stem, leaf sheath, stamen and endosperm. *OsMADS34* was highly expressed in the leaf sheath, sterile lemma, pistil and panicle, with low levels of expression in the stem, stamen and endosperm (Figure 6B). These results are consistent with the roles of *OsMADS1* and *OsMADS34* in rice sterile lemma development.

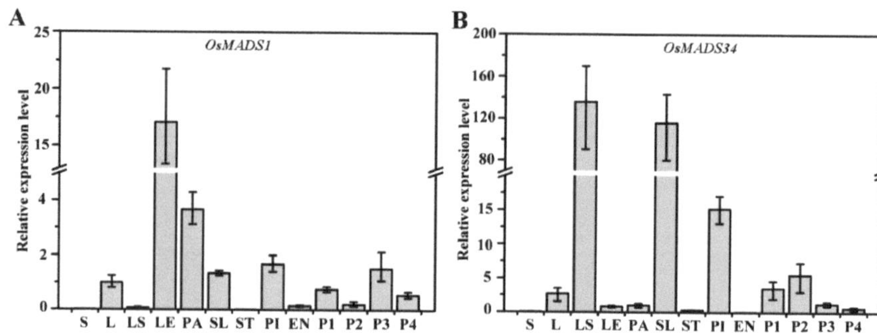

Figure 6. Expression patterns of *OsMADS1* and *OsMADS34* in rice. (**A**,**B**) Gene expression levels of *OsMADS1* (**A**) and *OsMADS34* (**B**) in various rice tissues. S: stem; L: leaf; LS: leaf sheath; LE: lemma, PA: palea, SL: sterile lemma, ST: stamen, PI: pistil. These tissues are taken from wild-type plants just undergoing heading. EN: endosperm at 9 days after flowering, P1: 0.5~2 cm panicles, P2: 3~5 cm panicles, P3: 6 cm panicles, P4: panicles before heading. Data represent means and standard errors (from at least three independent replicates).

2.7. Loss-of-Function Mutation of G1 Altered the Expression of OsMADS1

To explore the effect of G1 mutation on the expression of *G1*, *OsMADS1* and *OsMADS34*, the expression levels of three genes in the sterile lemmas of WT and M15 were analyzed using a qRT-PCR. The expression of *G1* in M15 was reduced by nearly half (Figure 7A), the expression of *OsMADS1* was significantly increased up to twofold (Figure 7B) and the expression of *OsMADS34* was almost unchanged compared with the WT (Figure 7C).

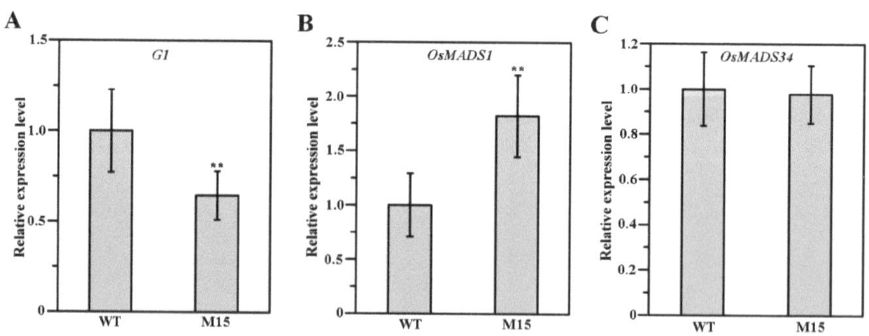

Figure 7. Expression levels of *G1* (**A**), *OsMADS1* (**B**) and *OsMADS34* (**C**) in the sterile lemmas of WT and M15. Data represent means and standard errors (from at least three independent replicates). Data were analyzed using a *t*-test; ** showed extremely significant difference ($p < 0.01$).

3. Discussion

3.1. G1 Controls the Specification of the Sterile Lemma

Rice spikelets are composed of one fertile floret, two sterile lemmas and two rudimentary glumes, and the sterile lemmas are located between the fertile floret and the rudimentary glumes. Some genes not only affect the development of the sterile lemma but also regulate the development of other flower organs. The AP2-family gene *SUPERNUMERARY BRACT* (*SNB*) regulates the transition from spikelet meristem to floral meristem. In *SNB* mutant plants, the sterile lemmas in some spikelets are transformed into lemma/palea-like organs, and the number of stamens and carpels is also altered [28]. In addition to affecting the number of sterile lemmas, mutations in *EXTRA GLUME1* (*EG1*) also lead to the formation of ectopic floral organs in each organ whorl [29]. In *nonstop glumes 1* (*nsg1*) mutants, the sterile lemma, palea, rudimentary glume and lodicule were elongated and transformed into lemma-like and/or a marginal region of the palea-like organs [30]. In this study, we isolated a long sterile lemma mutant, M15. Positional cloning and transgene complementation revealed that the mutant phenotype was controlled by *G1*. Except for the sterile lemma, the spikelet structure of M15 was complete, and the number and structure of other flower organs had not changed (Figure 1). Our SEM results further showed that the elongated sterile lemmas in M15 mimic lemmas (Figure 1). *G1* in M15 contained a missense mutation, and a substitution of G to A changed the glycine in the DUF640 domain to aspartate (Figure 2B,C). An amino acid sequence analysis showed that glycine (G) residue at amino acid position 107 was highly conserved in G1-related proteins among different plant species (Figure S3), suggesting that a single amino acid substitution in the DUF640 domain is critical for G1 protein function. Yoshida et al. [22] have reported four *G1* mutants with different mutations and named them *g1-1*, *g1–2*, *g1–3* and *g1–4*. The whole *G1* gene was deleted in *g1-1*, and missense and nonsense mutations were examined in *g1–2* (G82V, R117H), *g1–3* (G82V, A118V) and *g1–4* (68 bp deletion). *ele* is another *G1* mutant containing a missense mutation (G68D) [25]. Therefore, M15 is a new *G1/ELE* allelic mutant compared with the mutation sites that have been reported [22,25]. Though the mutations of these mutants are different, they all result in the same mutant phenotype, indicating that G1 plays an important and conservative role in the development of the sterile lemma. The ancestor of *Oryza* has three florets, and the three-floret spikelet hypothesis was confirmed using the *lateral florets 1* mutant [31]. Modern cultivated rice only bears a single fertile floret within a spikelet, and the two lateral florets degenerated and left only the lemma during evolution. Here we support the proposition that the rice sterile lemma belongs to a kind of lemma and palea and has the ability to transform into a lemma when G1 function is lost.

3.2. G1 and OsMADS1 Jointly Regulate Sterile Lemma Development in Rice

OsMADS1 plays a critical role in rice floral organ identity specification and floral meristem determinacy through interacting with B-, C- and D-class proteins. Previous studies show that OsMADS1 interacts with two A-class proteins, OsMADS14 and OsMADS15, and two E-class proteins, OsMADS7 and OsMADS8 [16,32]. In addition, OsMADS1 interacts with OsMADS3 and OsMADS58 to control floral meristem activity maintenance and regulate floral meristem determinacy, respectively [33]. In the present study, our yeast two-hybrid assay showed that G1 interacted with OsMADS1, and the interaction region was at the G1 C-terminus (Figure 5A). The BiFC assay revealed that G1 interacted with OsMADS1 in nuclei (Figure 5C). A temporal and spatial expression pattern analysis showed that both *G1* and *OsMADS1* were expressed abundantly in sterile lemmas and young panicles (Figures 3B and 6A). In rice, the overexpression of *OsMADS1* results in an elongated sterile lemma [23]. Interestingly, the expression of *OsMADS1* also increased in the M15 sterile lemma (Figure 7A). In addition, the functional interaction between G1 and OsMADS1 is also strongly supported by phenotypic impacts on the sterile lemma. Although OsMADS34 is also involved in sterile lemma development [21,24], our yeast two-hybrid assay (Figure 5A) and qRT-PCR (Figure 7C) showed that there is no direct relationship between G1 and OsMADS34. These results imply that G1 and OsMADS1 play simultaneous roles in regulating the development of the sterile lemma.

3.3. Characteristics of G1

Domains of unknown function (DUFs) represent a large number of uncharacterized proteins [34]. Most DUFs are highly conserved in plants and are critical for plant growth and development. Rice and Arabidopsis contain 10 genes encoding proteins with DUF640, also named the ALOG (*Arabidopsis LSH1* and *Oryza G1*) domain [22]. The genes functionally identified in this family contain rice *BEAK-SHAPED GRAIN1* (*BSG1*), which is involved in controlling lemma and palea development. The loss of function of BSG1 leads to defects in lemma and palea expansion, but with a normal sterile lemma [35]. *G1* encoded a protein composed of a conserved DUF640 and specifically regulated the development of the sterile lemma (Figures 1 and 2C). G1 is a homolog of rice BLS1 [35], while the changes in flower organs caused by the mutations of the two genes are completely different. Except for Arabidopsis, a BLASTP search showed that the homologs of G1 were found in other plants, including dicot and monocot plants (Figure S4). Moreover, a phylogenetic analysis indicated that these proteins were distinctly classified into two clades corresponding to proteins from monocots and dicots. Based on the functions of G1 homologs in Arabidopsis and rice, as well as the regulation of the sterile lemma and grain length in this study, we think that G1 is relatively conservative, while its roles in regulating plant development are diverse.

In this study, G1 was located in the nucleus and can form homodimers (Figures 3D and 4). In addition, *G1* was highly expressed in the sterile lemma and the early developmental stage of the panicle (Figure 3B), implying that G1 may affect the growth of the sterile lemma by controlling the expression of other genes. Our qRT-PCR assay showed that the expression level of *OsMADS1* was significantly increased in the M15 mutant (Figure 7B). Previous studies have shown that overexpression of *OsMADS1* also leads to a long sterile lemma phenotype [23]. Therefore, we speculate that G1 may act as a suppressor of OsMADS1 to regulate sterile lemma development. Arabidopsis *LSH1*, a homolog of G1, has been reported to function as a transcription regulator protein [36]. The G1 protein structure was predicted using the I-TASSER database (https://zhanggroup.org/I-TASSER, accessed on 4 February 2024), and the predicted result was further analyzed using PyMOL software (2.5.7). The predicted ligand of G1 was a double-stranded nucleic acid, and the key amino acid sites of G1 binding to DNA are A105, V113, N153, P154, F155, G256, G257, G258, G260 and F261 (Figure S5). Collectively, these results support our interpretation that G1 is involved in transcriptional regulation for sterile lemma growth. In this study, we fused a GFP tag at the C terminal of G1 and obtained transgenic plants with G1-GFP. Therefore, in future work, we will use Chip-seq and IP-MS to identify the target genes and other interacting proteins of G1, respectively, to gain a more comprehensive understanding of the regulatory network involved in sterile lemma development.

4. Materials and Methods

4.1. Plant Materials

The long sterile lemma mutant M15 was derived from a mutant pool of the rice variety (*Oryza sativa* L. ssp. japonica) Kitaake (wild-type, WT) [37]. The mutant was self-crossed for over 5 generations with a stable phenotype and is considered homozygous. All rice materials were planted in the experimental field of Yangzhou University under natural conditions.

4.2. Microscopy

The developing spikelets of WT and M15 rice were separated using tweezers and an anatomical needle. The lemma, lodicule, palea, pistil, sterile lemma and stamen were photographed under a Leica EZ4W stereomicroscope. Mature and dry WT and M15 grains were glued onto aluminum specimen stubs to observe the lemma and sterile lemma using a Hitachi S-4800 scanning electron microscope (SEM), following the method described previously [37].

4.3. Positional Cloning and Transgene Complementation

The M15 mutant was crossed with a wide-compatibility indica variety Dular to obtain an F_2 population. Twenty individuals with long sterile lemmas were selected from the F_2 population for positional cloning. A bulked segregant analysis (BSA) was used to identify the markers linked to the M15 long sterile lemma phenotype [37].

For a complementation test, the *G1* coding sequence without a stop codon was cloned and inserted into vector pCAMBIA1305-GFP to produce the recombinant plasmid p35s:G1-GFP. p35s:G1-GFP and pCAMBIA1305-GFP were introduced into M15 *calli* using *Agrobacterium tumefaciens*-mediated transformation. Twenty-two independent transgenic lines harboring p35s:G1-GFP and 15 independent transgenic lines carrying pCAMBIA1305-GFP were successfully obtained. The transgenic plants of free GFP were used as a control.

4.4. RNA Isolation, cDNA Synthesis and Quantitative Real-Time PCR Analysis

Total RNA from different rice tissues was extracted using an RNAprep pure plant kit (Tiangen, Beijing, China) and reverse-transcribed to cDNA using a HiScript III 1st Strand cDNA Synthesis Kit (Vazyme, Nanjing, China). A quantitative real-time PCR (qRT-PCR) was performed on a CFX Connect real-time PCR system (Bio-Rad, Hercules, CA, USA) using the AceQ Universal SYBR qPCR Master Mix (Vazyme, Nanjing, China). The rice *ACTIN* gene was used as an endogenous control. The qRT-PCR primers are listed in Table S1.

4.5. Protein Extraction and Western Blot

Proteins were extracted in an extraction buffer (50 mM Tris-HCl (pH 8.0), 0.25 M Sucrose, 2 mM EDTA, 2 mM DTT, 1 mM PMSF) at 4 °C. The Western blot was prepared as described by Zhang et al. [37].

4.6. Subcellular Localization of G1

To determine the subcellular localization of G1, the empty plasmid pCAMBIA1305-GFP, p35s:D53-mCherry and p35s:G1-GFP were expressed in *Nicotiana. benthamiana* leaves using an *Agrobacterium tumefaciens*-mediated transformation. The infected tobacco leaves were cultured at 28 °C for 2–3 days, and the fluorescence signals were then observed using a Zeiss LSM880 confocal laser microscope.

4.7. Yeast Two-Hybrid Assay

The full-length coding sequences of *G1*, *OsMADS1* and *OsMADS34* and truncations of *G1* were cloned into a pGBKT7 vector. The coding sequences of *G1* and *OsMADS1* were cloned into a pGADT7 vector (Clontech, Palo Alto, CA, USA). Various combinations of plasmids were transformed into yeast AH109 competent cells according to the manufacturer's instructions (Clontech, Palo Alto, CA, USA). The vectors pGBKT7 and pGADT7 were expressed in the yeast as negative controls. The PCR primers are listed in Supplementary Table S1.

4.8. Pull-Down Assay

The coding sequences of *G1* and *OsMADS1* were cloned into a pGEX4T-1 vector to generate fusion with glutathione S-transferase (GST), and the coding sequence of *G1* was then cloned into the pET-32a vector to generate fusion with a His tag. GST-G1, GST-OsMADS1, GST and His-G1 were transformed into BL21 Rosetta cells to induce protein expression using 0.5 mM isopropyl-β-d-thiogalactoside. The total protein concentration was quantified using a protein purification kit (Beyotime, Shanghai, China). The pull-down assay was performed as described previously [38]. The proteins were separated on a 10% SDS-PAGE gel and immunoblotted with anti-GST or anti-His antibodies (ABclonal, Wuhan, China).

4.9. Bimolecular Fluorescence Complementation (BiFC) Assay

The coding sequences of *G1* and *OsMADS1* were inserted into a pSAT1-cEYFP vector to create cYFP-G1 and cYFP-OsMADS1. The coding sequences of *G1* and *OsMADS1* were also inserted into a pSAT1-nEYFP vector to create nYFP-G1 and nYFP-OsMADS1. Various combinations of plasmids were then expressed in *Nicotiana. benthamiana* leaves as previously described [39]. Yellow fluorescent protein (YFP) fluorescence was observed using a Zeiss LSM880 confocal laser microscope. The relevant PCR primers are listed in Supplemental Table S1.

5. Conclusions

In this study, a new *G1* allelic mutant with long sterile lemma was identified. *G1* was highly expressed in the sterile lemma and young panicle, and its encoding protein was located in the nucleus. G1 functioned as a homomer and interacted with OsMADS1. The mutation of *G1* led to a decrease in its own expression level and an increase in the *OsMADS1* expression level. This study can provide useful information for better understanding the functions of *G1* and *OsMADS1* in flower development.

Supplementary Materials: The following supporting information can be downloaded at https://www.mdpi.com/article/10.3390/plants13040505/s1, Figure S1: Phenotypes of WT and M15 seeds; Figure S2: The transcriptional activation assay of G1 and truncated G1 in yeast cells; Figure S3: Alignment of amino acid sequences of G1 protein mutation regions in different plant species; Figure S4: Phylogenetic analysis of G1 and its homologs; Figure S5: The protein structure analysis of G1; Table S1: Primers used in this study.

Author Contributions: Methodology, H.F., H.C. and L.Z.; investigation, J.W. and N.L.; resources, H.C. and L.Z.; data curation, J.W. and N.L.; writing—original draft preparation, H.F. and L.Z.; writing—review and editing, L.Z. and C.W.; project administration, L.Z. and C.W.; funding acquisition, H.F., H.C., L.Z. and C.W. All authors have read and agreed to the published version of the manuscript.

Funding: This research was funded by grants from the National Natural Science Foundation of China (32001450), the Natural Science Foundation of Jiangsu Province (BK20221283), the Natural Science Foundation of the Jiangsu Higher Education Institutions of China (19KJB210006), the Innovation Program for Graduates of Jiangsu Province (KYCX22_3470), the Qinglan Project of Jiangsu (Su (2022) no.2) to Huimin Fang, and the Qinglan Project of Yangzhou University to Long Zhang.

Data Availability Statement: Data are contained within the article and supplementary materials.

Conflicts of Interest: The authors declare no conflicts of interest.

References

1. Coen, E.S.; Meyerowitz, E.M. The War of the Whorls: Genetic Interactions Controlling Flower Development. *Nature* **1991**, *353*, 31–37. [CrossRef]
2. Ditta, G.; Pinyopich, A.; Robles, P.; Pelaz, S.; Yanofsky, M.F. The *SEP4* gene of *Arabidopsis thaliana* functions in floral organ and meristem identity. *Curr. Biol.* **2004**, *14*, 1935–1940. [CrossRef]
3. Ferrario, S.; Immink, R.G.; Angenent, G.C. Conservation and diversity in flower land. *Curr. Opin. Plant Biol.* **2004**, *7*, 84–91. [CrossRef] [PubMed]
4. Theissen, G. Development of floral organ identity: Stories from the MADS house. *Curr. Opin. Plant Biol.* **2001**, *4*, 75–85. [CrossRef] [PubMed]
5. Theissen, G.; Saedler, H. Plant biology—Floral quartets. *Nature* **2001**, *409*, 469–471. [CrossRef] [PubMed]
6. Yoshida, H.; Nagato, Y. Flower development in rice. *J. Exp. Bot.* **2011**, *62*, 4719–4730. [CrossRef]
7. Yamaguchi, T.; Hirano, H.Y. Function and diversification of MADS-box genes in rice. *Sci. World J.* **2006**, *6*, 1923–1932. [CrossRef] [PubMed]
8. Lydia, G.; Günter, T.e. Phylogenomics of MADS-Box Genes in Plants—Two Opposing Life Styles in One Gene Family. *Biology* **2013**, *2*, 1150–1164.
9. Wang, K.J.; Tang, D.; Hong, L.L.; Xu, W.Y.; Huang, J.; Li, M.; Gu, M.H.; Xue, Y.B.; Cheng, Z.K. DEP and AFO Regulate Reproductive Habit in Rice. *PLoS Genet.* **2010**, *6*, e1000818. [CrossRef] [PubMed]
10. Nagasawa, N.; Miyoshi, M.; Sano, Y.; Satoh, H.; Hirano, H.; Sakai, H.; Nagato, Y. *SUPERWOMAN1* and *DROOPING LEAF* genes control floral organ identity in rice. *Development* **2003**, *130*, 705–718. [CrossRef]
11. Yao, S.G.; Ohmori, S.; Kimizu, M.; Yoshida, H. Unequal genetic redundancy of rice orthologs, *OsMADS2* and *OsMADS4*, in lodicule and stamen development. *Plant Cell Physiol.* **2008**, *49*, 853–857. [CrossRef] [PubMed]
12. Dreni, L.; Pilatone, A.; Yun, D.P.; Erreni, S.; Pajoro, A.; Caporali, E.; Zhang, D.B.; Kater, M.M. Functional Analysis of All *AGAMOUS* Subfamily Members in Rice Reveals Their Roles in Reproductive Organ Identity Determination and Meristem Determinacy. *Plant Cell* **2011**, *23*, 2850–2863. [CrossRef] [PubMed]
13. Dreni, L.; Jacchia, S.; Fornara, F.; Fornari, M.; Ouwerkerk, P.B.F.; An, G.H.; Colombo, L.; Kater, M.M. The D-lineage MADS-box gene *OsMADS13* controls ovule identity in rice. *Plant J.* **2007**, *52*, 690–699. [CrossRef]
14. Li, H.F.; Liang, W.Q.; Yin, C.S.; Zhu, L.; Zhang, D.B. Genetic Interaction of *OsMADS3*, *DROOPING LEAF*, and *OsMADS13* in Specifying Rice Floral Organ Identities and Meristem Determinacy. *Plant Physiol.* **2011**, *156*, 263–274. [CrossRef]

15. Arora, R.; Agarwal, P.; Ray, S.; Singh, A.K.; Singh, V.P.; Tyagi, A.K.; Kapoor, S. MADS-box gene family in rice: Genome-wide identification, organization and expression profiling during reproductive development and stress. *BMC Genom.* **2007**, *8*, 242. [CrossRef]
16. Cui, R.F.; Han, J.K.; Zhao, S.Z.; Su, K.M.; Wu, F.; Du, X.Q.; Xu, Q.J.; Chong, K.; Theissen, G.; Meng, Z. Functional conservation and diversification of class E floral homeotic genes in rice (*Oryza sativa*). *Plant J.* **2010**, *61*, 767–781. [CrossRef]
17. Malcomber, S.T.; Kellogg, E.A. Heterogeneous expression patterns and separate roles of the *SEPALLATA* gene *LEAFY HULL STERILE1* in grasses. *Plant Cell* **2004**, *16*, 1692–1706. [CrossRef]
18. Zahn, L.M.; King, H.Z.; Leebens-Mack, J.H.; Kim, S.; Soltis, P.S.; Landherr, L.L.; Soltis, D.E.; dePamphilis, C.W.; Ma, H. The evolution of the *SEPALLATA* subfamily of MADS-Box genes: A preangiosperm origin with multiple duplications throughout angiosperm history. *Genetics* **2005**, *169*, 2209–2223. [CrossRef]
19. Jeon, J.S.; Jang, S.; Lee, S.; Nam, J.; Kim, C.; Lee, S.H.; Chung, Y.Y.; Kim, S.R.; Lee, Y.H.; Cho, Y.G.; et al. *leafy hull sterile1* is a homeotic mutation in a rice MADS box gene affecting rice flower development. *Plant Cell* **2000**, *12*, 871–884. [CrossRef]
20. Khanday, I.; Yadav, S.R.; Vijayraghavan, U. Rice LHS1/OsMADS1 Controls Floret Meristem Specification by Coordinated Regulation of Transcription Factors and Hormone Signaling Pathways. *Plant Physiol.* **2013**, *161*, 1970–1983. [CrossRef] [PubMed]
21. Gao, X.C.; Liang, W.Q.; Yin, C.S.; Ji, S.M.; Wang, H.M.; Su, X.A.; Guo, C.C.; Kong, H.Z.; Xue, H.W.; Zhang, D.B. The *SEPALLATA*-like Gene *OsMADS34* Is Required for Rice Inflorescence and Spikelet Development. *Plant Physiol.* **2010**, *153*, 728–740. [CrossRef] [PubMed]
22. Yoshida, A.; Suzaki, T.; Tanaka, W.; Hirano, H.Y. The homeotic gene *long sterile lemma* (*G1*) specifies sterile lemma identity in the rice spikelet. *Proc. Natl. Acad. Sci. USA* **2009**, *106*, 20103–20108. [CrossRef] [PubMed]
23. Wang, L.; Zeng, X.Q.; Zhuang, H.; Shen, Y.L.; Chen, H.; Wang, Z.W.; Long, J.C.; Ling, Y.H.; He, G.H.; Li, Y.F. Ectopic expression of *OsMADS1* caused dwarfism and spikelet alteration in rice. *Plant Growth Regul.* **2017**, *81*, 433–442. [CrossRef]
24. Zhang, Y.; Yu, H.P.; Liu, J.; Wang, W.; Sun, J.; Gao, Q.; Zhang, Y.H.; Ma, D.R.; Wang, J.Y.; Xu, Z.J.; et al. Loss of function of *OsMADS34* leads to large sterile lemma and low grain yield in rice (*Oryza sativa* L.). *Mol. Breed.* **2016**, *36*, 147. [CrossRef]
25. Hong, L.L.; Qian, Q.; Zhu, K.M.; Tang, D.; Huang, Z.J.; Gao, L.; Li, M.; Gu, M.H.; Cheng, Z.K. ELE restrains empty glumes from developing into lemmas. *J. Genet. Genom.* **2010**, *37*, 101–115. [CrossRef] [PubMed]
26. Nakai, K.; Horton, P. Computational Prediction of Subcellular Localization. *Methods Mol. Biol.* **2007**, *390*, 429–466. [PubMed]
27. Zhang, J.; Cai, Y.; Yan, H.G.; Jin, J.; You, X.M.; Wang, L.; Kong, F.; Zheng, M.; Wang, G.X.; Jiang, L.; et al. A Critical Role of *OsMADS1* in the Development of the Body of the Palea in Rice. *J. Plant Biol.* **2018**, *61*, 11–24. [CrossRef]
28. Lee, D.Y.; Lee, J.; Moon, S.; Park, S.Y.; An, G. The rice heterochronic gene *SUPERNUMERARY BRACT* regulates the transition from spikelet meristem to floral meristem. *Plant J.* **2007**, *49*, 64–78. [CrossRef]
29. Li, H.G.; Xue, D.W.; Gao, Z.Y.; Yan, M.X.; Xu, W.Y.; Xing, Z.; Huang, D.N.; Qian, Q.; Xue, Y.B. A putative lipase gene *EXTRA GLUME1* regulates both empty-glume fate and spikelet development in rice. *Plant J.* **2009**, *57*, 593–605. [CrossRef]
30. Zhuang, H.; Wang, H.L.; Zhang, T.; Zeng, X.Q.; Chen, H.; Wang, Z.W.; Zhang, J.; Zheng, H.; Tang, J.; Ling, Y.H.; et al. *NONSTOP GLUMES1* Encodes a C2H2 Zinc Finger Protein That Regulates Spikelet Development in Rice. *Plant Cell* **2020**, *32*, 392–413. [CrossRef]
31. Zhang, T.; Li, Y.F.; Ma, L.; Sang, X.C.; Ling, Y.H.; Wang, Y.T.; Yu, P.; Zhuang, H.; Huang, J.Y.; Wang, N.; et al. *LATERAL FLORET 1* induced the three-florets spikelet in rice. *Proc. Natl. Acad. Sci. USA* **2017**, *114*, 9984–9989. [CrossRef]
32. Lim, J.; Moon, Y.H.; An, G.; Jang, S.K. Two rice MADS domain proteins interact with OsMADS1. *Plant Mol. Biol.* **2000**, *44*, 513–527. [CrossRef]
33. Hu, Y.; Liang, W.Q.; Yin, C.S.; Yang, X.L.; Ping, B.Z.; Li, A.X.; Jia, R.; Chen, M.J.; Luo, Z.J.; Cai, Q.; et al. Interactions of OsMADS1 with Floral Homeotic Genes in Rice Flower Development. *Mol. Plant* **2015**, *8*, 1366–1384. [CrossRef] [PubMed]
34. Bateman, A.; Coggill, P.; Finn, R.D. DUFs: Families in search of function. *Acta Crystallogr. F* **2010**, *66*, 1148–1152. [CrossRef] [PubMed]
35. Ma, X.D.; Cheng, Z.J.; Wu, F.Q.; Jin, M.N.; Zhang, L.G.; Zhou, F.; Wang, J.L.; Zhou, K.N.; Ma, J.; Lin, Q.B.; et al. *BEAK LIKE SPIKELET1* is Required for Lateral Development of Lemma and Palea in Rice. *Plant Mol. Biol. Rep.* **2013**, *31*, 98–108. [CrossRef]
36. Zhao, L.; Nakazawa, M.; Takase, T.; Manabe, K.; Kobayashi, M.; Seki, M.; Shinozaki, K.; Matsui, M. Overexpression of *LSH1*, a member of an uncharacterised gene family, causes enhanced light regulation of seedling development. *Plant J.* **2004**, *37*, 694–706. [CrossRef] [PubMed]
37. Zhang, L.; Zhao, L.L.; Lin, L.S.; Zhao, L.X.; Liu, Q.Q.; Wei, C.X. A novel mutation of *OsPPDKB*, encoding pyruvate orthophosphate dikinase, affects metabolism and structure of starch in the rice endosperm. *Int. J. Mol. Sci.* **2018**, *19*, 2268. [CrossRef] [PubMed]
38. Miernyk, J.A.; Thelen, J.J. Biochemical approaches for discovering protein-protein interactions. *Plant J.* **2008**, *53*, 597–609. [CrossRef]
39. Waadt, R.; Kudla, J. In planta visualization of protein interactions using bimolecular fluorescence complementation (BiFC). *Cold Spring Harb. Protoc.* **2008**, *4*, t4995. [CrossRef]

plants

MDPI

Article

Identification of the CNGC Gene Family in Rice and Mining of Alleles for Application in Rice Improvement

Xinchen Wang †, Fengcai Wu †, Jinguo Zhang, Yaling Bao, Nansheng Wang, Guohui Dou, Dezhuang Meng, Xingmeng Wang, Jianfeng Li and Yingyao Shi *

College of Agronomy, Anhui Agricultural University, Hefei 230036, China; 22721752@stu.ahau.edu.cn (X.W.); 22720535@stu.ahau.edu.cn (F.W.); zhangjinguo@stu.ahau.edu.cn (J.Z.); valleybao2019@126.com (Y.B.); 1341489953@stu.ahau.edu.cn (N.W.); douguohui@stu.ahau.edu.cn (G.D.); 22721677@stu.ahau.edu.cn (D.M.) wangxingmeng@stu.ahau.edu.cn (X.W.); 22720528@stu.ahau.edu.cn (J.L.)

* Correspondence: shiyy123@163.com

† These authors contributed equally to this work.

Abstract: Cyclic nucleotide-gated ion channel (CNGC) gene regulation plays important roles in plant immune and abiotic stress response. Here, we identified 16 CNGC genes in rice (Oryza sativa). Then, we analyzed their chromosomal location, physicochemical properties, subcellular localization, gene functional interaction network, cis-acting elements, phylogenetic relationships, collinearity, expression in tissues under normal conditions and abiotic stresses, and geng-cds-haplotype (gcHap) diversity in 3010 gcHaps. As a result, *OsCNGC3* (*Os06g0527300*) was identified as a gene different from previous report, and *OsCNGC* genes were found to play important roles in rice population differentiation and rice improvement. Our results revealed their very strong differentiation between subspecies and populations, important roles in response to abiotic stresses, as well as strong genetic bottleneck effects and artificial selection of gcHap diversity in the modern breeding process of Xian (indica) and Geng (japonica) populations. The results also suggested that natural variations in most rice CNGC loci are potentially valuable for improving rice productivity and tolerance to abiotic stresses. The favorable alleles at the CNGC loci should be explored to facilitate their application in future rice improvement.

Keywords: rice; cyclic nucleotide-gated ion channel gene; geng-cds-haplotype (gcHap) diversity; modification

Citation: Wang, X.; Wu, F.; Zhang, J.; Bao, Y.; Wang, N.; Dou, G.; Meng, D.; Wang, X.; Li, J.; Shi, Y. Identification of the CNGC Gene Family in Rice and Mining of Alleles for Application in Rice Improvement. *Plants* **2023**, *12*, 4089. https://doi.org/10.3390/plants12244089

Academic Editor: Tika Adhikari

Received: 31 October 2023
Revised: 2 December 2023
Accepted: 5 December 2023
Published: 6 December 2023

1. Introduction

Cyclic nucleotide-gated channels (CNGC) are widely present in plants and animals, playing important roles in their growth and development and response to stress. Cyclic nucleotides (CNMPs) are an important class of signaling molecules, such as 3′,5′-cyclic adenosine monophosphate (cAMP), which is an important component in signaling pathways in plants and animal life activities [1]. The CNGC family was first discovered in plants during the screening of a barley-pasteurized layer expression library for calmodulin (CaM) [2]. Sixteen members of the CNGC family have been identified in rice [3,4], among which four members have been cloned and verified. SSS1-D encodes *OsCNGC13*, a member of the CNGC family, which enhances fruit setting by promoting the growth of pollen tubes in the stylar tissue [5]. In Arabidopsis thaliana, *AtCNGC7* and *AtCNGC8* proteins are closely related to 74% amino acid sequence identity and play certain roles in pollen germination and male fertility [6]. *AtCNGC5*, *AtCNGC6*, and *AtCNGC9* are required for the structural growth of root hairs in *A. thaliana* [7]. *OsCNGC9* is a positive regulator of immune response (PTI), and *OsCNGC9* overexpression in rice transgenic lines showed significant enhancement of path-ogen-associated molecular patterns (PAMPs)-triggered immune response (PTI) and blast resistance include stronger calcium ion flux, higher burst of reactive oxygen species, and PTI-related gene expression levels [8], whereas in *A. thaliana*,

AtCNGC2 and *AtCNGC4* are positive regulators of PTI only at specific calcium concentrations (i.e., 1.5 mmol-L-1) [9]. In apple, *MdCNGC1* is probably a negative regulator of plant resistance to bacterial and fungal pathogens [10]; and in tomato, *SlCNGC1* and *SlCNGC14* inhibit rice stripe disease [11]. Moreover, overexpression of *OsCNGC9* was also found to significantly enhance rice resistance to low-temperature stress [12]. *OsCNGC14* and *OsCNGC16* positively regulate both heat and cold tolerance in rice, and their loss of function reduces or eliminates cytoplasmic calcium signaling induced by high or low-temperature stresses [13]. In Chinese jujube, a *ZjCNGC2*-mediated *ZjMAPK* cascade is involved in cold stress response [14]. In addition, *OsCNGC4, OsCNGC5*, and *OsCNGC8* may be associated with pollen development [4]. However, it remains to be explored how to target these CNGC genes for the improvement of crop yield and sustainability. Grain shape has an important impact on rice yield and its appearance, processing, cooking, and eating quality, thereby directly affecting the commercial value of rice. In addition, thousand-grain weight (TGW), a genetically stable trait, is another important factor affecting rice yield [8]. The factors that determine the TWG include grain length, grain width, and grain thickness [9]. Globally, different countries and regions have different preferences for rice quality traits. For example, people in South and Southeast Asia, southern China, the USA, and Latin America prefer long, fine grains with a fluffy and firm texture and medium amylose content. However, people in northern China, Korea, Japan, and parts of the Mediterranean area prefer short, round, soft, and sticky rice grains with a low amylose content [10,11].

Rice is one of the most important cereal crops in the world as well as a cereal model plant with high intraspecific genetic diversity [15,16]. In rice, more than 4600 genes have been cloned and functionally characterized at the molecular level (https://funricegenes.github.io/ (accessed on 26 September 2023)) [17]. The primary gene pool was found to have extremely high genomic diversity by sequencing 3010 different rice materials (3KRG) from 89 countries worldwide [18,19]. However, the rapid progress in functional and population genomic studies in rice has not yet been widely applied to the development of more efficient breeding techniques. This is due to the fact that there is incomplete information on the phenotypic effect(s) of cloned genes because most experiments were conducted in the laboratory instead of breeding target environments, and therefore, the effects of environmentally interacting genotypes and genetic backgrounds of most cloned genes on some important agronomic traits are largely uncertain [17]. Moreover, before spending much effort and funds to study a specific gene of interest, researchers should check whether similar alleles have been identified or fixed in commercial crop varieties; and if a gene has been studied for decades, it is unlikely to contribute to a sudden significant increase in yield [20]. Finally, due to the abundance of natural allelic variations at most loci in rice populations [19], it remains a great challenge to identify and mine desirable alleles from rice germplasm resources to improve specific target traits. Among *OsCNGC* genes, the functions of many genes still remain unknown. Therefore, it is time-consuming to understand the function of these genes by gene cloning, and a great challenge is to obtain the information about these genes needed for breeding without gene cloning.

In this study, we integrated a series of approaches including gene family identification and gene expression analysis with population genetic analysis of geng-cds-haplotype (gcHap) diversity in rice populations, developing an efficient strategy to provide important information on a large number of gene loci needed for future development of new breeding techniques of rice.

2. Results

2.1. Genome-Wide Identification and Characterization of OsCNGC Genes

OsCNGC genes in the whole rice genome were identified using BLAST, HMM search, and literature analysis [3,4]. Finally, a total of 16 *OsCNGC* genes were identified in the genome. To characterize these *OsCNGC* genes, we further identified their physical and chemical features, including their protein length, molecular weight, instability coefficient, lipolysis index, hydrophilicity index, and subcellular localization. The 16 genes showed protein sequence lengths ranging from 243 to 773 amino acids, molecular weights ranging from 27,613.65 to 88,353.96 Da, and PI (theoretical) ranging from 8.02 to 10.04 with an average of 9.36 (Table S1). Subcellular localization revealed that all 16 *OsCNGC* genes were localized to the plasma membrane or chloroplasts (Table S1). The *OsCNGC* genes showed different physicochemical properties, indicating their different biological functions. Moreover, the 16 genes were distributed on all chromosomes except for chromosomes 7, 8, 10, and 11 (Figure 1a). To further reveal the evolutionary and structural characteristics of *OsCNGC* genes, we constructed a phylogenetic tree with 16 CNGC genes in rice (japonica), 20 CNGC genes in *A. thaliana*, 12 CNGC genes in maize, and 19 CNGC genes in Maoyamoya fruit, which could be classified into groups I, II, III, and IV according to the distance of the evolutionary relationship, among which group IV could be subdivided into IV-A and IV-B subgroups (Figure 1b).

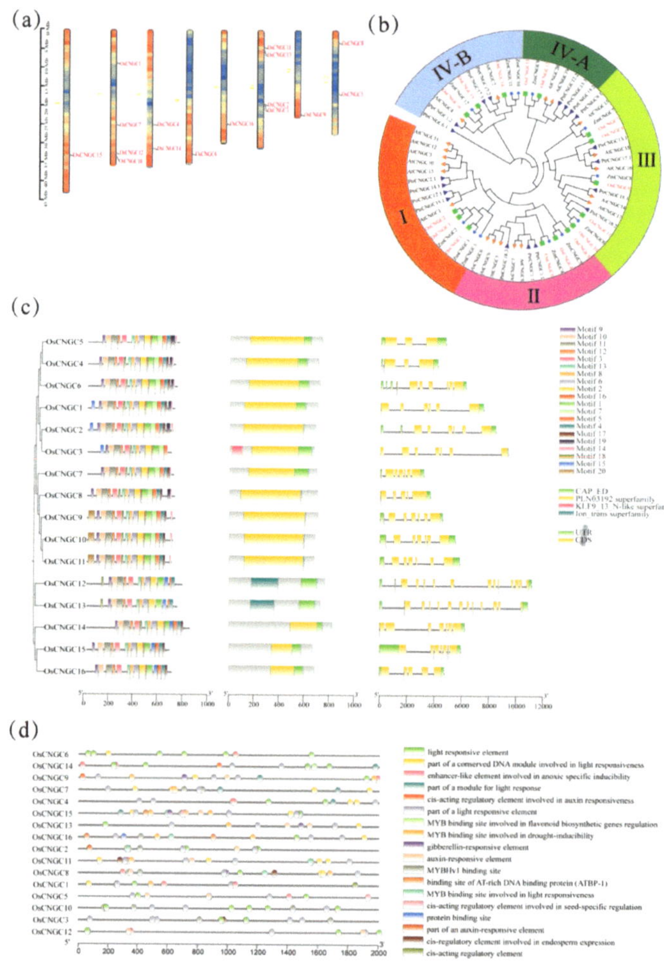

Figure 1. Characterization of *OsCNGC* genes. (**a**) Chromosomal localization of *OsCNGC* genes; (**b**) Phylogenetic tree of *OsCNGCs* in rice, Arabidopsis, maize, and Populus trichocarpa; (**c**) Phylogenetic tree, motif prediction, structural domains, and exon-intron structural distributions of *Os-CNGC* genes, from the left to the right, respectively; and (**d**) cis-acting elements of *OsCNGC* genes.

OsCNGCs had 14 to 19 motifs, and all contained CAP_ED structural domains, while the structural domain KLF9_13_N-like superfamily was only found in *OsCNGC3* (Figure 1c). The structure of *OsCNGCs* was analyzed, and the number of UTRs and CDS was identified for each gene. The CDS number ranged from 3 to 13, and *OsCNGC12* and *OsCNGC13* had more CDSs and a unique structural domain (Figure 1a).

In terms of cis-acting elements, most *OsCNGC* family members were related to light response. They also contained elements related to stress response, such as *OsCNGC3*, *OsCNGC7*, *OsCNGC8*, *OsCNGC9*, *OsCNGC13*, *OsCNGC14*, *OsCNGC15*, and *OsCNGC16*, which had drought-induced and -related MYB-binding elements (Figure 1d).

2.2. Ka/Ks Ratio Analysis of OsCNGC Homologous Genes

Covariance analysis of the *OsCNGC* genes showed three pairs of genes, including *OsCNGC2* and *OsCNGC4*, *OsCNGC12* and *OsCNGC13*, and *OsCNGC15* and OsCNGC16, (Figure 2), indicating their homology to each other, and each pair was formed due to gene duplication events. To elucidate the effect of selection pressure on the evolution of the *OsCNGC* family, Ka values, Ks values, and Ka/Ks ratio of homologous genes in the *OsCNGC* family were calculated. The results showed that the Ka/Ks ratios of the three pairs of homologous genes were all much lower than 1 (Table 1), indicating that the *OsCNGC* family has experienced strong purifying selection during evolution [21].

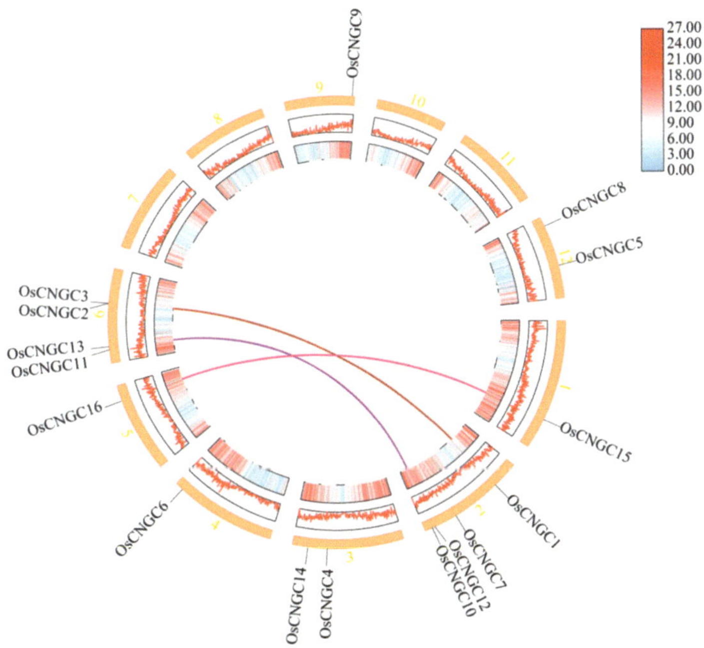

Figure 2. Homologous relationship and chromosomal localization of *OsCNGC* genes. The connected lines represent a homologous relationship.

Table 1. Ka/Ks ratio of *OsCNGC* homologous genes.

Gene 1	Gene 2	Ka	Ks	Ka/Ks
OsCNGC2	*OsCNGC4*	0.14	0.9	0.16
OsCNGC12	*OsCNGC13*	0.14	0.82	0.17
OsCNGC15	*OsCNGC16*	0.14	0.39	0.35

2.3. Functional Interaction Network Analysis of OsCNGC Genes

Functional interaction network prediction of *OsCNGC* genes showed that they had the same interacting proteins, including TPC1 as biportal calcium channel protein 1, and most of the remaining proteins are members of the protein kinase superfamily (Figure 3).

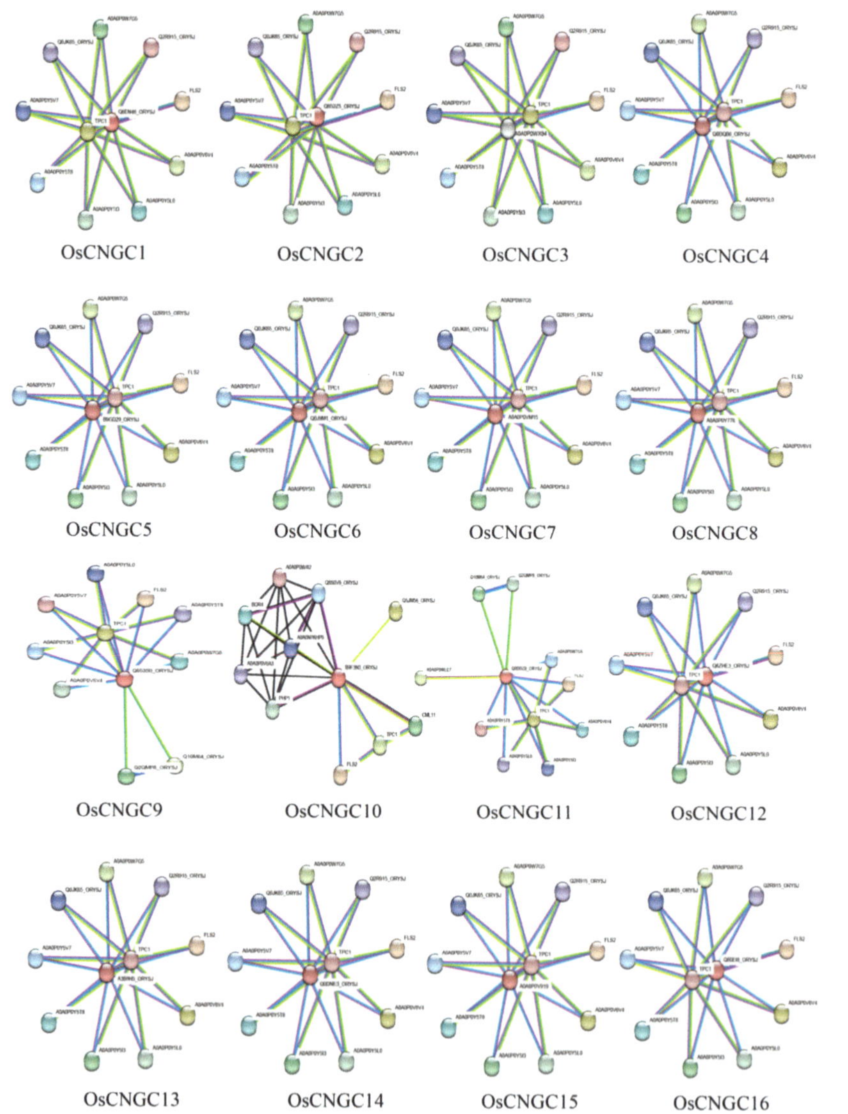

Figure 3. Functional interaction network of the 16 *OsCNGC* genes. The network summarizes the predicted association network for a set of specific proteins. Network nodes are proteins. Edges indicate predicted functional associations. The thickness of the line indicates the confidence prediction of the interaction. Red proteins in the center represent rice *OsCNGC* genes or their homologs. The red line indicates the presence of fusion evidence; the green line: neighborhood evidence; the blue line: co-expression evidence; the purple line: experimental evidence; the yellow line: text mining evidence; the light blue line: database evidence; the black line: co-expression evidence.

2.4. Expression Profiles of OsCNGC Genes in Tissues under Normal Conditions and Abiotic Stresses

OsCNGC gene expression in tissues under normal conditions was analyzed. *OsCNGC4*, *OsCNGC5*, and *OsCNGC8* showed high expression in flowers; *OsCNGC7*, *OsCNGC1*, and *OsCNGC13* were highly expressed in leaves; and *OsCNGC10* and *Os-CNGC9* showed high expression in roots and stems, respectively (Figure 4a). Detailed data are shown in (Figure S1). These results suggest that tissue specificity is an important aspect of functional differentiation for *OsCNGC* genes, and the mechanism for the differential expression patterns of most *OsCNGC* genes remains unclear. Under different abiotic stresses, *OsCNGC12*, *OsCNGC11*, and *OsCNGC15* showed high expression under drought, high-temperature stress, and low-temperature stress, respectively (Figure 4b), indicating that these genes play an important role in abiotic stress response.

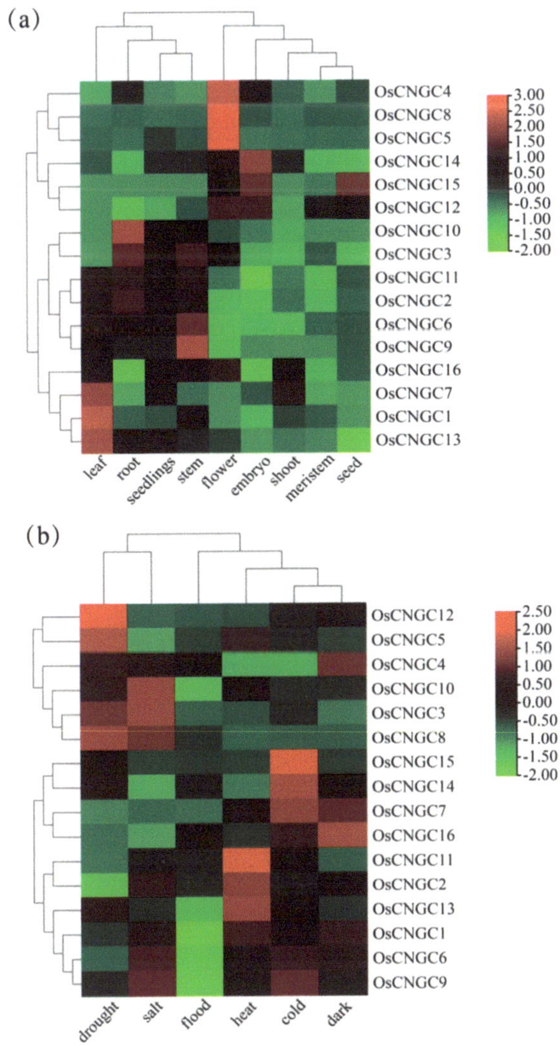

Figure 4. Expression analysis of *OsCNGC* genes. (**a**) *OsCNGC* gene expression in different tissues including the leaf, root, seedlings, stem, flower, embryo, shoot, meristem, and seed. (**b**) Expression of *OsCNGC* genes under abiotic stresses including drought, salt, flood, heat, cold, and dark.

2.5. Genetic and Allelic Diversity of OsCNCG Loci in Rice Populations

From the CDS haplotype (gcHap) data from the Rice Genome Project (3KRG), we obtained the Shannon fairness EH (Table S3). The average gcHaps, main gcHaps, and EH of the 16 *OsCNGC* genes were 134.4, 8.4, and 0.222, respectively. However, there were large variations in genetic diversity among different *OsCNGC* genes. *OsCNGC15* had an EH of 0.491 and 484 gcHaps (Table S3), followed by *OsCNGC5* with an EH of 0.438 and 569 gcHaps. *OsCNGC6* had the lowest EH of 0.021 and eight gcHaps (Table S3). The CNGC gene diversity also varied considerably across rice populations. The mean EH of the 16 *OsCNGCs* was 0.187 in the Xian population, 0.155 in the Geng population, 0.206 in the Aus population, and 0.267 in the Bas population. The average and major gcHaps were 81.7 and 5.5, 39.5 and 4.4, 6.8 and 4.3, and 12.3 and 4.1 in Xian, Geng, Aus, and Bas populations, respectively (Table S3), indicating that the frequency of gcHaps in *OsCNGC* genes is low and rare in most populations.

To understand the variations of *OsCNGC* genes among major rice populations, we estimated the genetic identity of Nei (INei) and gcHap data for the 16 polymorphic *OsCNGC* genes among all pairs of groups. *OsCNGC2*, 3, 4, 5, 8, 12, 15 and 16 showed strong XI-GJ differentiation (INei ≤ 0.2) specificity (Figure 5, Table S4). In addition, the 16 *OsCNGC* genes showed strong differentiation in other populations. For example, *OsCNGC2* also showed strong differentiation in *Aus-XI*, *Aus-GJ*, *Aus-Bas*, and *GJ-Bas* (Figure 5, Table S4). It can be inferred that allelic variation at the *OsCNGC* loci contributes significantly to the differentiation of major rice populations and the adaptation of different populations to the environment.

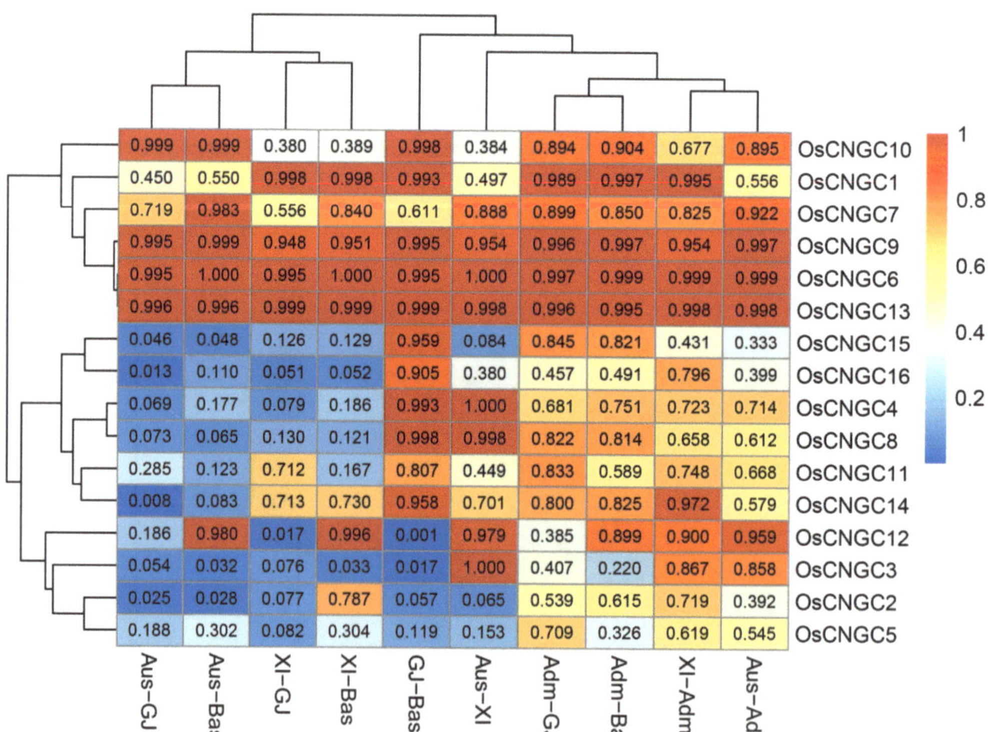

Figure 5. Nei's genetic identity (INei) of *OsCNGC* genes between all paired populations calculated from gcHap data.

2.6. Impact of Modern Breeding on gcHap Diversity of OsCNGC Genes

To understand the impact of modern breeding on the gcHap diversity of *OsCNGC* genes in recent decades, we conducted a study of modern varieties (MVs) and local varieties (LANs), including 732 Xian local varieties (LANs-Xian), 358 Xian modern varieties (MVs-Xian), 328 Geng local varieties (LANs-Geng), and 139 modern varieties of Geng (MVs-Geng) (Table S5). In the Xian population, the mean EH of the 16 *OsCNGC* genes in MVs-Xian was 0.249, which was increased by 14.1% compared with the 0.188 of LANs-Xian (Table 2). Nine *OsCNGC* genes (*OsCNGC3, 5, 7, 8, 9, 11, 14, 15,* and *16*) showed increased diversity in MVs-Xian genes (Table S5). Interestingly, MVs-Xian had an average of 31.6 gcHaps per locus, which was lower than 40.5 gcHaps per locus of LANs-Xian. A closer inspection revealed that on average, 26 gcHaps/per locus were missed in MVs-Xian relative to that in LANs-Xian (Table 2, Figure 6a), which is apparently caused by genetic bottlenecks during modern breeding (Table 2, Figure 6a). To understand the variations of *OsCNGC* genes among major rice populations, we estimated the genetic identity of Nei (INei) and gcHap data for the 16 polymorphic *Os-CNGC* genes among all pairs of groups. *OsCNGC2, 3, 4, 5, 8, 12, 15* and *16* showed strong *XI-GJ* differentiation (INei < 0.35) specificity (Figure 5, Table S4). In addition, the 16 *OsCNGC* genes showed strong differentiation in other populations. For example, *OsCNGC2* also showed strong differentiation in *Aus-XI, Aus-GJ, Aus-Bas,* and *GJ-Bas* (Figure 5, Table S4). It can be inferred that allelic variation at the OsCNGC loci contributes significantly to the differentiation of major rice populations and the adaptation of different populations to the environment.

Table 2. Comparison of genetic diversity of the 16 *OsCNGC* genes in Geng between local and modern varieties.

Gene Name	Xian (Indica)						
	LANs		MVs		Change of gcHapN (Major gcHapN)		
	E_H	gcHapN (Major gcHapN)	E_H	gcHapN (Major gcHapN)	Lost	New	Retained
OsCNGC1	0.13	16 (3)	0.168	14 (5)	7	5	9
OsCNGC2	0.196	39 (4)	0.248	28 (5)	26	15	13
OsCNGC3	0.153	25 (3)	0.215	18 (5)	12	5	13
OsCNGC4	0.111	27 (4)	0.101	21 (2)	11	5	23
OsCNGC5	0.38	143 (8)	0.511	93 (9)	123	73	20
OsCNGC6	0.003	3 (1)	0.004	2 (1)	2	1	1
OsCNGC7	0.307	47 (9)	0.391	42 (9)	23	18	24
OsCNGC8	0.07	4 (3)	0.12	4 (4)	0	0	4
OsCNGC9	0.21	39 (5)	0.301	33 (8)	21	15	18
OsCNGC10	0.143	15 (3)	0.183	13 (5)	4	2	11
OsCNGC11	0.157	13 (5)	0.226	13 (6)	3	3	10
OsCNGC12	0.056	18 (3)	0.082	11 (3)	9	2	9
OsCNGC13	0.056	14 (2)	0.061	10 (3)	9	5	5
OsCNGC14	0.26	50 (8)	0.354	48 (7)	31	29	19
OsCNGC15	0.502	150 (13)	0.639	111 (16)	109	70	41
OsCNGC16	0.271	45 (7)	0.385	44 (8)	26	25	19
Mean	0.188	40.5(5.1)	0.249	31.6 (6)	26	17.1	14.9

Note: EH and gcHapN (major gcHapN) are the fairness of the Shannon and the number of gcHaps (≥1% of varieties) identified, respectively.

In addition, the proportion of newly emerged gcHaps observed in modern varieties was significantly lower than that of lost gcHaps (Figure 6b,c). MVs-Xian gained an average of 17.1 new gcHaps per locus absent in LANs-Xian. The new gcHaps were evidently generated by intragenic recombination during breeding. The increase in diversity and decrease in gcHap number observed at many loci suggest that the Xian population underwent significant frequency changes in the major gcHaps during the breeding process. In fact, significant changes in the frequency of major gcHaps, F(P) were observed at nine out of the sixteen *OsCNGC* loci, including a significant decrease at eight *Os-CNGC* loci and a significant increase at one *OsCNGC* locus (Table S6).

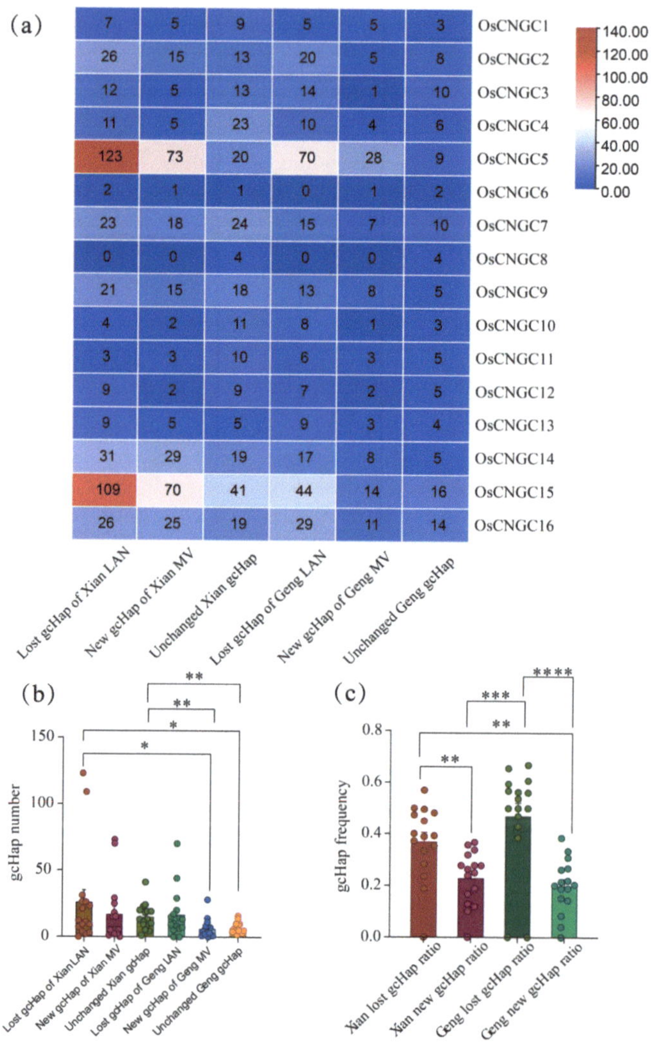

Figure 6. Lost, emerging, and unchanged gcHaps of the 16 *OsCNGC* genes during modern breeding. (**a**) Heat map of the number of lost, emerging, and unchanged gcHaps. (**b**) Comparative analysis of the number of lost, emerging, and unchanged gcHaps. (**c**) Frequency comparison of lost, emerging, and unchanged gcHaps. Tukey's multiple comparison method was used. **** $p < 0.0001$, *** $p < 0.001$, ** $p < 0.01$, * $p < 0.05$. Only significant differences are shown.

In the Geng population, the average EH of the 16 *OsCNGC* genes was 0.19 in MVs-Geng and 0.17 in LANs-Geng (Table 3). However, significant increases in diversity were observed only at two *OsCNGC* loci (*OsCNGC4* and *OsCNGC9*) in MVs-Geng (Table S5). MVs-Geng had a lower mean value (13.1 gcHaps per locus) than that of LANs-Geng (23.5 gcHaps per locus) (Table 3). In fact, on average, 6.3 new gcHaps per locus were newly gained, and 16.7 gcHaps per locus were lost in MVs-Geng relative to those in LANs-Xian (Table 3). These new gcHaps were apparently generated by intragenic recombination during breeding. Significant shifts in F(P) were observed at 6 out of the 16 *OsCNGC* loci, including significant decreases in F(P) at *OsCNGC4*, *OsCNGC7*, *OsCNGC9*, and *OsCNGC15* and significant increases in F(P) at *OsCNGC6* and *OsCNGC8* (Table S6). Notably, the same major gcHaps were found on all OsCNGC genes except for *OsCNGC2* and *OsCNGC16*, suggesting that the major gcHaps function simultaneously in Xian and Geng.

Table 3. Comparison of genetic diversity of the 16 *OsCNGC* genes in Geng between local and modern varieties.

Gene Name	Geng (Japonica)						
	LANs		MVs		Change of gcHapN (Major gcHapN)		
	E_H	gcHapN (Major gcHapN)	E_H	gcHapN (Major gcHapN)	Lost	New	Retained
OsCNGC1	0.102	8(3)	0.129	8(3)	5	5	3
OsCNGC2	0.171	28(6)	0.189	13(8)	20	5	8
OsCNGC3	0.174	24(7)	0.186	11(7)	14	1	10
OsCNGC4	0.07	16(3)	0.13	10(5)	10	4	6
OsCNGC5	0.436	79(9)	0.435	37(7)	70	28	9
OsCNGC6	0.045	2(2)	0.03	3(2)	0	1	2
OsCNGC7	0.271	25(6)	0.325	17(6)	15	7	10
OsCNGC8	0.084	4(4)	0.055	4(3)	0	0	4
OsCNGC9	0.09	18(3)	0.156	13(2)	13	8	5
OsCNGC10	0.068	11(2)	0.045	4(3)	8	1	3
OsCNGC11	0.152	11(3)	0.207	8(6)	6	3	5
OsCNGC12	0.046	12(1)	0.071	7(3)	7	2	5
OsCNGC13	0.063	13(3)	0.058	7(2)	9	3	4
OsCNGC14	0.177	22(4)	0.17	13(2)	17	8	5
OsCNGC15	0.346	60(11)	0.417	30(9)	44	14	16
OsCNGC16	0.421	43(11)	0.434	25(10)	29	11	14
Mean	0.170	23.5 (4.9)	0.19	13.1 (4.9)	16.7	6.3	6.8

Note: EH and gcHapN (major gcHapN) are the fairness of the Shannon and the number of gcHaps (≥1% of varieties) identified, respectively.

2.7. Association Analysis of Major gcHaps of OsCNGC Genes with Important Agronomic Traits

To further demonstrate the functional importance of the 16 CNGC genes, we constructed a gcHap network of the primary alleles of the 16 genes in 5 rice populations and analyzed their associations with four agronomic traits, including number of spikes per plant (PN), spike length (PL), plant height (PH), and thousand-grain weight (TGW) (Figures 7 and S2–S4). Strong ($p < 10^{-7}$) associations were observed in 39 (60.9%) out of 64 (16 × 4) instances, and many of the major alleles of *OsCNGC* genes were strongly associated with more than one trait. We further analyzed four cloned *OsCNGC* genes, among which *OsCNGC9* is a conserved gene with six major gcHaps. Hap4 had the highest frequency in LANs-Xian (Figure 7). Hap2 differed from Hap4 by two non-synonymous mutations. Hap2 has become the major gcHap in LANs-Geng and has been significantly reduced by selection during breeding, which significantly reduced the PN. *OsCNGC13* was conserved and had two major gcHaps. Hap2 had eight non-synonymous mutations from Hap1, and selection during breeding reduced the PN, PL, PH, and TGW. *OsCNGC14* belonged to "other genes" and had six major gcHaps. Hap2 had the highest frequency in LANs-Xian, and Hap4 had the highest frequency in LANs-Geng. Hap1 showed a decline relative to Hap4 for all three traits except for PL (no change), while Hap2 showed an increase in PN. *OsCNGC16* was a divergent gene between Xian and Geng and had a staggering number of major haplotypes (11). Figure 7 clearly shows that significant differentiation occurred after four non-synonymous mutations between Hap10 and Hap4. Among them, Hap6 showed a clear dominance in PN. Hap2 and Hap9 had the highest frequency in LANs-Xian and LANs-Geng, respectively. Hap2 showed an increase only in PN compared with Hap9. In addition, the proportion of newly emerged gcHaps observed in modern varieties was significantly lower than that of lost gcHaps (Figure 6b,c). MVs-Xian gained an average of 17.1 new gcHaps per locus absent in LANs-Xian. The new gcHaps were evidently generated by intragenic recombination during breeding. The increase in diversity and decrease in gcHap number observed at many loci suggest that the Xian population underwent significant frequency changes in the major gcHaps during the breeding process.

In fact, significant changes in the frequency of major gcHaps, F(P) were observed at nine out of the sixteen *OsCNGC* loci, including a significant decrease at eight *OsCNGC* loci and a significant increase at one *OsCNGC* locus (Table S6).

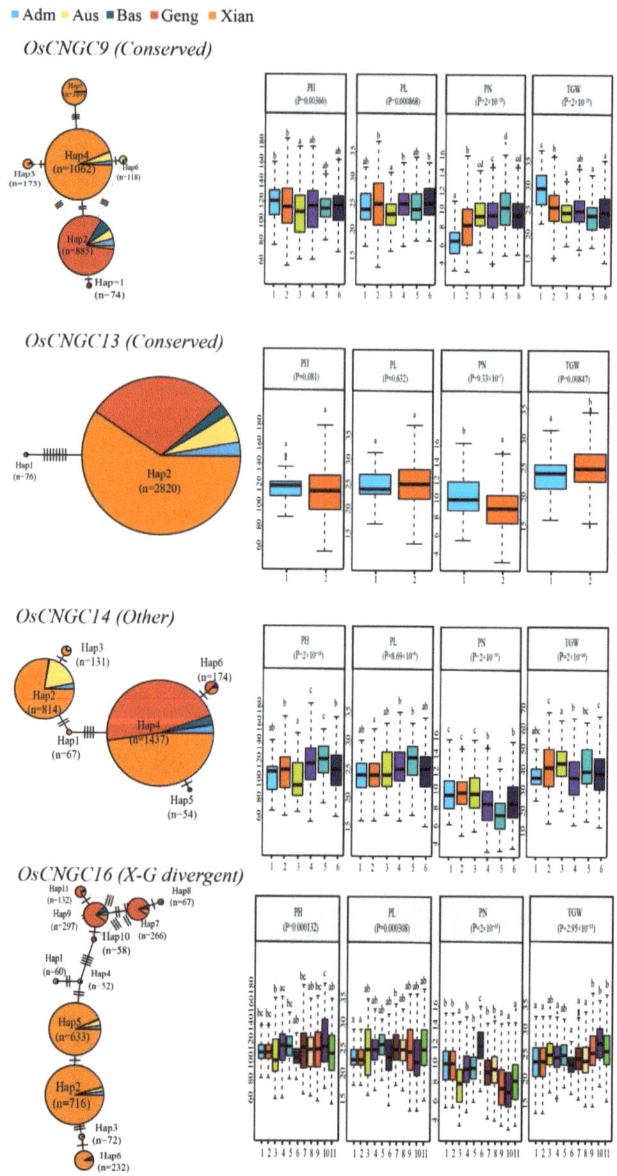

Figure 7. Haplotype networks of the four cloned CNGC genes and their association traits with four agronomic traits in 3KRG. *p*-Values indicate differences among haplotypes assessed by two-factor ANOVA, where different letters on the box-and-line plots indicate statistically significant differences based on the Duncan's Multiple Range Test at $p < 0.05$. The bars on the right show the frequency differences in dominant gcHaps between local varieties (LANs) and modern varieties (MVs) in Xian and Geng. The chi-square test was used to determine significant differences in the proportions of the same gcHap between groups.

2.8. Comparison of Trait Values for OsCNGC Gene Dominance and Unfavorable gcHaps in Rice

The major gcHaps (with the highest frequency) of the *OsCNGC* loci in a rice population are thought to have been favored by natural selection during evolution. On the contrary, the major gcHaps with the lowest frequency in a population are likely unfavorable gcHaps [9]. We compared the phenotypic differences between major gcHaps and unfavorable gcHaps at each of the 16 CNGC loci for 15 agronomic traits. Out of the 240 (16 × 15) comparisons, significant differences between major gcHaps and unfavorable gcHaps at all *OsCNGC* loci were detected in 100 (41.7%) comparisons (Figures S5–S20; Table S7). TGW was detected at 10 (62.5%) *OsCNGC* loci, and it was hypothesized that these genes might have a greater effect on grain weight (Table S7). These ten genes, except for *OsCNGC8*, also had certain expression in seeds under normal conditions (Figure S1). Among the 16 *OsCNGC* loci, significant differences in mean trait values between major gcHaps and unfavorable gcHaps were detected for *OsCNGC1* (8 traits), *OsCNGC4* (8 traits), *OsCNGC5* (8 traits), *OsCNGC6* (9 traits), and *OsCNGC9* (8 traits), and for more than half of the traits, significant differences were observed in the mean values between major gcHaps and unfavorable gcHaps. Interestingly, *OsCNGC4*, *OsCNGC5*, and *OsCNGC6* were all located on branch II of the phylogenetic tree (Figure 1b), suggesting that these five *OsCNGC* genes probably affect more traits than other *OsCNGC* loci (Table S7).

2.9. Mining of Favorable Alleles in OsCNGC Loci for Yield Improvement

We found significant differences between major gcHaps in their effects on important agronomic traits at most rice loci. Therefore, we hypothesized that favorable gcHap(s) for traits to increase productivity are present at most *OsCNGC* loci, even though the favorable gcHap(s) may differ in different target environments. Therefore, there should be one or more favorable alleles defined as gcHap(s) associated with high productivity traits. The frequency of these most favorable gcHaps in MVs, and different rice populations would be of particular interest to rice breeders. Figure 8 shows the frequencies of favorable gcHaps for five traits, including PL, CN, GL, GW, and TGW, in the cloned *OsCNGC* genes in MVs-Xian and MVs-Geng as well as 3 KRG rice populations. The frequencies of favorable alleles varied greatly across different yield traits, *OsCNGC* loci, MVs-Xian and MVs-Geng, and rice populations. For example, Hap7 and Hap5 at *OsCNGC16* were associated with the highest GW and TGW, respectively, in the 3KRG germplasm, which was fixed in MVs-Xian but present in low frequency in MVs-Geng. This is consistent with the empirical observation in rice breeding that very high values for any particular yield component do not necessarily lead to the highest productivity, as yield-related traits are quantitative, not qualitative. In addition, for specific traits, new potential alleles can be identified from natural germplasm from natural habitat-oriented species or species grown in unfavorable or isolated environments, and wild infiltration can also be helpful. This result also suggests that the favorable alleles for different yield traits may be very different in the two rice subspecies or genetic backgrounds and may also differ in different environments. We also found that *OsCNGC16*-favorable gcHaps were absent in MVs-Geng for PL and PN, which were present in all four traits. These results suggested that some favorable gcHaps function only in specific subpopulations. Therefore, there is a great potential to improve yield traits and productivity of MVs by aggregating the missing favorable gcHaps.

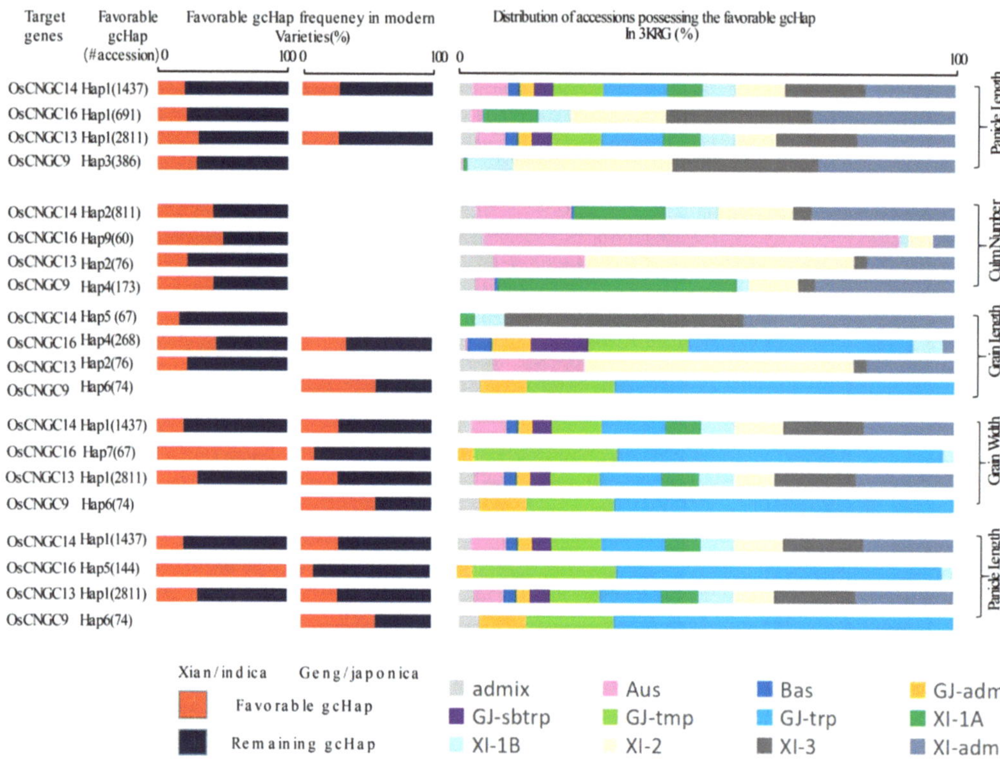

Figure 8. Frequency of favorable gcHaps of four cloned *OsCNGC* genes affecting five yield traits (GWT, GL, GW, PL, and CN) in modern varieties of Xian/indica (XI) and Geng/japonica (*GJ*) and different rice subgroups. "#accession" indicates the number of accessions that possess the favorable gcHap. Five subpopulations of XI (XI-1A, XI-1B, *XI-2*, *XI-3*, and *XI-adm*) and four subpopulations of *GJ* (temperate *GJ*: *GJ-tmp*, *subtropical GJ*: *GJ-sbtrp*, *tropical GJ*: *GJ-trp*, and *GJ-adm*) [18].

3. Discussion

With reference to previous studies [3,4], sixteen *OsCNGC* genes were identified in this study, which are distributed on eight chromosomes except for chromosomes 7, 8, 10, and 11. Evolutionary analysis classified these genes into four major groups (I, II, III, and IV) and two subgroups (IV-A and IV-B). Gene structure analysis showed that the number of introns in *OsCNGCs* ranged from 3 to 13, and the number of exons varied considerably among *OsCNGC* genes. Structural domain analysis showed that all 16 *OsCNGC* genes contained the CAP_ED structural domain. *Os06g0527300*, which was considered to contain no CAP_ED structural domain [3], was retained because this domain is present in this gene as indicated by our conserved domain analysis (Figure 1c). Moreover, we identified no *LOC_Os06g33610* gene. Hence, we excluded it and renamed Os06g0527300 as *OsCNGC3*. This result is consistent with the recently published gene family visualization results [22]. We hypothesize that the reason for this phenomenon may be the different sources of the reference genome. According to the cis-acting element analysis, the *OsCNGC* genes play important roles in normal plant development and response to adverse stress, among which *OsCNGC9*, *OsCNGC13*, *OsCNGC14*, and *OsCNGC16* have been validated [5,8,12,13]. Functional interaction network analysis showed that all *OsCNGCs* contain TPC1. AtTPC1 is the first TPC channel cloned from plants, which is localized to the vesicular membrane and responsible for generating slow vesicular (SV) currents. AtTPC1 is a non-selective cation channel that can permeate through a wide range of monovalent cations and Ca2+ and potentially plays an important role in regulating the cytosol ion concentration [23]. Protein kinases are involved in stress response in plants [24]. Considering the high expression of some *OsCNGCs* in tissues under normal conditions and stress, we hypothesized that these *OsCNGCs* may play a key role in the corresponding phenotype and stress response (Figure 4).

OsCNGC genes may have played important roles in rice subspecificity and population differentiation during rice evolution, and there are strong subspecificity and population differentiation at most rice *OsCNGC* loci (Figure 5, Table S4). Many *OsCNGC* genes may play an important role in rice improvement by significantly changing the diversity of major gcHaps at *OsCNGC* loci (Figures 5 and 6, Tables 2 and 3), as well as the agronomic traits associated with major gcHaps (Figures 7 and 8, Table S7). Because *OsCNGC* genes act as signaling receptors, they are expected to play important regulatory roles in many processes of rice growth and development and response to environmental cues. However, how different *OsCNGC* genes are involved in these important processes at the molecular level remains to be further elucidated.

Unexpectedly, we observed that most *OsCNGC* loci have increased diversity in MVs. However, due to manual selection, it may result in a lower MV. For example, in both selective breeding and breeding programs, selections are mostly associated with high productivity, including more compact plant types, larger or more panicles, appropriate grain weight or more tillers, and greater tolerance to biotic and abiotic stresses. In addition, genetic bottlenecks due to the relative fixity of germplasm resources used for breeding have reduced genetic diversity at most loci. Indeed, we observed an average loss of 26 (45%) gcHaps at the 16 *OsCNGC* loci during Xian breeding, while an average of 17.1 (29%) new gcHap loci appeared in MVs-Xian (Figure 6). We also found no increase or deletion of gcHaps related to *OsCNGC8* in both Xian and Geng, and it remains to be investigated whether this is an allele fixed in the variety. During the breeding in Geng, the average loss of gcHaps at the 16 CNGC loci was larger, which reached about 56%. At the same time, an average of 6.3 new gcHaps per locus (21%) appeared in MVs-Geng. Thus, the loss of gcHaps due to the genetic bottleneck effect is more significant in MVs-Geng than in MVs-Xian during modern breeding (Tables 2 and 3). We hypothesize that this is because interline crosses are generally performed in the breeding process, which greatly increases the recombination of genes relative to the low heterosis rate of local varieties. We found that the frequency of major gcHaps in MVs-Xian and MVs-Geng was significantly lower at specific *OsCNGC* loci, suggesting that the gcHaps favored by natural selection were not artificially selected by modern breeding. The reason for this phenomenon may be that the modern breeding process is mainly targeted at yield and some alleles for adaption to natural environmental changes are abandoned; however, it will be costly to mine and use these environment-adapting genes. This also means that many rare gcHaps may be of great value in rice improvement. In summary, previous modern breeding activities have had a greater impact on *OsCNGC* loci than previously estimated [11], suggesting that *OsCNGC* genes are important in rice improvement.

In conclusion, our results suggest that natural variations at most *OsCNGC* loci have potential value for improving the productivity and tolerance of rice to abiotic stresses. All *OsCNGC* genes have potential value for rice trait improvement. Among them, *OsCNGC9*, *OsCNGC13*, *OsCNGC14* and *OsCNGC16* have been cloned [5–7]. Also, *OsCNGC11*, *OsCNGC12*, and *OsCNGC15* may be used to improve rice tolerance to abiotic stresses, while other genes may be used to improve rice yield traits. However, since most of the *OsCNGC* genes identified in this study had significant interactive effects on yield traits, it remains a challenge to identify favorable alleles at specific *OsCNGC* loci and compare the dominant and unfavorable allelic traits to improve specific yield traits of rice. Therefore, more efforts are needed to obtain the required information, and it is expected that some alleles that are already present or fixed in certain varieties will be identified for future application to improve rice productivity.

4. Materials and Methods

4.1. Identification and Physicochemical Characterization of OsCNGC Genes

To identify all CNGC genes in the whole genome of rice, we first downloaded the whole genome data of rice (Oryza sativa Geng/japonica), the protein sequences of Arabidopsis CNGCs in the database (https://www.ncbi.nlm.nih.gov/, (accessed on 26 September 2023)), and the Hidden Markov Models of CNGCs in the pfam (http://pfam.xfam.org/ (accessed on 26 September 2023)). We searched (Perl script) for NBD (Cyclic Nucleotide Binding Domain, pfamID:PF00027) and TM/ITP (Transmembrane or Ion Transport Domain, pfamID:PF00520) using Hm [25], which were compared in Tbtools to take the intersection, and genes containing these two structural domains were considered as members of the CNGC candidate gene family [26], and were published in the pfam (http://pfam.xfam.org/ (accessed on 26 September 2023)), NCBI (http://www.ncbi.nlm.nih.gov/cdd/ (accessed on 26 September 2023)), and SMART website (http:smartbl de/(accessed on 26 September 2023)), as well as the rice database (https://ricedata.cn/gene/ (accessed on 26 September 2023)) to validate the candidate genes of *OsCNGCs*. Moreover, by analyzing the CNGC-related literature for naming [3,4], we finally identified all the *OsCNGC* genes. Physicochemical properties such as molecular weight and isoelectric point of *OsCNGC* genes were predicted using the online tools ExPASy (https://www.expasy.org/ (accessed on 26 September 2023)) and Rice Database (https://ricedata.cn/gene/ (accessed on 26 September 2023)). Subcellular localization prediction was performed using the online tool Cell-PLoc 2.0 (http://www.csbio.sjtu.edu.cn/bioinf/euk-multi-2/ (accessed on 26 September 2023)).

4.2. Gene Structure, Conserved Motif, Conserved Structural Domain, and Cis-Acting Element Analysis of OsCNGC Proteins

After downloading the gene structure annotation files from the NCBI website, the gene structure view (advanced) function in TBtools-II v2.025 software was used for visualization. The conserved analysis of CNGC protein sequences was achieved using the TBtools-II v2.025 software Simple MEME Wrapper Function Implementation Function, which set 20 motifs. The number of occurrences of motifs on each sequence was unlimited, and the other parameters were the default parameters. The conserved structural domains were analyzed using the CD search function of NCBI and visualized using Tbtools [26]. Gtf/Gff3 sequence extracts from TBtools were utilized for extracting the upstream 2000 bp of CDS and extracting the promoter sequences of CNGC genes, which were submitted to the PlantCARE database (http://bioinformatics.psb.ugent.be/webtools/plantcare/html/ (accessed on 26 September 2023)) to analyze their promoter region cis-acting elements [27]. The resultant files were filtered based on the information in the table, retained for viewing, and then presented visually using the Simple BioSequence Viewer feature of TBtools-II v2.025 software.

4.3. Chromosomal Localization, Covariance Analysis, and Functional Interaction Network Analysis of OsCNGC Genes

Rice DNA files and annotation files were downloaded from the NCBI website. Interspecific covariates were mapped using the one-step MCScanX and advanced Circos functions of TBtools [26]. Gene positions were visualized from the GTF/GFF function to map the distribution of OsCNGC genes on chromosomes. Functional interaction network analysis was performed according to the instructions on the online website (https://www.stringdb.org/, (accessed on 26 September 2023)). STRING used a spring model to generate network images. Nodes were modeled as masses and edges as springs. The nodes were used to compute the image by minimizing the "energy" of the system.

4.4. Phylogenetic Tree Construction and Gene Expression Specificity Analysis of OsCNGC Genes

Phylogenetic trees of CNGC genes of Arabidopsis, rice, maize, and Populus trichocarpa were constructed using MEGA11 software (https://www.megasoftware.net/

accessed on 30 October 2023). The full-length sequences of the CNGC proteins were matched to each other and analyzed phylogenetically. The protein sequences required for the construction of the evolutionary tree are presented in Table S2. The phylogenetic tree was then embellished using the online mapping site iTOL (https://itol.embl.de/ (accessed on 26 September 2023)). The newly published gene family visualization website (https://bis.zju.edu.cn/cropgf/analysis/gene-info (accessed on 26 September 2023)) was used for heatmap visualization using Tbtools after downloading the expression data of *OsCNGC* genes in different tissues and under abiotic stresses [22,26].

4.5. gcHaps and Their Diversity in Modern and Local Varieties

Shannon's fairness (EH) was used to assess the level of gcHap diversity at all *OsCNGC* loci in a given rice population or in the species as a whole [28]. Nei's genetic identity (INei) was used to measure the genetic similarity between two populations or individuals based on their gcHap frequencies at different *OsCNGC* loci [29]. The formulas used to calculate EH and INei referred to the methodology described in a previous study [11], where *OsCNGC* genes were classified into conserved genes based on gcHap diversity and subspecies/population differentiation (EH < 0.3 in 3KRG and INei ≥ 0.8 between pairwise populations), and Xian-Geng differentiated genes (*X-G*) (EH < 0.3 in both Xian and Geng and INei ≤ 0.2 between Xian and Geng), and other genes [17]. To understand how modern breeding over the past decades has affected the gcHap diversity of *OsCNGC* genes, we collected detailed information on a total of 3010 3KRG rice materials. Of these, 732 were local varieties of Xian (indica) (LANs-Xian), 358 were local varieties of Geng (japonica) (LANs-Geng), 328 were modern varieties of Xian (indica) (MVs-Xian), and 139 were modern varieties of Geng (japonica) (MVs-Geng). First, we downloaded the major gcHaps (highest frequency in 3KRG) of each *OsCNGC* gene from RFGB and 3K webpage (https://www.rmbreeding.cn/Index (accessed on 26 September 2023)). The frequency difference of major gcHaps for each OsCNGC gene between modern varieties (MVs) and local varieties (LAN) was then calculated based on R-4.2.2 scripts [17]. Finally, the distribution of gcHaps in modern Xian and Geng varieties and their respective local varieties were compared. Missing and newly occurring gcHaps in modern Xian/Geng varieties were also analyzed. Finally, GraphPad Prism 8 software was used to plot the above data.

4.6. Determination of Major gcHap Phenotypes of OsCNGC Genes

First, we collected phenotypic data on 3010 Asian-cultivated rice germplasm for 15 agronomic traits. This study examined 15 agronomic traits including cays to heading (DTH, day), plant height (PH, cm), flag leaf length (FLL, cm), flag leaf width (FLW, cm), panicle number (PN, count), panicle length (PL, cm), culm number (CN, count), culm length (CL, cm), grain length (GL, mm), grain width (GW, mm), grain length/width ratio (GLWR, ratio), thousand-grain weight (TGW, g), leaf rolling index (LRI,%), seedling height (SH, cm), and ligule length(LL, mm). Phenotypic data for 15 rice traits were downloaded from the RFGB website (https://www.rmbreeding.cn/Index/ (accessed on 26 September 2023)). Next, the major gcHaps (with frequency $\geq$1% in 3KRG) of all rice CNGC genes were obtained using R scripts [17]. Finally, associations of major gcHaps with these agronomic traits in different 3010 rice germplasm were realized using R scripts. Significance was calculated using one-way ANOVA and compared between major gcHaps using Tukey's multiple comparisons. The layout of the images was conducted in Adobe Illustrator 2023 software, version 28.0.

4.7. Construction of gcHap Network of OsCNGC Genes and Association Analysis between Major gcHaps and Yield Traits in 3KRG Materials

First, haplotypes (gcHaps) of rice CNGC genes were constructed using the R package pegas [14]. A network of gcHaps for each *OsCNGC* gene was generated using a statistical parsimony algorithm, which was performed with a method that first connects the most closely related haplotypes by the smallest number of mutations [11]. More detailed steps are presented in a previous study [18]. The layout of the images was performed in Adobe Illustrator 2023 software, version 28.0.

Supplementary Materials: The following supporting information can be downloaded at https://www.mdpi.com/article/10.3390/plants12244089/s1, Figure S1: Expression of *OsCNGC* genes in tissues under normal conditions; Figures S2–S4: Haplotype network of the remaining 12 *OsCNGC* genes and their associations with four agronomic traits in 3KRG. *p*-values under the names of the traits indicate differences between haplotypes as assessed by a two-factor ANOVA, where different letters on the box and line plots indicate statistically significant differences based on Duncan's multiple range test at $p < 0.05$. The bars on the right side show the frequency differences of dominant gcHaps between local varieties (LANs) and modern varieties (MVs) in Xian and Geng. A chi-square test was used to determine significant differences in the proportions of the same gcHaps between groups **** $p < 0.0001$, *** $p < 0.001$, ** $p < 0.01$, * $p < 0.05$ and N.S., not significant; Figures S5–S20: Comparative analysis of 15 agronomic traits of dominant gcHaps, unfavorable gcHaps and major gcHaps of *OsCNGC* genes; Table S1: Basic information of rice CNGC gene family; Table S2: Protein sequences used to construct the evolutionary tree; Table S3: Genetic diversity of 16 CNGC genes in four major rice populations; Table S4: Genetic identity of Nei in five major populations calculated from gcHap data of rice *OsCNGC* genes; Table S5: Comparison of genetic diversity of 16 *OsCNGC* genes between local and modern varieties; Table S6: Frequency differences in major gcHaps of 16 CNGC genes between local and modern varieties; Table S7: Phenotypic comparison of major gcHaps and unfavorable gcHaps of 16 polymorphic *OsCNGC* genes in 3KRG germplasm.

Author Contributions: X.W. (Xinchen Wang), F.W., Y.B. and N.W. performed data collection and graphing; X.W. (Xinchen Wang), J.Z., X.W. (Xingmeng Wang), J.L., G.D. and D.M., data organization and iconography; Y.S. and X.W. (Xinchen Wang) wrote the paper; and Y.S. designed the experiments, provided instructions, and wrote and reviewed the paper. All authors have read and agreed to the published version of the manuscript.

Funding: This research was funded by the Scientific Research Plan Major Projects of Anhui Province (grant number 2022AH040126), the Improved Varieties Joint Research (Rice) Project of Anhui Province (the 14th five-year plan), the Science and Technology Major Project of Anhui Province (grant number 2021d06050002), and the National Natural Science Foundation of China [grant numbers U21A20214].

Data Availability Statement: Data is contained within the article and Supplementary Materials.

Conflicts of Interest: The authors declare no conflict of interest.

References

1. Beavo, J.A.; Brunton, L.L. Cyclic nucleotide research—Still expanding after half a century. *Nat. Rev. Mol. Cell Biol.* **2002**, *3*, 710–718. [CrossRef]
2. Schuurink, R.C.; Shartzer, S.F.; Fath, A.; Jones, R.L. Characterization of a calmodulin-binding transporter from the plasma membrane of barley aleurone. *Proc. Natl. Acad. Sci. USA* **1998**, *95*, 1944–1949. [CrossRef] [PubMed]
3. Nawaz, Z.; Kakar, K.; Saand, M.A.; Shu, Q.Y. Cyclic nucleotide-gated ion channel gene family in rice, identification, characterization and experimental analysis of expression response to plant hormones, biotic and abiotic stresses. *BMC Genom.* **2014**, *15*, 853. [CrossRef] [PubMed]
4. Lee, S.K.; Lee, S.M.; Kim, M.H.; Park, S.K.; Jung, K.H. Genome-Wide Analysis of Cyclic Nucleotide-Gated Channel Genes Related to Pollen Development in Rice. *Plants* **2022**, *11*, 3145. [CrossRef] [PubMed]
5. Xu, Y.; Yang, J.; Wang, Y.; Wang, J.; Wan, J. OsCNGC13 promotes seed-setting rate by facilitating pollen tube growth in stylar tissues. *PLoS Genet.* **2017**, *13*, e1006906. [CrossRef] [PubMed]
6. Pan, Y.; Chai, X.; Gao, Q.; Zhou, L.; Zhang, S.; Li, L.; Luan, S. Dynamic Interactions of Plant CNGC Subunits and Calmodulins Drive Oscillatory Ca2+ Channel Activities. *Dev. Cell* **2019**, *48*, 710–725. [CrossRef] [PubMed]
7. Tan, Y.Q.; Yang, Y.; Zhang, A.; Fei, C.F.; Wang, Y.F. Three CNGC Family Members, CNGC5, CNGC6, and CNGC9, Are Required for Constitutive Growth of Arabidopsis Root Hairs as Ca2+-Permeable Channels. *Plant Commun.* **2019**, *1*, 100001. [CrossRef]
8. Wang, J.; Liu, X.; Zhang, A.; Ren, Y.; Wan, J. A cyclic nucleotide-gated channel mediates cytoplasmic calcium elevation and disease resistance in rice. *Cell Res.* **2019**, *29*, 820–829. [CrossRef]
9. Tian, W.; Hou, C.; Ren, Z.; Wang, C.; Zhao, F.; Dahlbeck, D.; Hu, S.; Zhang, L.; Niu, Q.; Li, L. A calmodulin-gated calcium channel links pathogen patterns to plant immunity. *Nature* **2019**, *572*, 131–135. [CrossRef] [PubMed]

10. Zhang, W.; Dong, C.; Zhang, Y.; Zhu, J.; Dai, H.; Bai, S. An apple cyclic nucleotide-gated ion channel gene highly responsive to Botryosphaeria dothidea infection enhances the susceptibility of Nicotiana benthamiana to bacterial and fungal pathogens. *Plant Sci.* **2018**, *269*, 94–105. [CrossRef]
11. Xuan-Rui, Z.; You-Ping, X.; Xin-Zhong, C. SlCNGC1 and SlCNGC14 Suppress Xanthomonas oryzae pv. oryzicola-Induced Hypersensitive Response and Non-host Resistance in Tomato. *Front. Plant Sci.* **2018**, *9*, 285.
12. Wang, J.; Ren, Y.; Liu, X.; Luo, S.; Wan, J. Transcriptional Activation and Phosphorylation of OsCNGC9 Confer Enhanced Chilling Tolerance in Rice. *Mol. Plants Engl. Ed.* **2021**, *14*, 15. [CrossRef] [PubMed]
13. Cui, Y.; Lu, S.; Li, Z.; Cheng, J.; Hu, P.; Zhu, T.; Wang, X.; Jin, M.; Wang, X.; Li, L. Cyclic Nucleotide-Gated Ion Channels 14 and 16 Promote Tolerance to Heat and Chilling in Rice. *Plant Physiol.* **2020**, *183*, 1794–1808. [CrossRef]
14. Wang, L.; Li, M.; Liu, Z.; Dai, L.; Zhang, M.; Wang, L.; Zhao, J.; Liu, M. Genome-wide identification of CNGC genes in Chinese jujube (Ziziphus jujuba Mill.) and ZjCNGC2 mediated signalling cascades in response to cold stress. *BMC Genom.* **2020**, *21*, 191. [CrossRef] [PubMed]
15. Li, Z.; Rutger, J.N. Geographic distribution and multilocus organization of isozyme variation of rice (*Oryza sativa* L.). *TAG Theor. Appl. Genet.* **2000**, *101*, 379–387. [CrossRef]
16. Yu, S.B.; Xu, W.J.; Vijayakumar, C.H.; Ali, J.; Fu, B.Y.; Xu, J.L.; Jiang, Y.Z.; Marghirang, R.; Domingo, J.; Aquino, C.; et al. Molecular diversity and multilocus organization of the parental lines used in the International Rice Molecular Breeding Program. *Theor. Appl. Genet.* **2003**, *108*, 131–140. [CrossRef]
17. Zeng, W.; Li, H.; Zhang, F.L.; Wang, X.C.; Rehman, S.; Huang, S.J.; Zhang, C.Y.; Wu, F.C.; Li, J.F.; Lv, Y.M.; et al. Functional characterization and allelic mining of OsGLR genes for potential uses in rice improvement. *Front. Plant Sci.* **2023**, *14*, 20. [CrossRef]
18. Wang, W.; Mauleon, R.; Hu, Z.; Chebotarov, D.; Leung, H. Genomic variation in 3010 diverse accessions of Asian cultivated rice. *Nature* **2018**, *557*, 43–49. [CrossRef]
19. Zhang, F.; Wang, C.; Li, M.; Cui, Y.; Shi, Y.; Wu, Z.; Hu, Z.; Wang, W.; Xu, J.; Li, Z. The landscape of gene-CDS-haplotype diversity in rice (*Oryza sativa* L.): Properties, population organization, footprints of domestication and breeding, and implications in genetic improvement—ScienceDirect. *Mol. Plants* **2021**, *14*, 18.
20. Khaipho-Burch, M.; Cooper, M.; Crossa, J.; de Leon, N.; Holland, J.; Lewis, R.; McCouch, S.; Murray, S.C.; Rabbi, I.; Ronald, P.; et al. Genetic modification can improve crop yields—But stop overselling it. *Nature* **2023**, *621*, 470–473. [CrossRef]
21. Kong, H.; Landherr, L.L.; Frohlich, M.W.; Leebens-Mack, J.; Depamphilis, C.W. Patterns of gene duplication in the plant SKP1 gene family in angiosperms: Evidence for multiple mechanisms of rapid gene birth. *Plant J.* **2010**, *50*, 873–885. [CrossRef] [PubMed]
22. Xu, J.T.; Zhu, C.; Su, M.Z.; Li, S.D.; Chao, H.Y.; Chen, M. CropGF: A comprehensive visual platform for crop gene family mining and analysis. *Database* **2023**, *2023*, 7. [CrossRef] [PubMed]
23. Guo, J.T.; Zeng, W.Z.; Chen, Q.F.; Lee, C.; Chen, L.P.; Yang, Y.; Cang, C.L.; Ren, D.J.; Jiang, Y.X. Structure of the voltage-gated two-pore channel TPC1 from *Arabidopsis thaliana*. *Nature* **2016**, *531*, 196–201. [CrossRef] [PubMed]
24. Wang, P.C.; Hsu, C.C.; Du, Y.Y.; Zhu, P.P.; Zhao, C.Z.; Fu, X.; Zhang, C.G.; Paez, J.S.; Macho, A.P.; Tao, W.A.; et al. Mapping proteome-wide targets of protein kinases in plant stress responses. *Proc. Natl. Acad. Sci. USA* **2020**, *117*, 3270–3280. [CrossRef]
25. Finn, R.D.; Clements, J.; Eddy, S.R. HMMER web server: Interactive sequence similarity searching. *Nucleic Acids Res.* **2011**, *39*, W29–W37. [CrossRef] [PubMed]
26. Chen, C.; Chen, H.; Zhang, Y.; Thomas, H.R.; Xia, R. TBtools: An Integrative Toolkit Developed for Interactive Analyses of Big Biological Data. *Mol. Plant.* **2020**, *13*, W29–W37. [CrossRef]
27. Letunic, I.; Bork, P. Interactive tree of life (iTOL) v3: An online tool for the display and annotation of phylogenetic and other trees. *Nucleic Acids Res.* **2016**, *49*, W242–W245. [CrossRef]
28. Jelassi, R.; Khemaissia, H.; Zimmer, M.; Garbe-Schönberg, D.; Nasri-Ammar, K. Biodiversity of Talitridae family (Crustacea, Amphipoda) in some Tunisian coastal lagoons. *Zool. Stud.* **2015**, *54*, e17. [CrossRef]
29. Nei, M. Genetic distance between populations. *Am. Nat.* **1972**, *106*, 283–292. [CrossRef]

plants

Article

OsLEA1b Modulates Starch Biosynthesis at High Temperatures in Rice

Gang Li [1,†], Ruijie Cao [1,†], Liuyang Ma [1], Guiai Jiao [1], Pengfei Chen [1], Nannan Dong [1], Xinwei Li [1], Yingqing Duan [1], Xiaoxue Li [1], Mingdong Zhu [2], Gaoneng Shao [1], Zhonghua Sheng [1], Shikai Hu [1], Shaoqing Tang [1], Xiangjin Wei [1,*], Yinghong Yu [2,*] and Peisong Hu [1,*]

1 State Key Laboratory of Rice Biology, China National Center for Rice Improvement, China National Rice Research Institute, Hangzhou 310006, China; Ganglee0102@163.com (G.L.); caoruijie@caas.cn (R.C.); maliuyang2018@163.com (L.M.); jiaoguiai@caas.cn (G.J.); 82101215140@caas.cn (P.C.); 13126531292@163.com (N.D.); lixinwei162013@163.com (X.L.); 18355093180@163.com (Y.D.); xx5990302@126.com (X.L.); shaogaoneng@caas.cn (G.S.); shengzhonghua@caas.cn (Z.S.); hushikai@caas.cn (S.H.); tangshaoqing@caas.cn (S.T.)

2 Hunan Rice Research Institute, Changsha 410125, China; uhz_uhz@hotmail.com

* Correspondence: weixiangjin@caas.cn (X.W.); yyh30678@163.com (Y.Y.); hupeisong@caas.cn (P.H.)

† These authors contributed equally to this work.

Abstract: High temperatures accelerate the accumulation of storage material in seeds, often leading to defects in grain filling. However, the mechanisms regulating grain filling at high temperatures remain unknown. Here, we want to explore the quality factors influenced by the environment and have identified a *LATE EMBROYGENESIS ABUNDANT* gene, *OsLEA1b*, a heat-stress-responsive gene in rice grain filling. *OsLEA1b* is highly expressed in the endosperm, and its coding protein localizes to the nucleus and cytoplasm. Knock-out mutants of *OsLEA1b* had abnormal compound starch granules in endosperm cells and chalky endosperm with significantly decreased grain weight and grain number per panicle. The *oslea1b* mutants exhibited a lower proportion of short starch chains with degrees of polymerization values from 6 to 13 and a higher proportion of chains with degrees from 14 to 48, as well as significantly lower contents of starch, protein, and lipid compared to the wild type. The difference was exacerbated under high temperature conditions. Moreover, *OsLEA1b* was induced by drought stress. The survival rate of *oslea1b* mutants decreased significantly under drought stress treatment, with significant increase in ROS levels. These results indicate that *OsLEA1b* regulates starch biosynthesis and influences rice grain quality, especially under high temperatures. This provides a valuable resource for genetic improvement in rice grain quality.

Keywords: high temperature; *OsLEA1b*; rice grain quality; starch biosynthesis

check for updates

Citation: Li, G.; Cao, R.; Ma, L.; Jiao, G.; Chen, P.; Dong, N.; Li, X.; Duan, Y.; Li, X.; Zhu, M.; et al. *OsLEA1b* Modulates Starch Biosynthesis at High Temperatures in Rice. *Plants* **2023**, *12*, 4070. https://doi.org/10.3390/plants12234070

Academic Editor: Diego F. Gomez-Casati

Received: 10 November 2023
Revised: 28 November 2023
Accepted: 29 November 2023
Published: 4 December 2023

1. Introduction

Rice (*Oryza sativa*) is an important crop consumed as a staple food by more than half of the world's population. The rice endosperm is the main storage organ for nutrients, including starch, proteins, and lipids [1]. Starch is the main endosperm component, and its content, composition, and structure greatly affect grain quality and yield [2]. Sucrose is efficiently converted into starch through a series of reactions during rice endosperm development [3].There are three main types of enzymes involved in starch biosynthesis. Starch synthase (SS) lengthens the non-reducing end of glucose chains; branching enzymes (BEs) then generate branches from the existing chain; and debranching enzymes (DBEs) hydrolyze branches [4]. In addition, several non-enzymatic proteins participate in starch biosynthesis in the cereal endosperm, such as PROTEIN TARGETING TO STARCH (PTST) and the starch-binding protein FLOURY ENDOSPERM 6 (FLO6) [5,6].

The environment also affects endosperm starch biosynthesis. Heat-induced crop damage is becoming increasingly common as a result of global warming, severely limiting high and stable rice yields [7]. Rice can tolerate certain temperature ranges during different developmental stages [8]. High temperatures perturb the expression of starch biosynthesis-related genes, hindering starch biosynthesis and adversely affecting rice during the grain-filling period [9]. Suppression of *OsMADS7*, encoding a transcription factor, in rice endosperm stabilizes amylose content under high-temperature stress [10]. Inhibiting the expression of α-amylase genes *Amy1A*, *Amy1C*, *Amy3A*, *Amy3D*, and *Amy3E* reduces starch degradation and results in chalky grains under high-temperature conditions [11]. The high pre-mRNA splicing efficiency of *Waxy* (*Wx*) likely stabilizes amylose content in rice seeds under high temperatures [12]. Moreover, genes specifically expressed in the endosperm play an important role in crop growth and development.

There are several endosperm-specific genes that were identified through a genomic survey [13]. These genes encode grain storage proteins and proteins related to starch biosynthesis, such as the drought-stress-responsive LATE EMBROYGENESIS ABUNDANT (LEA) proteins [14]. In Arabidopsis (*Arabidopsis thaliana*), LEA proteins are associated with seed dehydration tolerance [15]. Four NAC transcription factors bind the *OsLEA* promoter to activate its expression to high levels in embryonic tissues during endosperm development [16]. However, the effects of LEA proteins on starch biosynthesis under high-temperature conditions remain unknown.

Here, we want to explore the quality factors influenced by environment and have identified an embryonic protein homologous to OsLEA3-2 [17], which we named OsLEA1b. The *oslea1b* mutants were more sensitive to high-temperature stress and drought stress than the wild type. Milled rice from *oslea1b* mutants showed significantly lower starch content and abnormal structure under a high-temperature environment, which resulted in significantly greater chalkiness and chalky grain rate than that from the wild type. A series of assays were utilized to determine the effects of OsLEA1b on starch biosynthesis under high-temperature conditions. These findings enhance our understanding of OsLEA proteins and grain filling in rice, providing useful information for genetic improvement of rice grain yield and quality.

2. Results

2.1. *OsLEA1b* Encodes an Embryonic Protein Mainly Expressed in the Endosperm

To isolate genes that regulate rice grain filling under high-temperature conditions, we searched the expression of genes that are induced under high temperatures in our RNA-Seq data (Supplementary Figure S1). We identified *OsLEA1b* (LOC_Os01g16920), which is predominantly expressed during the grain-filling stage and is homologous to *OsLEA3-2* (Supplementary Figure S1). Reverse transcription quantitative PCR (RT-qPCR) analysis validated that *OsLEA1b* is highly expressed in developing seeds, particularly during the late stages of seed development (Figure 1A). In contrast, *OsLEA1b* was rarely expressed in leaf and stem tissues. We transformed a vector harboring the *β-glucuronidase* (*GUS*) reporter gene driven by the *OsLEA1b* promoter into rice. Histochemical analysis corroborated that GUS activity in transgenic plants was strongest in the endosperm (Figure 1B), consistent with the RT-qPCR results. Transient expression analysis of a 35S:OsLEA1b–GFP construct in rice protoplasts revealed that the OsLEA1b–GFP fusion protein was widely present in the cytoplasm and nucleus, similar to the GFP control (Figure 1C). We therefore concluded that *OsLEA1b* encodes an embryonic protein.

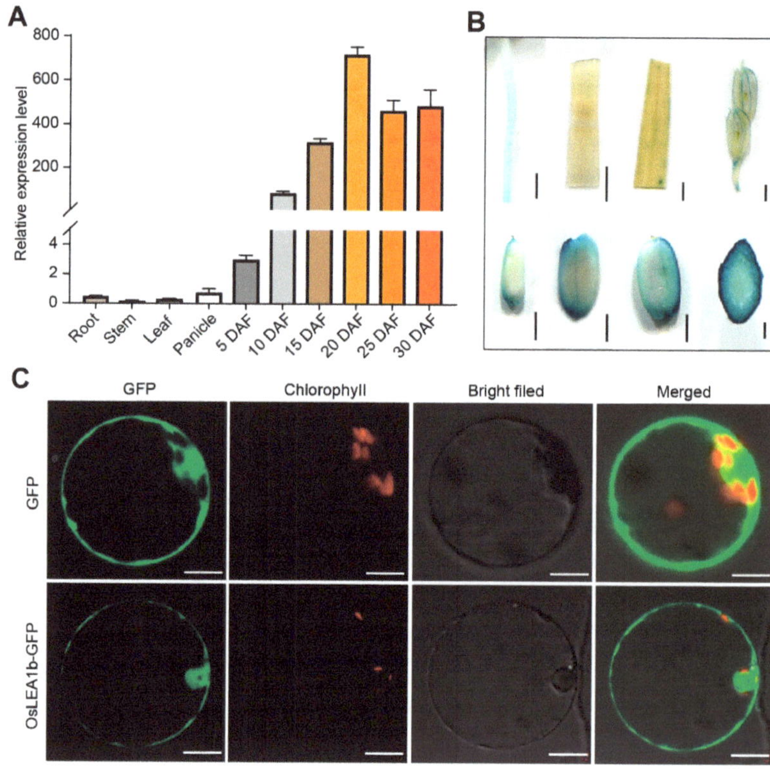

Figure 1. Expression pattern and subcellular localization of OsLEA1b. (**A**) Relative *OsLEA1b* expression levels in various tissues and in developing endosperms at 5, 10, 15, 20, 25, and 30 days after fertilization (DAF). Values are means ± SD from three biological replicates. (**B**) GUS staining in root, stem, leaf, spikelet, and developing grains expressing *ProOsLEA1b:GUS*. (**C**) Subcellular localization of OsLEA1b in rice protoplasts. Confocal microscopy images show OsLEA1b–GFP localized in the nucleus and cytoplasm. GFP signaling, chlorophyll autofluorescence, bright field, and merged images are shown for each construct. Scale bars, 2 mm in (**B**) and 10 μm in (**C**).

2.2. The oslea1b Mutants Exhibit Chalky Endosperm and Decreased Yield

The *OsLEA1b* gene contains two exons and one intron and does not harbor any known domains aside from four low-complexity regions (determined using SMART, http://smart.embl-heidelberg.de, accessed on 1 January 2018, Figure S2). To investigate the function of *OsLEA1b* in rice endosperm development and plant growth, we knocked out *OsLEA1b* in the *japonica* rice cultivar 'ZH11' using clustered regularly interspaced short palindromic repeat (CRISPR)/CRISPR-associated nuclease 9 (Cas9)-mediated genome editing. We successfully obtained two independent homozygous mutants, *oslea1b-1* and *oslea1b-2*. RT-qPCR analysis showed that *OsLEA1b* expression in the two mutants was significantly lower than that in the wild type (Figure S3A–C). Compared to the wild type, the grain width of *oslea1b* mutants was significantly lower, and this resulted in a lower 1000-grain weight even though the grains were longer in length (Figure 2A–C,H–K). Moreover, we observed that most mutant grains showed a semi-opaque endosperm, and the grain chalkiness rate in the mutants reached 95% (Figure 2D–G,O). A dramatically lower grain number per panicle in the mutants resulted in lower yields, only 49% relative to the wild-type control. But panicle length, tiller number, and seed setting rate were not significantly different in the mutants compared to the wild type (Figures 2L–N and S4). Besides the shorter plant height, the grain-filling rate in the *oslea1b* mutants was slower on the 20th and 30th days after fertilization (DAF) than that of the wild type (Figures S4 and S5). Thus, the *OsLEA1b* gene plays an important role in grain quality and yield.

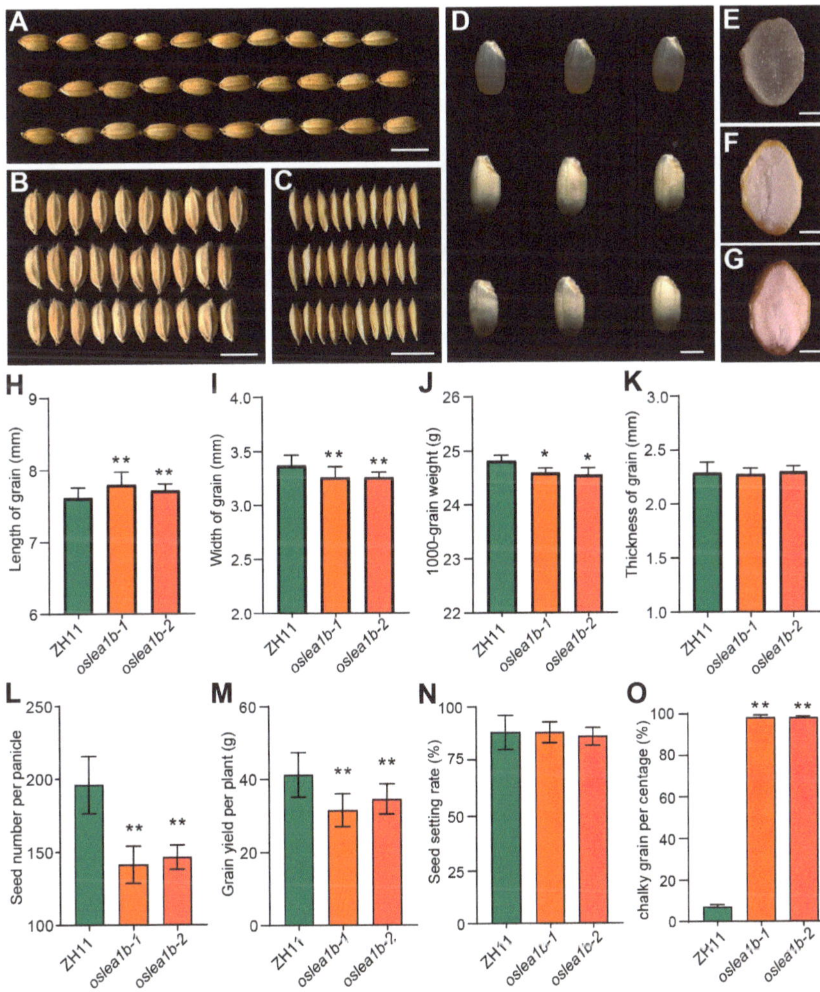

Figure 2. Analysis of agronomic traits of *oslea1b* mutants compared to the wild type. (**A–C**) Grain length, width, and thickness of wild-type (WT) and *oslea1b* grains. Scale bars, 6 mm. (**D**) Appearance of brown rice grains from WT and *oslea1b* mutant plants. Scale bar, 2 mm. (**E–G**) Transverse sections of WT (**E**) and *oslea1b* (**F**,**G**) endosperm. Scale bars, 2 mm. (**H–O**) Quantification of grain length (**H**), grain width (**I**), 1000-grain weight (**J**), grain thickness (**K**), seed number per panicle (**L**), grain yield per plant (**M**), seed setting rate (**N**), and chalky grain percentage (**O**) for WT plants and *oslea1b* mutants. Data are means ± SD from at least three biological replicates. Statistically significant differences were determined using Student's *t*-test (*, $p < 0.05$, **, $p < 0.01$).

2.3. *OsLEA1b Affects Starch Granule Development in the Endosperm*

We next observed the starch grain morphology in endosperm cells of *oslea1b* mutants and the wild type. Scanning electron microscopy (SEM) revealed that wild-type starch granules were polygonal in shape and densely arranged. In contrast, starch granules in the chalky endosperm of the mutants were round and had numerous air spaces (Figure 3A–C). We prepared semi-thin sections of grains on the 10th DAF to observe the morphological differences in complex starch granules in the endosperm. Mutant grains had greater numbers of loosely arranged complex starch granules in peripheral and central endosperm cells than exhibited in wild-type grains (Figure 3D–G). In addition, gaps between amyloplasts in the center of the mutant endosperm were larger than those in wild-type endosperm, with many dispersed starch granules in the mutant (Figure 3H,I). Meanwhile, transmission electron microscopy (TEM) revealed many abnormal starch granules developed in the mutant endosperm (Figure 3J–O).

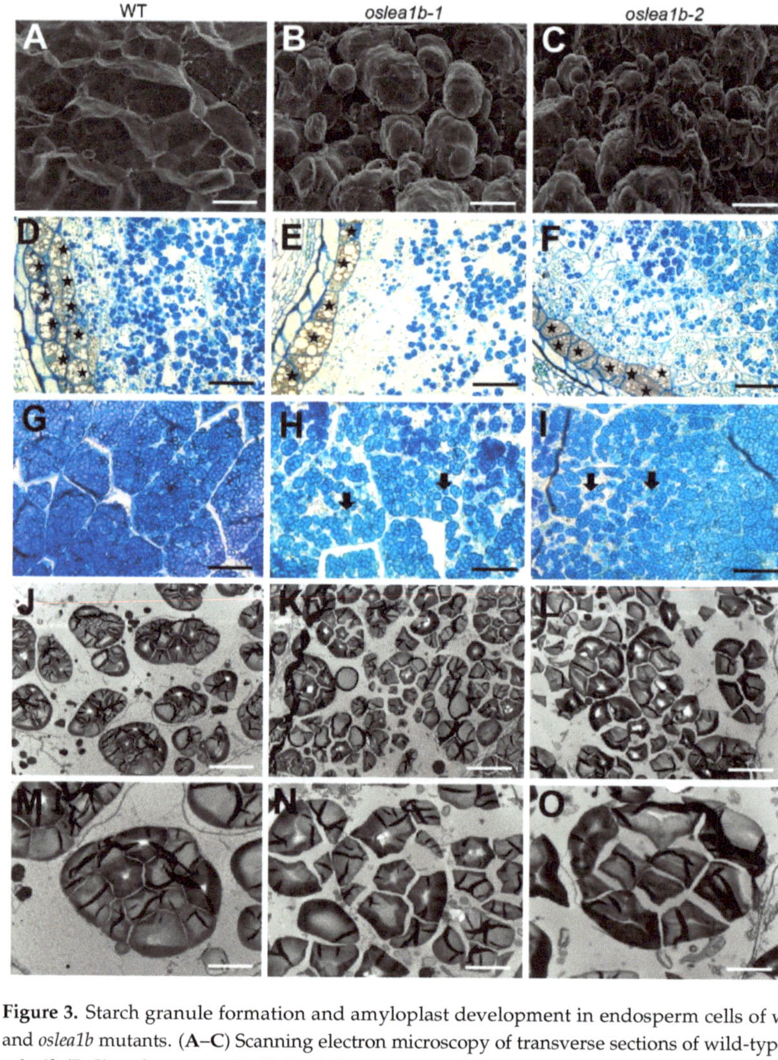

Figure 3. Starch granule formation and amyloplast development in endosperm cells of wild-type and *oslea1b* mutants. (**A–C**) Scanning electron microscopy of transverse sections of wild-type (**A**) and *oslea1b* (**B**,**C**) endosperm. (**D–I**) Semi-thin sections of WT and *oslea1b* endosperm at 10 days after fertilization (DAF). (**D–F**) Micrographs showing the peripheral region; (**G–I**) micrographs showing the central region. Black stars in (**D–F**) indicate aleurone cells. Arrows in (**H**,**I**) indicate broken amyloplasts. (**J–L**) Transmission electron microscopy shows well-developed amyloplasts in WT endosperm cells (**J**,**M**) and broken amyloplasts in *oslea1b* endosperm cells (**K**,**L**,**N**,**O**). Scale bars, 15 μm in (**A–C**), 40 μm in (**D–I**), 5 μm in (**J–L**), and 2 μm in (**M–O**).

We examined additional physicochemical properties of starch from wild-type and *oslea1b* endosperm. Amylose, total starch, and protein contents in the mutants were significantly lower than those in the wild type (Figure 4A–C). The gel consistency of starch in the wild type was 81 mm, while that of *oslea1b* mutant starch was 39.5–53 mm (Figure 4D). The distribution of amylopectin chain lengths suggested a lower proportion of short chains with degrees of polymerization values between 6 and 13 but a higher proportion of intermediate chains (degrees 14 to 48) in the *oslea1b* mutants than that in the wild type (Figure 4E). Starch gelatinization and viscosity analyses indicated that the mutant starch was more difficult to gelatinize (Figure 4F,G). These results show that the physicochemical properties of endosperm starch were altered in the *oslea1b* mutants.

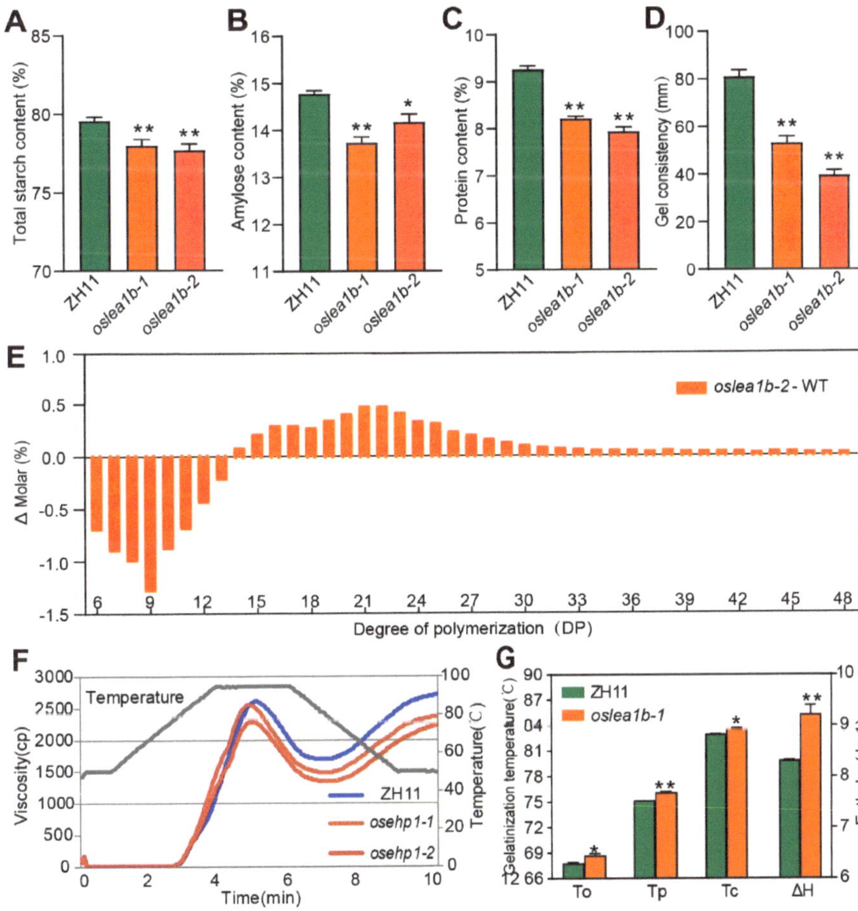

Figure 4. Starch physicochemical properties in the endosperm of *oslea1b* mutants. (**A**–**D**) Total starch content (**A**), amylose content (**B**), protein content (**C**), and gel consistency (**D**) of wild-type and *Osesp1* mutant endosperm. (**E**) Differences in amylopectin chain length distribution between WT and *oslea1b*. (**F**) Pasting properties of WT and *oslea1b* endosperm starch determined using a Rapid Visco Analyzer (RVA). The gray line indicates temperature changes during the measurements. (**G**) Gelatinization temperature of endosperm starch. To, Tp, and Tc represent the onset, peak, and concluding gelatinization temperatures, respectively. Data in (**A**–**D**,**G**) are presented as means ± SD from three replicates. Statistically significant differences were determined using Student's *t*-test (*, $p < 0.05$, **, $p < 0.01$).

2.4. The oslea1b Mutants Showed Lower Grain Quality under High Temperature Conditions

High temperature during rice grain filling often reduces grain quality [18]. We grew wild-type and *oslea1b* plants under field conditions until flowering, after which plants were subjected to artificial high temperatures (35 °C for 12 h in light and 28 °C for 12 h in darkness) and normal temperatures (28 °C for 12 h in light and 22 °C for 12 h in darkness). Under high-temperature conditions, mature wild-type grains were slightly chalky, but mature *oslea1b* grains were floury with a significantly lower 1000-grain weight (Figure 5A–H). Compared to those under normal temperature conditions, wild-type and *oslea1b* grains had significantly lower amylose and total protein contents under high-temperature conditions. However, the differences between conditions were greater for *oslea1b* grains than for wild-type grains (Figure 5I–K).

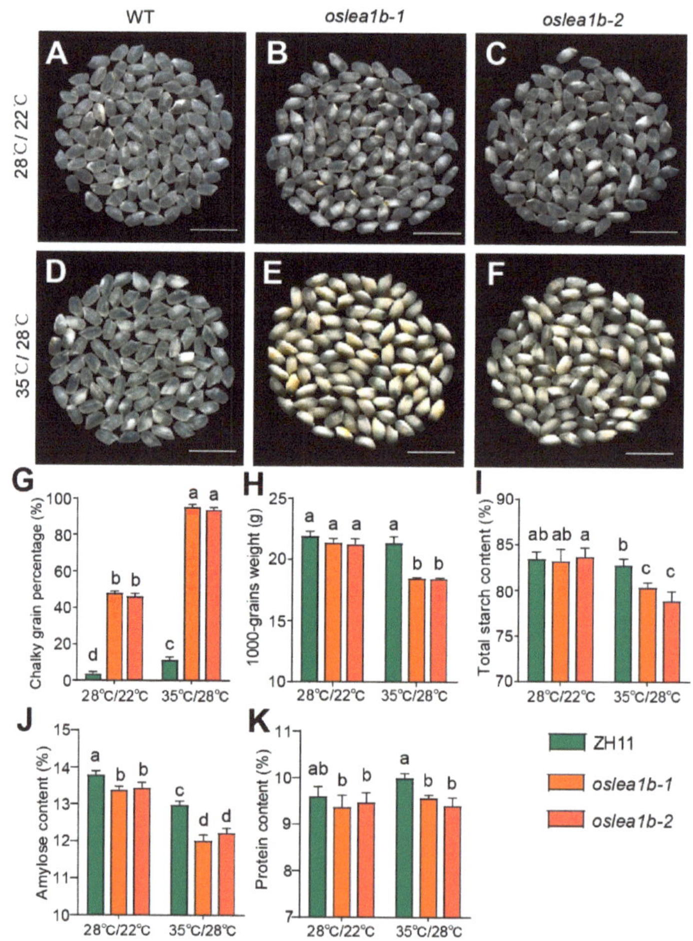

Figure 5. The *oslea1b* mutants were more sensitive to high temperatures during the grain-filling stage. (**A–F**) Appearance of wild-type (WT) and *oslea1b* mutant grains under high-temperature treatment (35 °C for 12 h in light and 28 °C for 12 h in darkness) and normal temperature treatment (28 °C for 12 h in light and 22 °C for 12 h in darkness). Scale bars, 1 cm. (**G–K**) Chalky grain percentage (**G**), 1000-grain weight (**H**), total starch content (**I**), amylose content (**J**), and total protein content (**K**) of WT and *oslea1b* mutant grains under normal and high temperatures. Different letters indicate significant differences at $p < 0.05$ by ANOVA and Duncan's test.

We analyzed the expression of genes related to starch biosynthesis under high and normal temperatures. High temperature induced *OsLEA1b* expression at the early grain-filling stage (Figure S6). The expression levels of *STARCH BRANCHING ENZYME 1* (*OsBE1*) and *BRITTLE 1* (*OsBT1*) were significantly lower in the mutant than in the wild type under high temperature conditions. Conversely, the expression levels of *SOLUBLE STARCH SYNTHASE IIA* (*OsSSIIa*), *PULLULANASE* (*PUL*), *ISOAMYLASE 3* (*ISA3*), *STARCH BRANCHING ENZYME IIb* (*OsBEIIb*), *GRANULE-BOUND STARCH SYNTHASE I* (*GBSS1*), *RAG2*, and *GLUTELIN TYPE-D 1* (*GLUD1*) were significantly higher in the mutant than in the wild type under high temperatures (Figure 6). These differences in gene expression likely led to the differences observed in amylopectin's fine structure in the mutants under high-temperature conditions.

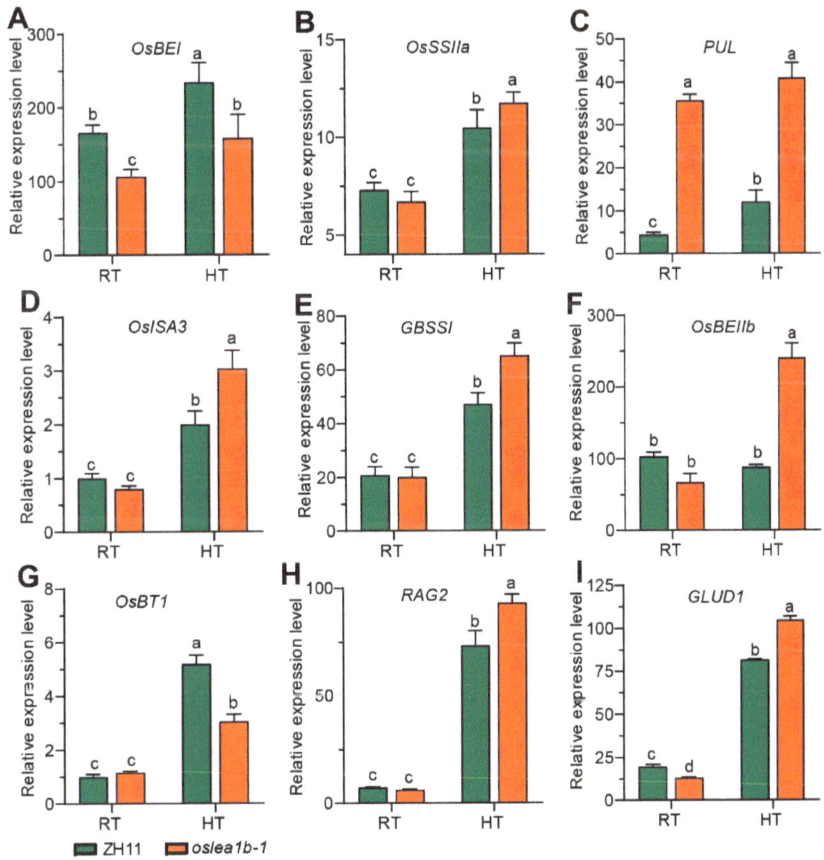

Figure 6. Transcription levels of genes related to regulate amylopectin synthesis in endosperm at 10 DAF. (**A–I**) Relative expression of genes related to amylopectin biosynthesis. Gene expression levels were measured in WT and *oslea1b-1* grains at 9 days after fertilization under high (HT) and normal (RT) temperature conditions. Different letters indicate significant differences at $p < 0.05$ by ANOVA and Duncan's test.

2.5. OsLEA1b Promotes Drought Tolerance

The *oslea1b* mutants were sensitive to high temperature. We also examined the sensitivity of the *oslea1b* mutants to abiotic stresses such as drought and salinity. Treatment with 20% PEG6000 or 200 mM NaCl for 1 week induced greater reductions of root and shoot length in the mutant than in the wild type. Moreover, the survival rates of *oslea1b* mutants were only 13% and 11% when subjected to 20% PEG6000 (Figures 7A–C and S7), indicating that the mutants were more sensitive to osmotic stress. When we exposed two-week-old seedlings to drought stress under soil conditions, the survival rate of *oslea1b* mutants was obviously reduced to 10% compared to that of the wild type (Figure 7D,E). *OsLEA1b* expression was induced after 12 h of 20% PEG6000 treatment in wild-type plants but not in the mutants (Figure 7F). These results indicate that the *oslea1b* mutants have reduced tolerance to drought stress.

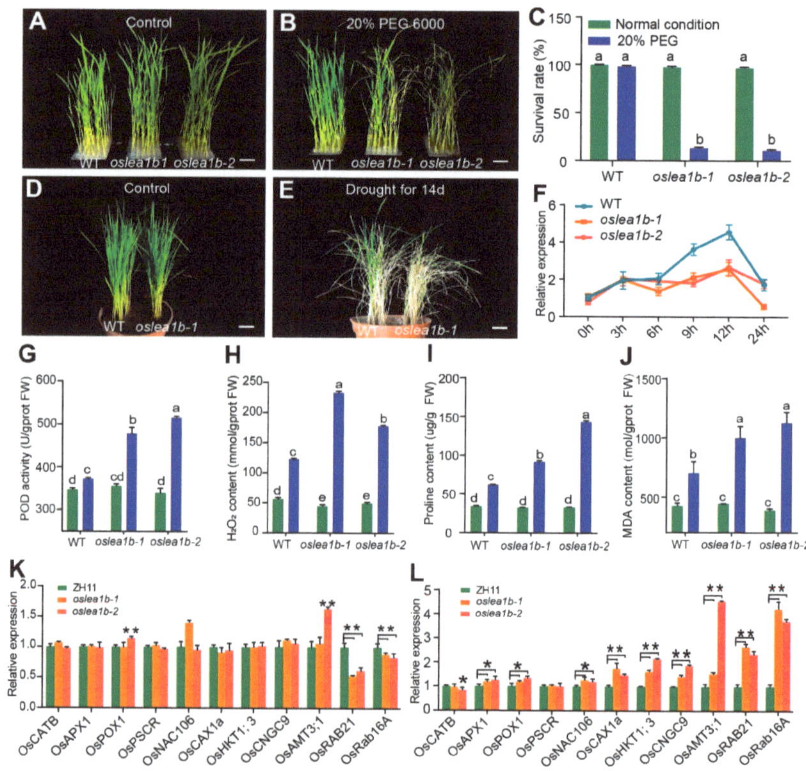

Figure 7. *OsLEA1b* gene positively regulates drought stress tolerance. (**A**,**B**) Phenotypes of seedlings of wild-type and *oslea1b* before and after 20% PEG treatment. (**C**) Survival rate of seedlings of WT and *oslea1b* after 20% PEG treatment. (**D**,**E**) Phenotypes of WT and *oslea1b* before and after drought stress. Water was withheld for 5 d followed by a 14-day recovery period. (**F**) Relative expression of *OsLEA1b* was induced by drought stress. (**G**–**J**) Peroxidase (POD) enzyme activity (**G**), and contents of hydrogen peroxide (H_2O_2; **H**), proline (**I**), and malondialdehyde (MDA; (**J**)) in WT and *oslea1b* mutant leaves. (**K**,**L**) Relative expression of genes related to reactive oxygen species scavenging and ABA signaling in wild type and mutants before (**K**) and after (**L**) 20% PEG treatment. Data are means ± SD from at least three biological replicates. Different letters in (**C**, **G**–**J**) indicate significant differences at $p < 0.05$ by ANOVA and Duncan's test. In (**K**, **L**), statistically significant differences were determined using Student's *t*-test (*, $p < 0.05$, **, $p < 0.01$).

Reactive oxygen species (ROS) levels were determined in leaves of the wild type and the *oslea1b* mutants. There were no differences in various ROS indexes between control and stress treatments in the wild type. But *oslea1b* mutants had significantly increased peroxidase (POD) activity, hydrogen peroxide (H_2O_2) content, proline content, and malondialdehyde (MDA) content after PEG treatment compared to the wild type (Figure 7G–J). Genes related to stress, ROS scavenging, and ABA signaling were significantly up-regulated in the *oslea1b* mutants after PEG treatment but not in the control (Figure 7K,L). These results indicate that the *oslea1b* mutants are more sensitive to drought stress than the wild type.

3. Discussion

3.1. The oslea1b Mutants Showed Greater Sensitivity to High Temperature

High temperature during the rice grain-filling stage often results in lower quantity and worse quality, such as higher chalkiness and lower grain-setting rate [19]. Grain chalkiness affects the appearance of rice. The imbalance of a source-to-sink under high temperature conditions leads to insufficient grain filling in endosperm cells and disordered enzyme activity, which bring a high grain chalkiness rate [20–24]. In this study, we observed that grains of *oslea1b* mutants exhibit high chalkiness rates under normal temperatures and chalky endosperm under high temperatures. The abnormally developed starch granules in the mutant endosperm cells were loosely arranged rather than whole complex granules. The *oslea1b* mutants were more sensitive to high temperature during endosperm development than wild-type plants. Moreover, the chalkiness rate and chalkiness degree of the mutants were dramatically higher than those of the wild type under both natural and artificial high temperatures. By contrast, the gelatinization and gel consistency of the two mutants were significantly lower than those in the wild type at high temperatures. Recent studies have shown that in addition to starch content, the arrangement of starch granules also affects rice grain cooking and eating qualities. It is likely that the abnormal starch granule development induced by high temperature in *oslea1b* mutants would affect the cooking and eating qualities. The higher proportion of short starch chain length and easier pasting properties of *oslea1b* may confer it greater sensitivity to high temperature tolerance.

Starch content and structure directly influence grain appearance such as chalkiness [25]. And many genes regulating starch biosynthesis were reported to influence rice chalkiness [26]. *Chalk5*, the major QTL control chalky grain rate of rice, two polymorphic nucleotides on the promoter of *Chalk5* affect its expression level. The lower expression level could decrease the chalkiness of rice and increase the percentage of milled rice. Total starch and amylose content in the *oslea1b* mutants were significantly lower than those in the wild type, especially under high temperatures. Moreover, a significantly different distribution of amylose chain lengths in the mutant led to an altered starch structure. The expression of *BE1*, *SSIIa*, *PUL*, *ISA3*, and *BEIIb* showed significant differences between the wild type and mutants under high-temperature treatment, which resulted in differences in the fine structure of starch [27,28]. The fine structure of starch affects starch grain expansion and gelatinization, which impacts the cooking performance and taste of rice [29]. Gelatinization temperature was higher in the mutants, causing hard, gluey consistency and reduced viscosity. The expression of *SSIIa* was significantly higher in the mutant than in the wild type under high temperatures, which may be responsible for the reduced viscosity. Although the effects of high temperature on rice quality are well known, additional research is needed to explore the underlying regulatory network to improve rice quality.

3.2. OsLEA1b Plays an Important Role in Rice Stress Responses

Phylogenetic analysis showed that the embryonic protein OsLEA1b is homologous to the late embryogenesis abundant protein OsLEA3-2. LEA proteins are widely distributed in rice, Arabidopsis, maize (*Zea mays*), cotton (*Gossypium hirsutum*), potato (*Solanum tuberosum*), and other plant species [30–34]. The role of plant LEA proteins in stress tolerance has been well documented. Plants maintain cell membrane stability in low temperature environments through LEA protein biosynthesis [35]. In vitro and in vivo experiments demonstrated that soybean (*Glycine max*) PM2 (LEA3) protects proteins and maintains enzyme activity under unfavorable temperatures [36]. In addition, LEA proteins bind to certain ions to maintain osmotic pressure in cells. Two soybean LEA4 proteins, GmPM1 and GmPM9, possess metal-binding properties that reduce abiotic stress-induced oxidative damage [37].

Furthermore, LEA proteins enhance plant salt and drought tolerance. LEA proteins also play an important role in the adaptive responses to water deficit in higher plants [38]. Transgenic tobacco (*Nicotiana tabacum*) plants heterologously expressing melon (*Cucumis melo*) *CmLEA-S* display greater tolerance to drought and salt compared to wild-type plants [39]. In this study, *oslea1b* mutants were more sensitive to drought stress than wild-type plants, showing extremely low survival rates and significantly increased cellular ROS levels. *CATALASE B (OsCATB), ASCORBATE PEROXIDASE 1 (OsAPX1), PEROXIDASE 1 (OsPOX1), Ca^{2+}/H^{+} EXCHANGER 1a (OsCAX1a), OsP5CR, OsNAC106, HIGH-AFFINITY K+ TRANSPORTER 1;3 (OsHKT1;3), CYCLIC NUCLEOTIDE-GATED CHANNEL 9 (OsCNGC9), RESPONSIVE TO ABA 21 (OsRAB21), OsRAB16a*, and *AMMONIUM TRANSPORTER 3 (OsAMT3)* are stress-related genes [40–51]. The stress-induced expression of these genes in *oslea1b* confers the sensitivity of the mutants to drought stress. This validates that *OsLEA1b* responds to drought stress by influencing the expression of those genes. Our results indicate that OsLEA1b regulates stress tolerance. We will further investigate the effects of overexpression of OsLEA1b on drought tolerance.

OsLEA1b is highly expressed in seeds yet barely detectable in other rice tissues. However, the drought sensitivity of *oslea1b* mutants was observed in other parts of the plant, such as the leaves. Additional studies are needed to explain how OsLEA1b, which is enriched in the endosperm, regulates plant stress tolerance. Overall, our research reveals that OsLEA1b regulates starch biosynthesis in the grain endosperm and plays an important role in rice drought tolerance. This study provides useful information for potential genetic improvement of rice yield and grain quality.

4. Materials and Methods

4.1. Plant Materials and Growth Conditions

The *oslea1b* mutants were generated by CRISPR/Cas9-mediated genome editing in the *japonica* rice (*Oryza sativa* ssp. *japonica*) of 'Zhonghua 11′ (ZH11) background. Homozygous T_2 lines of *oslea1b* were used in experiments. Rice plants were grown in a paddy field in Hangzhou, Zhejiang province, China, during the regular growing season. Wild-type and *oslea1b* plants (pre-flowering) were subjected to normal temperatures (12 h of light at 28 °C; 12 h of darkness at 22 °C) and high temperatures (12 h of light at 35 °C; 12 h of darkness at 28 °C) in plant growth chambers. The average outdoor daytime temperature in Hangzhou was 37 °C from 20 July 2020 to 20 August 2020.

4.2. Vector Construction and Plant Transformation

To knock out *OsLEA1b* using CRISPR/Cas9, single guide RNAs (sgRNA) targeting *OsLEA1b* (Os01g0276300) were cloned into the BGK03 vector (Biogle, Hangzhou, China; http://www.biogle.cn/index/excrispr, accessed on 1 October 2014). The CRISPR/Cas9–*OsLEA1b* construct was transformed into ZH11 calli via *Agrobacterium tumefaciens* (Agrobacterium)-mediated transformation. Primer sequences used for vector construction and validation are listed in Table S1.

4.3. RNA Extraction, RT-qPCR Analysis, and GUS Staining

Total RNA was isolated from roots, stems, leaves, flowers, and developing grains at 5, 10, 15, 20, 25, and 30 days after fertilization (DAF) using an RNA extraction kit (Trizol, Invitrogen, Carlsbad, CA, USA) following the manufacturer's instructions. Total RNA was reverse transcribed by priming with oligo (dT18) based on the instructions from the ReverTra Ace qPCR RT Kit (Toyobo, Osaka, Japan). Gene expression was quantified by reverse transcription quantitative PCR (RT-qPCR). RT-qPCR was performed using SYBR Green Real-time PCR Master Mix (Toyobo) and a Light Cycler 480 system (Roche, Basel, Switzerland). Relative gene expression was calculated using the $2^{-\Delta\Delta CT}$ method [52]. Primer sequences used in this analysis are listed in Table S1. The putative promoter region of *OsLEA1b* (~2 kb upstream of ATG) was amplified by PCR and cloned into the *Eco*RI and *Nco*I sites of pCAMBIA1305. This construct was transformed into ZH11 calli through Agrobacterium-mediated transformation. Positive T_0 transgenic progeny were used to detect GUS activity as previously described [52].

4.4. Subcellular Localization

The *OsLEA1b* coding sequence was cloned into the pAN580-GFP vector capable of generating an OsLEA1b-GFP fusion protein. The fusion construct and an empty vector control (PAN580-GFP) were separately transformed into rice protoplasts [53]. GFP fluorescence was detected using a confocal laser scanning microscope with 488 nm (Leica TCS SP5, Wetzlar, Germany).

4.5. Electron Microscopy

Dry and completely brown rice grains of *oslea1b* mutants and the wild type were randomly selected. Images were obtained using a HITACHI S3400N scanning electron microscope (Hi-tachi, Tokyo, Japan). Scanning electron microscopy was performed as described previously [54]. To examine the development of compound granules, transverse sections (approximately 1-mm thick) of wild-type and *oslea1b* endosperms at 10 DAF were used to prepare semi-thin sections (800 nm). Samples were stained with I_2-KI for 5 s and subsequently examined under a light microscope (Nikon Eclipse 80i, Tokyo, Japan; http://www.nikon.com accessed on 1 January 2018). All treatments were performed as described previously [55]. To observe amyloplast ultrastructure, developing seeds (10 DAF) of wild-type and *oslea1b* plants were analyzed using transmission electron microscopy (H7650; Hitachi). Samples were treated as described previously [56].

4.6. Analysis of Endosperm Starch Physicochemical Properties

Total starch and amylose contents were measured using the Megazyme K-TSTA and K-AMYL (Megazyme, Wicklow, Ireland) kits, respectively. Lipid and protein contents in *oslea1b* mutant and wild type were quantified following a previously established method [54]. Gel consistency was measured according to published methods [57]. To determine the pasting properties of endosperm starch, 3 g of milled rice flour (0.5 mm or less, 14% moisture) was transferred into a sample vessel containing 25 mL of distilled water. The sample was mixed and assessed using a Rapid Visco Analyzer (RVA Techmaster, Newport Scientific, Narrabeen, Australia). To determine the chain length distributions of amylopectin, 5 mg of rice powder was digested with *Pseudomonas amyloderamosa* isoamylase (Megazyme) and then analyzed by capillary electrophoresis (PA800 plus pharmaceutical analysis system, Beckman Coulter, Brea, CA, USA).

4.7. Measurement of POD Activity and H_2O_2, Proline, and MDA Levels

Peroxidase activity in leaves was measured using an Amplex™ Red Hydrogen Peroxide/Peroxidase Assay Kit (A22180). MDA, proline, and H_2O_2 levels were determined by visible spectrophotometry using a malondialdehyde (MDA) assay kit (A003-1,TBA method), proline assay kit (A084-3-1), and hydrogen peroxide assay kit (A064-1-1, Nanjing, Jiancheng biotechnology company), respectively.

4.8. Data Analysis

Data are presented as the means ± standard deviation (SD), shown by error bars. The chi-square (χ^2) test and independent samples *t*-test (* $p < 0.05$; ** $p < 0.01$; NS, not significant) were used for statistical analysis using IBM SPSS Statistics 26 software.

5. Conclusions

In summary, we identified *OsLEA1b*, a heat-stress-responsive gene in rice grain filling. *OsLEA1b* is highly expressed in the endosperm, and its coding protein localizes to the nucleus and cytoplasm. The *oslea1b* mutants had abnormal compound starch granules in the endosperm cells and chalky endosperm with significantly decreased grain weight and grain number per panicle, as well as significantly lower contents of starch, protein, and lipid compared to the wild type. Moreover, *OsLEA1b* was induced by high temperature and drought stress. The *oslea1b* mutants showed more sensitivity to high temperatures. These results indicate that *OsLEA1b* regulates starch biosynthesis and influences rice grain quality, especially under high temperatures. In conclusion, this research provides a valuable gene resource and strategy for genetic improvement of rice grain quality.

Supplementary Materials: The following supporting information can be downloaded at: https://www.mdpi.com/article/10.3390/plants12234070/s1, Figure S1. Phylogenetic analysis and amino acid sequence alignment of OsLEA1b. Figure S2. Structure and expression pattern prediction of OsLEA1b. Figure S3. CRISPR/Cas9 mediated editing lines of *OsLEA1b*. Figure S4. Appearance analysis of WT and *oslea1b* mutants. Figure S5. Grains filling of WT and *oslea1b* mutants. Figure S6. High temperature induces the expression of *OsLEA1b*. Figure S7. OsLEA1b is a positive regulator of drought tolerance. Table S1. Primers used in this study.

Author Contributions: X.W., Y.Y. and P.H. directed the project. G.L. performed the experiments. R.C., L.M., G.J., P.C., N.D., X.L. (Xinwei Li), Y.D., X.L. (Xiaoxue Li) and M.Z. participated in the experiments. G.L., R.C., G.J. and X.W. wrote the manuscript. G.S., Z.S., S.H., S.T., X.W., Y.Y. and P.H. revised and finalized the manuscript. All authors have read and agreed to the published version of the manuscript.

Funding: This work was supported by the National Natural Science Foundation of China (32372099, 32172080, and 32301821) and the International Science & Technology Innovation Program of the Chinese Academy of Agricultural Sciences, China (CAAS-ZDRW202109). The funding agencies had no role in the study design, data collection and analysis, decision to publish, or manuscript preparation.

Data Availability Statement: The datasets supporting the conclusions of this article are included within the article and Supplementary Materials.

Conflicts of Interest: The authors declare no conflict of interest.

References

1. Xiong, H.; Wang, W.; Sun, M. Endosperm development is an autonomously programmed process independent of embryogenesis. *Plant Cell* **2021**, *33*, 1151–1160. [CrossRef] [PubMed]
2. Matsukura, C.; Saitoh, T.; Hirose, T.; Ohsugi, R.; Perata, P.; Yamaguchi, J. Sugar uptake and transport in rice embryo. Expression of companion cell-specific sucrose transporter (OsSUT1) induced by sugar and light. *Plant Physiol.* **2000**, *124*, 85–93. [CrossRef] [PubMed]
3. Olsen, O. The Modular Control of Cereal Endosperm Development. *Trends Plant Sci.* **2020**, *25*, 279–290. [CrossRef] [PubMed]
4. Deschamps, P.; Haferkamp, I.; d'Hulst, C.; Neuhaus, H.E.; Ball, S.G. The relocation of starch metabolism to chloroplasts: When why and how. *Trends Plant Sci.* **2008**, *13*, 574–582. [CrossRef]
5. Zhang, L.; Li, N.; Zhang, J.; Zhao, L.; Qiu, J.; Wei, C. The CBM48 domain-containing protein FLO6 regulates starch synthesis by interacting with SSIVb and GBSS in rice. *Plant Mol. Biol.* **2022**, *108*, 343–361. [CrossRef] [PubMed]
6. Seung, D.; Soyk, S.; Coiro, M.; Maier, B.A.; Eicke, S.; Zeeman, S.C. PROTEIN TARGETING TO STARCH is required for localising GRANULE-BOUND STARCH SYNTHASE to starch granules and for normal amylose synthesis in Arabidopsis. *PLoS Biol.* **2015**, *13*, e1002080. [CrossRef]
7. Zhao, C.; Liu, B.; Piao, S.; Wang, X.; Lobell, D.B.; Huang, Y.; Huang, M.; Yao, Y.; Bassu, S.; Ciais, P.; et al. Temperature increase reduces global yields of major crops in four independent estimates. *Proc. Natl. Acad. Sci. USA* **2017**, *114*, 9326–9331. [CrossRef]
8. Pang, Y.; Hu, Y.; Bao, J. Comparative Phosphoproteomic Analysis Reveals the Response of Starch Metabolism to High-Temperature Stress in Rice Endosperm. *Int. J. Mol. Sci.* **2021**, *22*, 10546. [CrossRef]
9. Tabassum, R.; Dosaka, T.; Ichida, H.; Morita, R.; Ding, Y.; Abe, T.; Katsube Tanaka, T. FLOURY ENDOSPERM11-2 encodes plastid HSP70-2 involved with the temperature-dependent chalkiness of rice (*Oryza sativa* L.) grains. *Plant J. Cell Mol. Biol.* **2020**, *103*, 604–616. [CrossRef]
10. Zhang, H.; Xu, H.; Feng, M.; Zhu, Y. Suppression of OsMADS7 in rice endosperm stabilizes amylose content under high temperature stress. *Plant Biotechnol. J.* **2018**, *16*, 18–26. [CrossRef]
11. Hakata, M.; Kuroda, M.; Miyashita, T.; Yamaguchi, T.; Kojima, M.; Sakakibara, H.; Mitsui, T.; Yamakawa, H. Suppression of alpha-amylase genes improves quality of rice grain ripened under high temperature. *Plant Biotechnol. J.* **2012**, *10*, 1110–1117. [CrossRef] [PubMed]
12. Zhang, H.; Duan, L.; Dai, J.S.; Zhang, C.Q.; Li, J.; Gu, M.H.; Liu, Q.Q.; Zhu, Y. Major QTLs reduce the deleterious effects of high temperature on rice amylose content by increasing splicing efficiency of Wx pre-mRNA. *Theor. Appl. Genet.* **2014**, *127*, 273–282. [CrossRef] [PubMed]
13. Nie, D.M.; Ouyang, Y.D.; Wang, X.; Zhou, W.; Hu, C.G.; Yao, J. Genome-wide analysis of endosperm-specific genes in rice. *Gene* **2013**, *530*, 236–247. [CrossRef] [PubMed]
14. Yu, J.; Lai, Y.; Wu, X.; Wu, G.; Guo, C. Overexpression of OsEm1 encoding a group I LEA protein confers enhanced drought tolerance in rice. *Biochem. Biophys. Res. Commun.* **2016**, *478*, 703–709. [CrossRef] [PubMed]

15. Hundertmark, M.; Hincha, D.K. LEA (late embryogenesis abundant) proteins and their encoding genes in *Arabidopsis thaliana*. *BMC Genom.* **2008**, *9*, 118. [CrossRef] [PubMed]
16. Fang, Y.; You, J.; Xie, K.; Xie, W.; Xiong, L. Systematic sequence analysis and identification of tissue-specific or stress-responsive genes of NAC transcription factor family in rice. *Mol. Genet. Genom.* **2008**, *280*, 547–563. [CrossRef] [PubMed]
17. Duan, J.; Cai, W.; Park, S. OsLEA3-2, an abiotic stress induced gene of rice plays a key role in salt and drought tolerance. *PLoS ONE* **2012**, *7*, e45117. [CrossRef]
18. Lin, C.; Li, C.; Lin, S.; Yang, F.; Huang, J.; Liu, Y.; Lur, H. Influence of High Temperature during Grain Filling on the Accumulation of Storage Proteins and Grain Quality in Rice (*Oryza sativa* L.). *J. Agric. Food Chem.* **2010**, *58*, 10545–10552. [CrossRef]
19. Xia, Y.; Sun, Y.; Yuan, J.; Xing, C. Grain quality evaluation of japonica rice effected by cultivars, environment, and their interactions based on appearance and processing characteristics. *Food Sci. Nutr.* **2021**, *9*, 2129–2138. [CrossRef]
20. Misra, G.; Anacleto, R.; Badoni, S.; Butardo, V.; Molina, L.; Graner, A.; Demont, M.; Morell, M.K.; Sreenivasulu, N.; Rebetzke, G. Dissecting the genome-wide genetic variants of milling and appearance quality traits in rice. *J. Exp. Bot.* **2019**, *70*, 5115–5130. [CrossRef]
21. Gao, B.; Hu, S.; Jing, L.; Wang, Y.; Zhu, J.; Wang, K.; Li, H.; Sun, X.; Wang, Y.; Yang, L. Impact of Elevated CO_2 and Reducing the Source-Sink Ratio by Partial Defoliation on Rice Grain Quality–A 3-Year Free-Air CO_2 Enrichment Study. *Front. Plant Sci.* **2021**, *12*, 788104. [CrossRef] [PubMed]
22. Zhang, C.; Zhou, L.; Zhu, Z.; Lu, H.; Zhou, X.; Qian, Y.; Li, Q.; Lu, Y.; Gu, M.; Liu, Q. Characterization of Grain Quality and Starch Fine Structure of Two Japonica Rice (*Oryza sativa*) Cultivars with Good Sensory Properties at Different Temperatures during the Filling Stage. *J. Agric. Food Chem.* **2016**, *64*, 4048–4057. [CrossRef] [PubMed]
23. Xu, X.; Li, X.; Li, Z.; Li, Y.; Chen, K.; Wu, L.; Fa, Y.; Xu, Z.; Xu, Q. Effects of Genetic Background and Environmental Conditions on Amylopectin Chain-Length Distribution in a Recombinant Inbred Line of an Inter-subspecies Rice Cross. *J. Agric. Food Chem.* **2020**, *68*, 7444–7452. [CrossRef] [PubMed]
24. Xu, Y.; Chu, C.; Yao, S. The impact of high-temperature stress on rice: Challenges and solutions. *Crop J.* **2021**, *9*, 963–976. [CrossRef]
25. Liu, Z.; Jiang, S.; Jiang, L.; Li, W.; Tang, Y.; He, W.; Wang, M.; Xing, J.; Cui, Y.; Lin, Q.; et al. Transcription factor OsSGL is a regulator of starch synthesis and grain quality in rice. *J. Exp. Bot.* **2022**, *73*, 3417–3430. [CrossRef] [PubMed]
26. Huang, L.; Tan, H.; Zhang, C.; Li, Q.; Liu, Q. Starch biosynthesis in cereal endosperms: An updated review over the last decade. *Plant Commun.* **2021**, *2*, 100237. [CrossRef]
27. Zhang, Z.; Tappiban, P.; Ying, Y.; Hu, Y.; Bao, J. Functional Interactions between Enzymes Involved in Amylose and Amylopectin Biosynthesis in Rice Based on Mathematical Models. *Biomacromolecules* **2022**, *23*, 1443–1452. [CrossRef]
28. Crofts, N.; Iizuka, Y.; Abe, N.; Miura, S.; Kikuchi, K.; Matsushima, R.; Fujita, N. Rice Mutants Lacking Starch Synthase I or Branching Enzyme IIb Activity Altered Starch Biosynthetic Protein Complexes. *Front. Plant Sci.* **2018**, *9*, 1817. [CrossRef]
29. Peng, Y.; Mao, B.; Zhang, C.; Shao, Y.; Wu, T.; Hu, L.; Hu, Y.; Tang, L.; Li, Y.; Tang, W.; et al. Influence of physicochemical properties and starch fine structure on the eating quality of hybrid rice with similar apparent amylose content. *Food Chem.* **2021**, *353*, 129461. [CrossRef]
30. Magwanga, R.O.; Lu, P.; Kirungu, J.N.; Lu, H.; Wang, X.; Cai, X.; Zhou, Z.; Zhang, Z.; Salih, H.; Wang, K.; et al. Characterization of the late embryogenesis abundant (LEA) proteins family and their role in drought stress tolerance in upland cotton. *BMC Genet.* **2018**, *19*, 6. [CrossRef]
31. Charfeddine, S.; Saïdi, M.N.; Charfeddine, M.; Gargouri-Bouzid, R. Genome-wide identification and expression profiling of the late embryogenesis abundant genes in potato with emphasis on dehydrins. *Mol. Biol. Rep.* **2015**, *42*, 1163–1174. [CrossRef] [PubMed]
32. Amara, I.; Odena, A.; Oliveira, E.; Moreno, A.; Masmoudi, K.; Pagès, M.; Goday, A. Insights into Maize LEA Proteins: From Proteomics to Functional Approaches. *Plant Cell Physiol.* **2012**, *53*, 312–329. [CrossRef] [PubMed]
33. Kamarudin, Z.S.; Yusop, M.R.; Ismail, M.R.; Tengku, M.M.M.; Harun, A.R.; Yusuff, O.; Magaji, U.; Fatai, A. LEA Gene Expression Assessment in Advanced Mutant Rice Genotypes under Drought Stress. *Int. J. Genom.* **2019**, *2019*, 8406036. [CrossRef] [PubMed]
34. Knox-Brown, P.; Rindfleisch, T.; Gunther, A.; Balow, K.; Bremer, A.; Walther, D.; Miettinen, M.S.; Hincha, D.K.; Thalhammer, A. Similar Yet Different-Structural and Functional Diversity among Arabidopsis thaliana LEA_4 Proteins. *Int. J. Mol. Sci.* **2020**, *21*, 2794. [CrossRef] [PubMed]
35. Liu, Y.; Xie, L.; Liang, X.; Zhang, S. CpLEA5, the Late Embryogenesis Abundant Protein Gene from Chimonanthus praecox, Possesses Low Temperature and Osmotic Resistances in Prokaryote and Eukaryotes. *Int. J. Mol. Sci.* **2015**, *16*, 26978–26990. [CrossRef] [PubMed]
36. Liu, Y.; Zheng, Y.; Zhang, Y.; Wang, W.; Li, R. Soybean PM2 Protein (LEA3) Confers the Tolerance of Escherichia coli and Stabilization of Enzyme Activity Under Diverse Stresses. *Curr. Microbiol.* **2010**, *60*, 373–378. [CrossRef] [PubMed]
37. Liu, G.; Xu, H.; Zhang, L.; Zheng, Y. Fe Binding Properties of Two Soybean (*Glycine max* L.) LEA4 Proteins Associated with Antioxidant Activity. *Plant Cell Physiol.* **2011**, *52*, 994–1002. [CrossRef] [PubMed]
38. Olvera-Carrillo, Y.; Campos, F.; Reyes, J.L.; Garciarrubio, A.; Covarrubias, A.A. Functional Analysis of the Group 4 Late Embryogenesis Abundant Proteins Reveals Their Relevance in the Adaptive Response during Water Deficit in Arabidopsis. *Plant Physiol.* **2010**, *154*, 373–390. [CrossRef]
39. Aduse Poku, S.; Nkachukwu Chukwurah, P.; Aung, H.H.; Nakamura, I. Over-Expression of a Melon Y3SK2-Type LEA Gene Confers Drought and Salt Tolerance in Transgenic Tobacco Plants. *Plants* **2020**, *9*, 1749. [CrossRef]

40. Imran, S.; Tsuchiya, Y.; Tran, S.; Katsuhara, M. Identification and Characterization of Rice OsHKT1;3 Variants. *Plants* **2021**, *10*, 2006. [CrossRef]
41. Piao, W.; Kim, S.; Lee, B.; An, G.; Sakuraba, Y.; Paek, N. Rice transcription factor OsMYB102 delays leaf senescence by down-regulating abscisic acid accumulation and signaling. *J. Exp. Bot.* **2019**, *70*, 2699–2715. [CrossRef] [PubMed]
42. Kim, S.; Choi, H.; Cho, Y.; Kim, S. Cold-Responsive Regulation of a Flower-Preferential Class III Peroxidase Gene, OsPOX1, in Rice (*Oryza sativa* L.). *J. Plant Biol. Singmul Hakhoe Chi* **2012**, *55*, 123–131. [CrossRef]
43. Nagar, P.; Kumar, A.; Jain, M.; Kumari, S.; Mustafiz, A.; Prasad, M. Genome-wide analysis and transcript profiling of PSKR gene family members in *Oryza sativa*. *PLoS ONE* **2020**, *15*, e236349. [CrossRef] [PubMed]
44. Ganguly, M.; Roychoudhury, A.; Sengupta, D.N.; Datta, S.K.; Datta, K. Independent overexpression of OsRab16A and AtDREB1A exhibit enhanced drought tolerance in transgenic aromatic rice variety Pusa Sugandhi 2. *J. Plant Biochem. Biot.* **2020**, *29*, 503–517. [CrossRef]
45. Huang, S.; Hu, L.; Zhang, S.; Zhang, M.; Jiang, W.; Wu, T.; Du, X. Rice OsWRKY50 Mediates ABA-Dependent Seed Germination and Seedling Growth, and ABA-Independent Salt Stress Tolerance. *Int. J. Mol. Sci.* **2021**, *22*, 8625. [CrossRef] [PubMed]
46. Sakuraba, Y.; Piao, W.; Lim, J.; Han, S.; Kim, Y.; An, G.; Paek, N. Rice ONAC106 Inhibits Leaf Senescence and Increases Salt Tolerance and Tiller Angle. *Plant Cell Physiol.* **2015**, *56*, 2325–2339. [CrossRef]
47. Sheng, C.; Yu, D.; Li, X.; Yu, H.; Zhang, Y.; Saqib Bilal, M.; Ma, H.; Zhang, X.; Baig, A.; Nie, P.; et al. OsAPX1 Positively Contributes to Rice Blast Resistance. *Front. Plant Sci.* **2022**, *13*, 843271. [CrossRef]
48. Ye, N.; Zhu, G.; Liu, Y.; Li, Y.; Zhang, J. ABA controls H_2O_2 accumulation through the induction of OsCATB in rice leaves under water stress. *Plant Cell Physiol.* **2011**, *52*, 689–698. [CrossRef]
49. Kamiya, T.; Akahori, T.; Ashikari, M.; Maeshima, M. Expression of the vacuolar Ca^{2+}/H^{+} exchanger, OsCAX1a, in rice: Cell and age specificity of expression, and enhancement by Ca^{2+}. *Plant Cell Physiol.* **2006**, *47*, 96–106. [CrossRef]
50. Wang, J.; Ren, Y.; Liu, X.; Luo, S.; Zhang, X.; Liu, X.; Lin, Q.; Zhu, S.; Wan, H.; Yang, Y.; et al. Transcriptional activation and phosphorylation of OsCNGC9 confer enhanced chilling tolerance in rice. *Mol. Plant* **2021**, *14*, 315–329. [CrossRef]
51. Ma, C.; Li, Q.; Song, Z.; Su, L.; Tao, W.; Zhou, B.; Wang, Q. Irrigation with Magnetized Water Alleviates the Harmful Effect of Saline–Alkaline Stress on Rice Seedlings. *Int. J. Mol. Sci.* **2022**, *23*, 10048. [CrossRef] [PubMed]
52. Cao, R.; Zhao, S.; Jiao, G.; Duan, Y.; Ma, L.; Dong, N.; Lu, F.; Zhu, M.; Shao, G.; Hu, S.; et al. OPAQUE3, encoding a transmembrane bZIP transcription factor, regulates endosperm storage protein and starch biosynthesis in rice. *Plant Commun.* **2022**, *3*, 100463. [CrossRef] [PubMed]
53. Chen, S.; Tao, L.; Zeng, L.; Vega-Sanchez, M.E.; Umemura, K.; Wang, G.L. A highly efficient transient protoplast system for analyzing defence gene expression and protein-protein interactions in rice. *Mol. Plant Pathol.* **2006**, *7*, 417–427. [CrossRef] [PubMed]
54. Kang, H.; Park, S.; Matsuoka, M.; An, G. White-core endosperm floury endosperm-4 in rice is generated by knockout mutations in the C-type pyruvate orthophosphate dikinase gene (OsPPDKB). *Plant J.* **2005**, *42*, 901–911. [CrossRef] [PubMed]
55. Peng, C.; Wang, Y.; Liu, F.; Ren, Y.; Zhou, K.; Lv, J.; Zheng, M.; Zhao, S.; Zhang, L.; Wang, C.; et al. FLOURY ENDOSPERM6 encodes a CBM48 domain-containing protein involved in compound granule formation and starch synthesis in rice endosperm. *Plant J. Cell Mol. Biol.* **2014**, *77*, 917–930. [CrossRef]
56. Li, S.; Wei, X.; Ren, Y.; Qiu, J.; Jiao, G.; Guo, X.; Tang, S.; Wan, J.; Hu, P. OsBT1 encodes an ADP-glucose transporter involved in starch synthesis and compound granule formation in rice endosperm. *Sci. Rep.* **2017**, *7*, 40124. [CrossRef]
57. Tan, Y.F.; Li, J.X.; Yu, S.B.; Xing, Y.Z.; Xu, C.G.; Zhang, Q. Three important traits for cooking and eating quality of rice grains are controlled by a single locus in an elite rice hybrid, Shanyou 63. *Theor. Appl. Genet.* **1999**, *99*, 642–648. [CrossRef]

plants

Article

OsCIPK9 Interacts with OsSOS3 and Affects Salt-Related Transport to Improve Salt Tolerance

Zhenling Zhou [1,2], Weijie Tang [3,4], Zhiguang Sun [1], Jingfang Li [1], Bo Yang [1], Yan Liu [1], Baoxiang Wang [1,*], Dayong Xu [1], Jianchang Yang [2] and Yunhui Zhang [3,4,5,*]

1 Lianyungang Academy of Agricultural Sciences, Lianyungang 222000, China; zhouzl3716@163.com (Z.Z.); zhiguangsun@126.com (Z.S.); jingfangli@163.com (J.L.); nkykjc19@163.com (B.Y.); ly516.bester@163.com (Y.L.); xudayong3030@163.com (D.X.)
2 Jiangsu Key Laboratory of Crop Genetics and Physiology, Jiangsu Key Laboratory of Crop Cultivation and Physiology, Agricultural College of Yangzhou University, Yangzhou 225009, China; 003335@yzu.edu.cn
3 Provincial Key Laboratory of Agrobiology, Institute of Germplasm Resources and Biotechnology, Jiangsu Academy of Agricultural Sciences, Nanjing 210014, China; weijiet08@126.com
4 Zhongshan Biological Breeding Laboratory, No. 50 Zhongling Street, Nanjing 210014, China
5 Jiangsu Co-Innovation Center for Modern Production Technology of Grain Crops, Yangzhou University, Yangzhou 225009, China
* Correspondence: 20101404@jaas.ac.cn (B.W.); zyhrice@163.com (Y.Z.)

Abstract: Salt is harmful to crop production. Therefore, it is important to understand the mechanism of salt tolerance in rice. *CIPK* genes have various functions, including regulating salt tolerance and other types of stress and nitrogen use efficiency. In rice, OsCIPK24 is known to regulate salt tolerance, but other OsCIPKs could also function in salt tolerance. In this study, we identified another OsCIPK—OsCIPK9—that can regulate salt tolerance. Knockout of *OsCIPK9* in rice could improve salt tolerance. Through expression analyses, *OsCIPK9* was found to be mainly expressed in the roots and less expressed in mature leaves. Meanwhile, *OsCIPK9* had the highest expression 6 h after salt treatment. In addition, we proved the interaction between OsCIPK9 and OsSOS3. The RNA-seq data showed that *OsCIPK9* strongly responded to salt treatment, and the transporters related to salt tolerance may be downstream genes of *OsCIPK9*. Finally, haplotype analyses revealed that Hap6 and Hap8 mainly exist in *indica*, potentially providing a higher salt tolerance. Overall, a negative regulator of salt tolerance, OsCIPK9, which interacted with OsSOS3 similarly to OsCIPK24 and influenced salt-related transporters, was identified, and editing *OsCIPK9* potentially could be helpful for breeding salt-tolerant rice.

Keywords: *OsCIPK9*; *OsSOS3*; sat-related transporters; salt tolerance; rice

Citation: Zhou, Z.; Tang, W.; Sun, Z.; Li, J.; Yang, B.; Liu, Y.; Wang, B.; Xu, D.; Yang, J.; Zhang, Y. OsCIPK9 Interacts with OsSOS3 and Affects Salt-Related Transport to Improve Salt Tolerance. *Plants* **2023**, *12*, 3723. https://doi.org/10.3390/plants12213723

Academic Editor: Koji Mikami

Received: 11 September 2023
Revised: 17 October 2023
Accepted: 24 October 2023
Published: 30 October 2023

1. Introduction

An increase in the output of rice, as a staple food feeding almost half the world, will be required over the next 50 years. Therefore, either a higher yield per ha or more cultivated land is needed for a larger harvest. Making full use of the soil would make higher yields achievable. However, salt in saline–alkali soil damages crop production. Salinity includes osmotic stress, ionic toxicity, and nutritional deficiencies, which inhibit rice development [1].

As a major abiotic stressor, researchers have focused on the mechanism of salt tolerance using quantitative trait loci (QTL) mapping [2] and genome-wide association studies (GWASs) [3,4]. With progress in rice genome sequencing, related genes that could be used for rice breeding, such as *SKC1*, have been cloned [5]. *SKC1* is a high-affinity K^+ transporter which was cloned from Pokkali with extreme salt resistance. OsWRKY53 is a key regulator cloned from GWAS which can regulate the expression of *OsMKK10.2* in promoting ion homeostasis and trans-represses *SKC1* [4]. The superior alleles identified could be useful for breeding rice with salt tolerance. Meanwhile, greater yields would be obtained in saline–alkali soils with the use of salt-tolerant genes [6].

In addition to genes like *SKC1*, *CIPK* family genes have an important role in salt tolerance. CIPK family genes are plant-specific proteins which interact with CBL and serve as major downstream signaling components [7,8]. The CBL-CIPK network plays a vital role in salinity stress, disease defense, drought tolerance, and other stresses [9]. In Arabidopsis, the SOS (Salt Overly Sensitive) pathway was the first well-studied pathway, and this acts as an example for other networks. The AtCBL4–AtCIPK24 complex activates downstream AtSOS1, which affects Na^+ extrusion and long-distance Na^+ transport [10]. Also, AtCIPK24 can form an AtCIPK24–GI complex to delay the flowering time under saline conditions [11]. After that, another network, AtCBL10-AtCIPK8, was identified, which also regulates AtSOS1 [12]. In addition to salinity stress, the ATCBL4-AtCIPK6 pathway regulates K^+ allocation [13]. The AtCBL9-AtCIPK3 network can affect seed dormancy through activating ABR1 in the nucleus [14]. In short, CIPK family genes have multiple functions in Arabidopsis. In rice, CIPKs can phosphorylate many transporters that have multiple functions in various processes, such as nitrogen uptake [15], K^+ uptake [16], and microbe-associated molecular pattern-induced defense [17]. *OsCIPK9, 14, 15* regulates microbe-associated molecular pattern-induced hypersensitive cell death, phytoalexin production, and defense gene expression in cultured cells [18]. Twelve *OsCIPK* genes, including *OsCIPK9*, have been demonstrated to be induced by salinity stress [19]. In addition, *CIPKs* are upregulated during panicle development and abiotic stress [18]. In salt tolerance, *OsSOS2* (*OsCIPK24*) and *OsSOS3* (*OsCBL4*) play vital roles [20]. The calcium-binding protein OsSOS3/CBL4 can sense the cytosolic calcium signal elicited by salt stress, then interact with and activate OsSOS2 (OsCIPK24). Then, activated OsSOS2/OsCIPK24 phosphorylates and activates OsSOS1 to regulate Na^+ homeostasis and improve rice tolerance to salt stress [21]. In rice, the *CIPK* family has 33 members, with *OsCIPK24* regulating salt tolerance [18], but other *OsCIPKs* which may also have a role in salt tolerance are still unknown. A previous study showed that the mutant *Oscipk9* showed a mild salt tolerance [16], but the possible mechanisms of salt tolerance and the haplotypes of *OsCIPK9* are still unclear.

Here, we proved that OsCIPK9 is a negative regulator of salt tolerance in rice using CRISPR-Cas9. *OsCIPK9* knockouts of rice showed an increased salt tolerance, and overexpression lines were more sensitive. The Na^+ concentration changed significantly after salt treatment. *OsSOS3* showed a higher expression in the *OsCIPK9-cas* line, and OsSOS3 interacted with OsCIPK9. As determined through RNA-seq data, *OsCIPK9* strongly responded to salt treatment. We found obvious changes in the expression of transporters like *OsKAT1*, especially under saline conditions. Finally, we analyzed the haplotypes of *OsCIPK9* and showed that the haplotype of *OsCIPK9* had an obvious subpopulation classification.

2. Results

2.1. OsCIPK9 Negatively Regulated Salt Tolerance

To study the specific role of *OsCIPK9* in rice, we constructed knockout lines using CRISPR-Cas9. One knockout line (*OsCIPK9-cas*) with a base insertion was identified (Figure 1a). The insertion caused a frameshift, and translation stopped after 148 aa (Figure 1a). The protein sequence of the knockout line showed that the domain lacked 140 normal amino acids (143–283) and contained six mutated amino acids (from 143 to 148) compared to that of the wild type (Figure 1b). We also created overexpression lines for further function validation. The overexpression lines OsCIPK9-OE2 and OsCIPK9-OE3 had approximately 9- and 5-fold higher expressions than the wild type (Nipponbare), respectively (Figure 1c).

To verify whether *OsCIPK9* could function in salt tolerance, we treated the transgenic lines and wild type with 0 and 120 mM NaCl. As a result, the knockout line (*OsCIPK9-cas*) showed a higher tolerance to salt treatment, while the overexpression lines were more sensitive (Figure 2a–c). The OsCIPK9-OE2 and OsCIPK9-OE3 lines had a lower relative fresh weight compared with the WT, while that of the knockout line was higher (Figure 2d,e).

We also measured the Na^+ and K^+ concentrations of whole seedlings. In the shoots, the four lines showed no differences in Na^+ concentrations and Na^+/K^+ content ratios under no-salt conditions (Figure S1), but a discrepancy was observed under salt treatment, in which these values were higher in the OE lines and lower in the knockout lines (Figure 2f,h). In the roots, the Na^+ concentration of the OE lines was lower than the wild type and the knockout line under no-salt conditions, but displayed almost no difference under saline conditions (Figure S1). The Na^+/K^+ content ratio was higher in the OE lines than in the knockout lines (Figure 2g,i). Thus, *OsCIPK9* conferred salt tolerance in rice, and knocking out this gene improved salt tolerance through regulating the Na concentration.

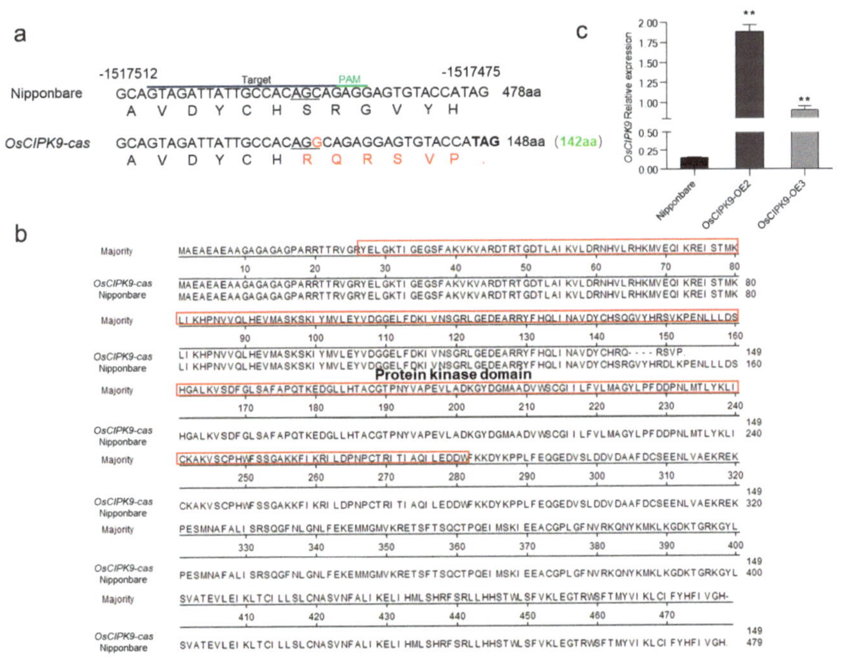

Figure 1. Transgenic line construction. (**a**) Targeted mutagenesis of *OsCIPK9* with CRISPR/Cas9. The mutant alleles have 1 nucleotide insertion in *OsCIPK9-cas*. The base shown in red represents the inserted nucleotide. The amino acids in red represent mutated amino acids. The line below in nucleotides represents the positions which translated the mutated amino acids. (**b**) The result of protein alignment between Nipponbare and *OsCIPK9-cas*. The domain was predicted in Pfam (pfam.xfam.org, accessed on 30 August 2022) presented in red square. (**c**) The expression of Nipponbare and overexpression lines (OsCIPK9-OE2, OsCIPK9-OE3). **: $p < 0.01$. Statistical significance (versus Nipponbare) was calculated using a Student's *t*-test.

2.2. Expression Pattern of OsCIPK9

To determine the expression pattern of *OsCIPK9*, various tissues were collected and the mRNA abundance of *OsCIPK9* was examined using RT-PCR analysis. The results showed that *OsCIPK9* was most expressed in the roots and least expressed in mature leaves (Figure 3a). Also, we measured the expression at different time points after salt treatment. The results show that *OsCIPK9* had the highest expression 6 h after salt treatment (Figure 3b). The knock-out line showed lower expression from 1 h to 7 days, with the largest difference observed after 6 h. These results indicate that *OsCIPK9* responded to salt treatment, and knocking out the gene reduced the expression under salt treatment.

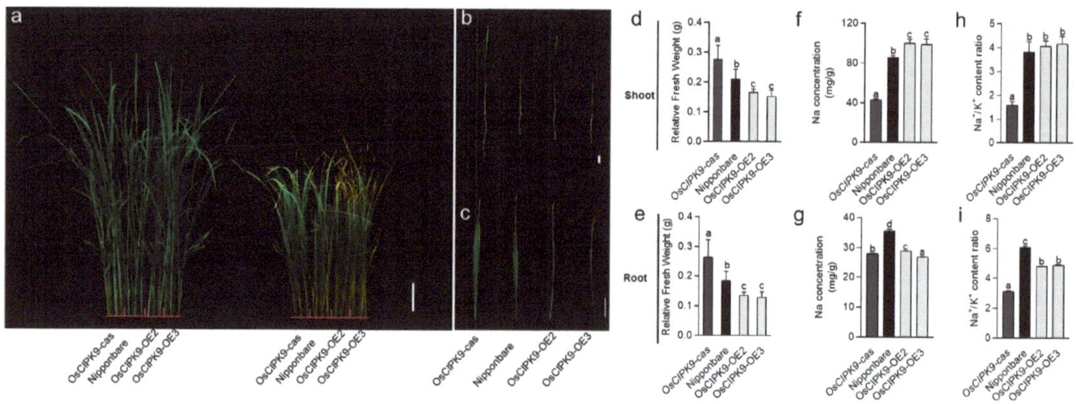

Figure 2. The phenotype of Nipponbare and transgenic lines under 0 and 120 mM NaCl conditions. (**a**–**c**) The phenotype of the wild type (Nipponbare), knockout line (*OsCIPK9-cas*) and overexpression lines (OsCIPK9-OE2, OsCIPK9-OE3) under 0 mM and 120 mM NaCl conditions, bar = 10, 2, 2 cm. (**d**,**f**,**h**) Comparison of Nipponbare, knockout line (*OsCIPK9-cas*) and overexpression lines (OsCIPK9-OE2, OsCIPK9-OE3) in terms of relative fresh weights, Na concentrations per plant and Na^+/K^+ content ratios in shoots under 120 mM NaCl conditions. $n = 3$. (**e**,**g**,**i**) Comparison of Nipponbare, the knockout line (*OsCIPK9-cas*) and overexpression lines (OsCIPK9-OE2, OsCIPK9-OE3) in terms of relative fresh weights, Na concentrations per plant and Na^+/K^+ content ratios in roots under 120 mM NaCl conditions. $n = 3$. The data are presented as means ± SDs. Statistical significance (versus Nipponbare) was calculated using a Student's *t*-test. Different letters indicate significant difference, $p < 0.05$.

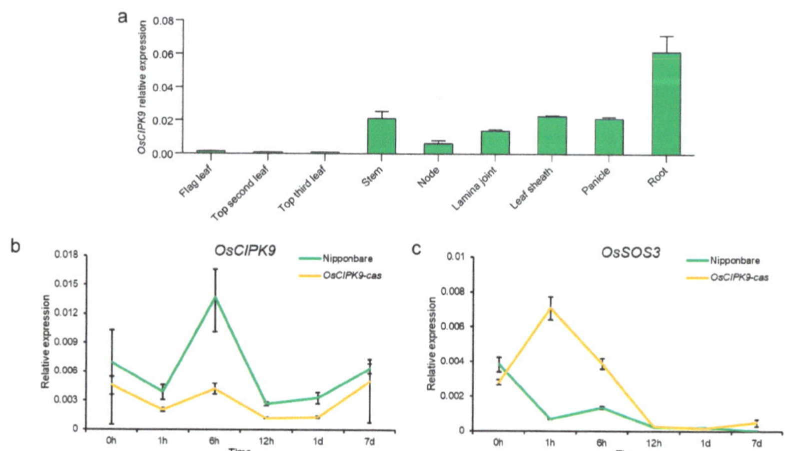

Figure 3. The expression of *OsCIPK9* and *OsSOS3*. (**a**) The expression of *OsCIPK9* in various tissues. (**b**,**c**) Comparison of the expression of *OsCIPK9* and *OsSOS3* between Nipponbare and the knockout line (*OsCIPK9-cas*) under 0 and 120 mM NaCl conditions. $n = 3$. The data are presented as means ± SDs.

2.3. *OsCIPK9 Interacted with OsSOS3*

According to the change in Na^+ concentrations in different transgenic lines, including the knock-out and overexpression lines, we examined whether OsSOS3 in the SOS pathway could interact with OsCIPK9 similar to OsSOS2/OsCIPK24. Firstly, *OsSOS3* showed a lower expression in Nipponbare compared with *OsCIPK9-cas*, while *OsCIPK9* showed a higher expression (Figure 3b,c). Thus, we supposed that OsSOS3 may interact with OsCIPK9, playing an opposing role in salt tolerance. To test this hypothesis, we analyzed the interaction between OsSOS3 and OsCIPK9 using a yeast two-hybrid (Y2H) assay. *OsSOS3* strongly interacted with OsCIPK9 (Figure 4a). Also, we verified the interaction between OsSOS3 and OsCIPK9 using an in vivo firefly luciferase complementation imaging (LCI) assay in Nicotiana benthamiana leaf epidermal cells (Figure 4b). Taken together, these results suggest that OsSOS3 could interact with OsCIPK9 in rice.

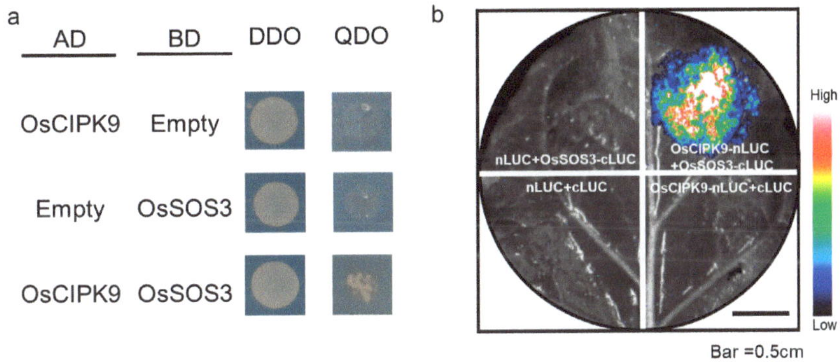

Figure 4. OsCIPK9 physically interacts with OsSOS3. (**a**) Y2H assay showing that OsCIPK9 can interact with OsSOS3 (A). AD, activation domain; BD, binding domain; DDO, SD/−Trp/−Leu; QDO, SD/−Trp/−Leu/−His/−Ade. (**b**) LCI assay showing that OsCIPK9 interacts with OsSOS3 in leaf cells of N. benthamiana. Colored scale bar indicates the luminescence intensity in counts per second (cps). CLUC, C terminus of LUC; NLUC, N terminus of LUC.

2.4. Transcriptome Analysis of OsCIPK9

To identify possible downstream genes influenced by *OsCIPK9*, we performed a transcriptome analysis using knockout lines with different treatments (0 and 120 mM salt). Many upregulated and downregulated DEGs (differential expression genes) were identified, but the number of up- or downregulated genes under NaCl treatment was much greater than under control conditions (Figure 5a). The DEGs included 2285 regulated genes without salt treatment and 9190 genes under salt treatment. Only 1338 genes were found for both treatments (Figure 5b). A GO analysis showed that genes in the plasma membrane of cellular components were identified, but the number of genes under salt treatment (1190) was more than three times that under no salt treatment (358) (Figures 5c and S2). In biological processes, more genes were clustered in cell wall organization under no salt treatment, while the ratio increased under salt treatment. Additionally, more genes were grouped into translation and carbohydrate metabolic processes under salt treatment (Figure S3). The GO molecular function analysis showed that the DNA-binding transcription faction activity process had a large number of genes with a lower Q value under no salt treatment. Under salt treatment, the process with the most regulated genes was structural constituents of the ribosome (Figure S4). Additionally, a KEGG analysis revealed almost no difference between Nipponbare and *OsCIPK9-cas* without NaCl treatment (Figure 5d). A large difference was identified under 120 mM NaCl treatment, as expected (Figure 5d). In the KEGG analysis, almost all processes differed, including transport and catabolism, signal transduction, and membrane transport. The processes consisted of the functions of CIPKs, which have been reported previously [22]. The GO and KEGG results indicated that *OsCIPK9* strongly reacted to salt treatment and played a vital role in salt tolerance, especially in membranes.

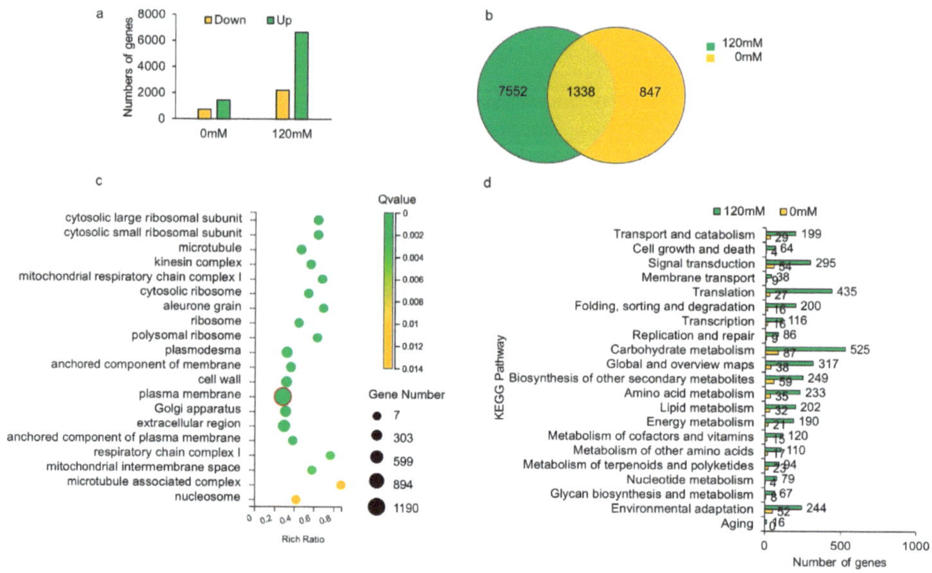

Figure 5. The analysis of RNA-seq data. (**a**) The number of DEGs after 0 mM and 120 mM NaCl treatment. (**b**) The overlap of genes identified after 0 mM and 120 mM NaCl treatment. (**c**) GO analysis of genes identified by RNA-seq. The red circle highlighted the gene number of the plasma membrane identified by GO analysis. (**d**) KEGG analysis of genes identified by RNA-seq.

The RNA-seq analysis showed that *OsCIPK9* affected membrane transport. To identify possible influencing genes in the membrane, we investigated the expression of previously reported transporters, such as *SKC1* and *OsKAT1*, in the RNA-seq data. Luckily, we found that *SKC1*, *OsKAT1*, and *OsNHX1* expression was higher in *OsCIPK9-cas* than in Nipponbare under salt treatment (Figure 6a–c). In particular, the expression of *OsKAT1* rose almost 15-fold under salt treatment, but only 2-fold under no salt treatment (Figure 6b). To verify the expression results, we used real-time PCR to measure the expression of the three transporters and obtained results similar to the RNA-seq data (Figure 6d,e). Therefore, we concluded that *OsCIPK9* conferred salt tolerance by affecting the expression of salt-related transporters expressed in the membrane, like OsKAT1.

2.5. The Haplotypes of OsCIPK9

OsCIPK9 plays a role in salt tolerance, and the haplotype of *OsCIPK9* may be helpful in breeding or for germplasm improvement in rice. To determine the haplotypes of *OsCIPK9*, we analyzed them using SNP data from the Rice3K database. *OsCIPK9* contains 21 SNPs with 4 missense variants and 12 SNPs in the promoter (Figure 7a). All nine haplotypes were identified based on SNP variants. Hap1 predominantly emerged in *japonica*, mostly in temperate *Japonica* (55.5%), while Hap6 and Hap8 (89.2%) were identified in *indica*. Hap2 and Hap9 were identified only in the *Aus* subpopulation. Hap3 and Hap7 were identified in the *Basmati* subpopulation. Hap5 was identified in various subpopulations but was rarely found in *Indica* (Figure 7b). A previous study demonstrated that the tolerance level of *INDICA* was higher than that of *japonica* at the seedling stage [23]. Therefore, varieties with Hap6 and Hap8 may have a higher salt tolerance than those with Hap1. Subsequently, we analyzed the haplotype network. The results showed that an unknown haplotype connected the haplotypes, mostly in *Japonica* and *Indica* (Figure 7c). Hap7, which was found in almost all *Aus* varieties, was the key haplotype connecting the present haplotypes in *Japonica* and *Indica* (Figure 7c). These results indicate that *OsCIPK9* has a subpopulation classification, and Hap7 is the key haplotype that connects the haplotypes existing in *Japonica* and *Indica*.

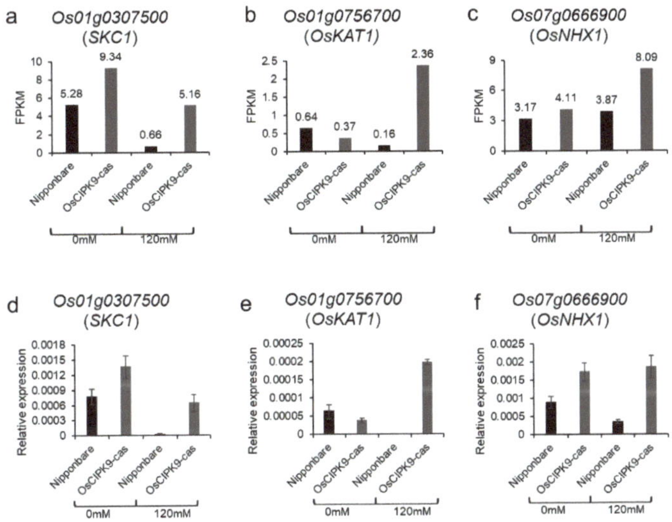

Figure 6. The expression of related transporters. (**a–c**) The FPKM of *SKC1*, *OsKAT1*, and *OsNHX1*. (**d–f**) The relative expressions of *SKC1*, *OsKAT1*, and *OsNHX1*.

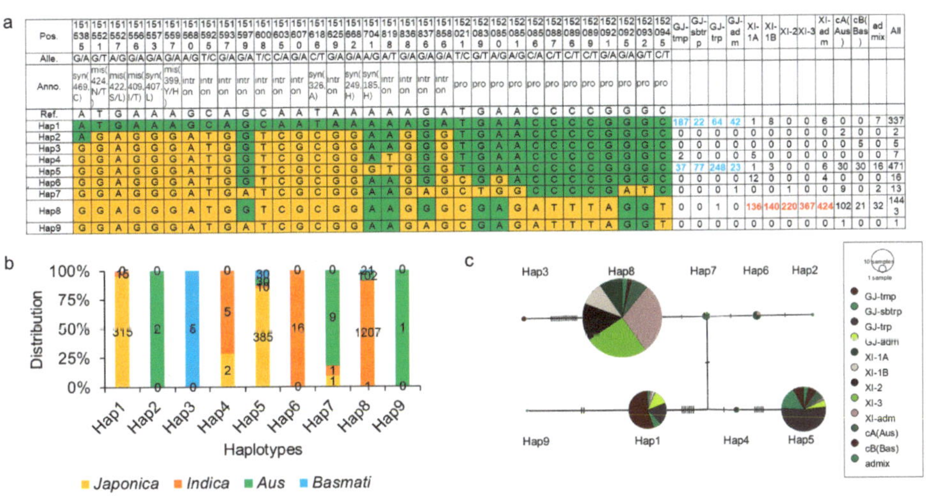

Figure 7. Analysis of haplotypes of *OsCIPK9*. (**a**) Haplotypes of *OsCIPK9* based on the Rice3K database. The green represented the same nucleotides with the reference. The yellow represented the variations compared with the reference. (**b**) The distribution of subpopulations in each haplotype. (**c**) The gene network of different haplotypes. Pos: position; Alle: allele; Anno: annotation; GJ: *japonica*, tmp: temperate, sbtrp: subtropical, trp: tropical, adm: admix. XI: *indica*, Bas: *basmati*, Pro: promoter variation, Syn: synonymous variation, Mis: missense variation, Intron: intron variation.

3. Discussion

In a previous study, *OsCIPK9* was found to regulate ammonium-dependent root growth [15]. Also, AtCIPK23 was found to function in salt tolerance and nitrogen use, including nitrate and ammonium in *Arabidopsis*. Here, knocking out *OsCIPK9* increased the tolerance to salt stress at the seedling stage. Thus, this gene has multiple functions in nitrogen use and salt tolerance. In our results, *OsCIPK9-cas* also showed higher expressions of *OsAMT1.1* and *OsAMT2.1* than Nipponbare under salt stress, especially *OsAMT1.1*, with a two-fold higher expression than the wild type (Figure S5), indicating that *OsCIPK9* plays a role not only in salt tolerance but also in ammonium transport. Overall, *OsCIPK9* could regulate nitrogen use and salt tolerance like *AtCPIK23*.

In this study, OsSOS3 was found to interact with OsCIPK9. In a previous study, OsSOS2 was found to be activated by OsSOS3 [24]. Secondly, *OsKAT1* and some other transporter genes were differentially expressed. In a previous study, researchers found that *CIPK* genes could phosphorylate transporters such as AtCIPK23 and CHL1 in *Arabidopsis* [25]. The phosphorylation of CHL1 has different functions in nitrate transport and sensing. In addition, AtCIPK23 can phosphorylate AtAMT1.1 and influence ammonium uptake [26]. OsSOS2 can phosphorylate OsSOS1, which is a Na^+/H^+ antiporter 1, and regulates salt tolerance in rice [27]. Further studies should be performed to reveal whether OsSOS3 can activate OsCIPK9 and whether OsCIPK9 can phosphorylate the transporters detected in our study, which would provide more powerful evidence for salt tolerance.

AtSOS3 (AtCBL4) could function in salt tolerance by interacting with AtSOS2 (AtCIPK24) [10]. In our study, we proved that OsSOS3 could also interact with OsCIPK9. This result indicates that the CBL-CIPK network plays a vitally important role in plant salt tolerance. Also, AtSOS3 (AtCBL4) regulates K^+ homeostasis through the CBL4-CIPK6-AKT2 pathway [13]. Moreover, AtSOS3 (AtCBL4) was found to be involved in auxin transport [28]. In future studies, more phenotypes may be studied in the OsSOS3-OsCIPK9 pathway.

Other OsCIPKs like *OsCIPK04* also responded to salt treatment according to a previous study [19]. With the development of knockout technology, transgenic lines can be produced more easily than before. Our study showed that *OsCIPK9* could improve salt tolerance using the knockout line. Future studies could focus on other OsCIPKs which may regulate salt tolerance, and rapidly validate the function using CRISP technology. With more OsCIPKs being identified in the future, we may identify a more comprehensive OsCIPK pathway in salt tolerance.

A haplotype analysis showed that lots of variations were present between *japonica* and *indica*. *Japonica* had a different evolutionary process compared to *indica* [29]. Thus, these variations may cause a change in *OsCIPK9*'s function in salt tolerance. Complementation experiments and functional studies of various haplotypes should be conducted in the future. More evidence about the function or causal SNPs of different haplotypes would help in molecular breeding, especially MAS (Marker Assistant Selection).

With progress in gene editing, more editing strategies, such as one-base editing and primer editing, have been used for crop improvements [30]. OsCIPK9 is a negative regulator of salt treatment; therefore, it is better to edit the genomic or promoter region of *OsCIPK9* to produce new alleles for breeding. A haplotype analysis showed that the promoter had more polymorphisms than the gene in our study. Therefore, editing the promoter and creating more lines with differential expressions [31] would be useful for breeding based on the evaluation of other agronomic traits.

4. Conclusions

In summary, we identified a negative regulator of salt tolerance in rice, OsCIPK9. In addition, we proposed the possible mechanism in which OsCIPK9 is involved through interaction with OsSOS3. The most affected cellular component was that of the plasma membrane, and the downstream genes of OsCIPK9 may be the transporters located in the plasma membrane, like OsKAT1, as observed through RNA-seq analyses. Finally, we suggest the possible pathway of salt tolerance in rice in which OsCIPK9 is involved (Figure 8). Our study shows that other OsCIPKs could also function in salt tolerance, like OsCIPK24. In addition, OsSOS3, which can interact not only with OsCIPK24 but also with OsCIPK9, regulates salt tolerance in rice. More importantly, our study investigated more OsCIPKs which may regulate salt tolerance in rice.

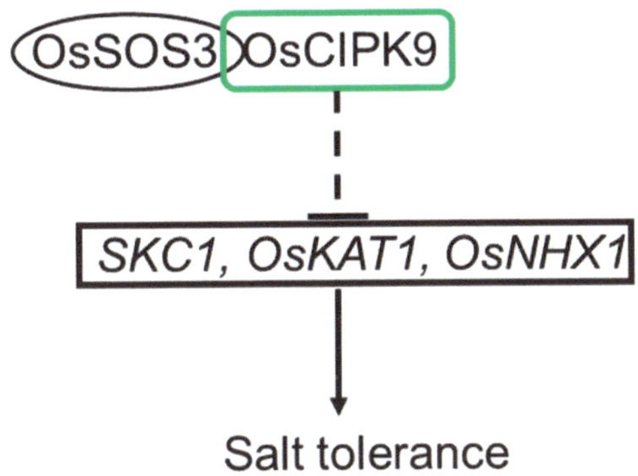

Figure 8. The pathway of OsCIPK9 in the salt tolerance of rice.

5. Materials and Methods

5.1. Plant Materials and Transgenic Lines Construction

The Line *japonica* Nipponbare was used as the wild type, and CRISPR/Cas9 technology was used to produce the knockout line (*OsCIPK9-cas*). The pam sequence is shown in Figure 1a. The full CDS of *OsCIPK9* was cloned from the cDNA of Nipponbare, and the pCUbi1390 recombination vector was constructed. Overexpression lines were developed using *Agrobacterium*-mediated genetic transformation. Two independent overexpression lines (OsCIPK9-OE2 and OsCIPK9-OE3) were isolated from transgenic lines. The expression of overexpression lines was determined using real-time PCR.

5.2. Salt Treatment and Phenotypic Identification

The seeds of all lines were geminated and grown in Yoshida nutrition solution for 14 days. NaCl concentrations of 120 mM and 0 mM (CK) were used for the treatment and control, respectively. The traits, including fresh weight and Na^+ and K^+ concentrations, were measured 7 days after NaCl treatment. Relative fresh weight = fresh weight under no salt conditions − fresh weight under salt treatment. Na^+ and K^+ concentrations were determined according to the method described by Wang et al. [32].

5.3. Y2H Assays

A Y2H assay was performed with a MatchMaker GAL4 Two-Hybrid System (Clontech, Mountain View, CA, USA, https://www.takarabio.com/, accessed on 14 August 2023). The full-length coding sequences of genes (*OsCIPK9* and *OsSOS3*) of interest were cloned from the cDNA of Nipponbare into pGADT7 and pGBKT7 (Clontech, Mountain View, CA, USA, https://www.takarabio.com/, accessed on 14 August 2023), and different combinations of constructs were transformed together into the yeast (*Saccharomyces cerevisiae*) AH109 strain. Positive transformants were selected on synthetic dropout (SD/−Leu/−Trp, DDO) nutrient media, while the interactions were screened in SD medium (SD/−Leu/−Trp/−His/−Ade, QDO).

5.4. Firefly LCI Assays

The coding sequences of *OsCIPK9* and *OsSOS3* were cloned into the pCAMBIA-nLUC or pCAMBIA-cLUC vectors. These constructed vectors were introduced into the Agrobacterium tumefaciens strain EHA105. Various combinations of EHA105 strains were used to infiltrate *N. benthamiana* leaves. The relative LUC activity was measured by using a Nightshade LB 985 system (Berthold Technologies, Baden, Germany, 10 August 2023), as described previously [33].

5.5. RNA-Seq

Additionally, the roots of different lines under 0 and 120 mM NaCl treatments were collected for RNA-seq analysis. The total RNA of three independent plants was isolated using a plant RNA purification reagent (Invitrogen, Shanghai, China, 10 November 2022). RNA-seq was performed using the BGI T7 platform, and analyses of DEGs (Differentially Expressed Genes), KEGG (Kyoto Encyclopedia of Genes and Genomes), and GO (Gene Ontology) were performed using Dr. Tom website produced by BGI, China (https://biosys.bgi.com, accessed on 20 January 2023). The expression of all genes produced via RNA-seq is listed in Table S1.

5.6. Real-Time PCR

The total RNA of different tissues and roots at different time points was extracted using an RNA prep Pure Plant Kit (Tiangen Biotech, Beijing, China). Then, ~1 μg of total RNA was reverse-transcribed into cDNA using a PrimeScipt™ Reverse Transcriptase kit (Takara, Shiga, Japan, www.takarabio.com, accessed on 5 March 2023). Quantitative RT-PCR assays were performed using an SYBR Premix Ex Taq™ kit (Takara, Shiga, Japan, www.takarabio.com, accessed on 5 March 2023). Real-time PCR was performed in a real-time PCR machine (I-Cycle, Bio-Rad, Hercules, CA, USA). The primers are listed in Table S2. The rice *UBQ* (*Os03g0234350*) gene was used as an internal control.

5.7. Haplotype Analysis

SNP data of Rice3K were downloaded from the Rice SNP-Seek Database (https://snp-seek.irri.org/, accessed on 5 June 2022). The haplotypes were separated based on all SNP variants of the *OsCIPK9* gene and promoter (~2K upstream sequence of *OsCIPK9*). The subpopulations of rice were divided into *japonica* (temperate, subtropical, tropical, and admix), *indica* (1A,1B,2,3, and admix), Aus, Basmati, and admix. The haplotype network was constructed using PopART software (https://popart.maths.otago.ac.nz/, accessed on 5 July 2022) [34].

Supplementary Materials: The following supporting information can be downloaded at: https://www.mdpi.com/article/10.3390/plants12213723/s1, Figure S1: Comparison of Nipponbare (Nip), knock-out line (*OsCIPK9-cas*) and overexpression lines (OsCIPK9-OE2, OsCIPK9-OE3) Na concentrations per plant and Na^+/K^+ content ratios in shoot and root under 0 mM NaCl conditions; Figure S2: The result of GO cellular component analysis between *OsCIPK9-cas* and Nipponbare under 0 mM NaCl treatment; Figure S3: The result of GO biological process analysis between *OsCIPK9-cas* and Nipponbare under 0 mM and 120 mM NaCl treatment; Figure S4: The result of GO molecular function analysis between *OsCIPK9-cas* and Nipponbare under 0 mM and 120 mM NaCl treatment; Figure S5: The FPKM of *OsAMT1.1* and *OsAMT2.1* in Nipponbare and knock-out line (*OsCIPK9-cas*) under 0 mM and 120 mM NaCl conditions; Table S1: The expression of all genes produced by RNA-seq; Table S2: The list of primers.

Author Contributions: Y.Z., B.W. and Z.Z. directed the project. Z.Z. and W.T. performed the experiments. Z.S., J.L., B.Y., Y.L., B.W., D.X. and J.Y. participated in the experiments. Z.Z. and Y.Z. wrote the manuscript. Y.Z. and B.W. revised and finalized the manuscript. All authors have read and agreed to the published version of the manuscript.

Funding: Lianyungang "521 Project" scientific research project, The Financial Grant Support Program of Lianyungang City (QNJJ1803; QNJJ2201), Special Project of Science and Technology in Northern Jiangsu Province (LYG-SZ201930), The Natural Science Foundation of Jiangsu Province (BK20221276) supported this study. The funding agencies had no role in the study design, data collection and analysis, decision to publish or manuscript preparation.

Institutional Review Board Statement: Not applicable.

Informed Consent Statement: Not applicable.

Data Availability Statement: The datasets supporting the conclusions of this article are included within the article and Supplementary Materials.

Conflicts of Interest: The authors declare no conflict of interest.

References

1. Qin, H.; Li, Y.X.; Huang, R.F. Advances and Challenges in the Breeding of Salt-Tolerant Rice. *Int. J. Mol. Sci.* **2020**, *21*, 8385. [CrossRef] [PubMed]
2. Singh, R.K.; Kota, S.; Flowers, T.J. Salt tolerance in rice: Seedling and reproductive stage QTL mapping come of age. *Theor. Appl. Genet.* **2021**, *134*, 3495–3533. [CrossRef] [PubMed]
3. Nayyeripasand, L.; Garoosi, G.A.; Ahmadikhah, A. Genome-Wide Association Study (GWAS) to Identify Salt-Tolerance QTLs Carrying Novel Candidate Genes in Rice During Early Vegetative Stage. *Rice* **2021**, *14*, 9. [CrossRef] [PubMed]
4. Yu, J.; Zhu, C.; Xuan, W.; An, H.; Tian, Y.; Wang, B.; Chi, W.; Chen, G.; Ge, Y.; Li, J.; et al. Genome-wide association studies identify OsWRKY53 as a key regulator of salt tolerance in rice. *Nat. Commun.* **2023**, *14*, 3550. [CrossRef] [PubMed]
5. Ren, Z.H.; Gao, J.P.; Li, L.G.; Cai, X.L.; Huang, W.; Chao, D.Y.; Zhu, M.Z.; Wang, Z.Y.; Luan, S.; Lin, H.X. A rice quantitative trait locus for salt tolerance encodes a sodium transporter. *Nat. Genet.* **2005**, *37*, 1141–1146. [CrossRef] [PubMed]
6. Ahmad, I.; Zhu, G.; Zhou, G.; Younas, M.U.; Suliman, M.S.E.; Liu, J.; Zhu, Y.M.; Salih, E.G.I. Integrated approaches for increasing plant yield under salt stress. *Front. Plant Sci.* **2023**, *14*, 1215343. [CrossRef] [PubMed]
7. Ding, X.; Liu, B.; Sun, X.; Sun, X.; Zheng, C. New functions of CIPK gene family are continue to emerging. *Trends Plant Sci.* **2022**, *49*, 6647–6658. [CrossRef]
8. Shi, J.; Kim, K.N.; Ritz, O.; Albrecht, V.; Gupta, R.; Harter, K.; Luan, S.; Kudla, J. Novel protein kinases associated with calcineurin B-like calcium sensors in Arabidopsis. *Plant Cell* **1999**, *11*, 2393–2405.
9. Mao, J.; Mo, Z.; Yuan, G.; Xiang, H.; Visser, R.G.F.; Bai, Y.; Liu, H.; Wang, Q.; van der Linden, C.G. The CBL-CIPK network is involved in the physiological crosstalk between plant growth and stress adaptation. *Plant Cell Environ.* **2022**, *46*, 3012–3022. [CrossRef]
10. Qiu, Q.S.; Guo, Y.; Dietrich, M.A.; Schumaker, K.S.; Zhu, J.K. Regulation of SOS1, a plasma membrane Na^+/H^+ exchanger in *Arabidopsis thaliana*, by SOS2 and SOS3. *Proc. Natl. Acad. Sci. USA* **2002**, *99*, 8436–8441. [CrossRef]
11. Kim, W.Y.; Ali, Z.; Park, H.J.; Park, S.J.; Cha, J.Y.; Perez-Hormaeche, J.; Quintero, F.J.; Shin, G.; Kim, M.R.; Qiang, Z.; et al. Release of SOS2 kinase from sequestration with GIGANTEA determines salt tolerance in *Arabidopsis*. *Nat. Commun.* **2013**, *4*, 1352. [CrossRef] [PubMed]
12. Yin, X.; Xia, Y.; Xie, Q.; Cao, Y.; Wang, Z.; Hao, G.; Song, J.; Zhou, Y.; Jiang, X. The protein kinase complex CBL10-CIPK8-SOS1 functions in Arabidopsis to regulate salt tolerance. *J. Exp. Bot.* **2020**, *71*, 1801–1814. [CrossRef] [PubMed]
13. Held, K.; Pascaud, F.; Eckert, C.; Gajdanowicz, P.; Hashimoto, K.; Corratgé-Faillie, C.; Offenborn, J.N.; Lacombe, B.; Dreyer, I.; Thibaud, J.B.; et al. Calcium-dependent modulation and plasma membrane targeting of the AKT2 potassium channel by the CBL4/CIPK6 calcium sensor/protein kinase complex. *Cell Res.* **2011**, *21*, 1116–1130. [CrossRef] [PubMed]
14. Sanyal, S.K.; Kanwar, P.; Yadav, A.K.; Sharma, C.; Kumar, A.; Pandey, G.K. Arabidopsis CBL interacting protein kinase 3 interacts with ABR1, an APETALA2 domain transcription factor, to regulate ABA responses. *Plant Sci.* **2017**, *254*, 48–59. [CrossRef] [PubMed]
15. Xuan, Y.H.; Kumar, V.; Han, X.; Kim, S.H.; Jeong, J.H.; Kim, C.M.; Gao, Y.; Han, C.D. CBL-INTERACTING PROTEIN KINASE 9 regulates ammonium-dependent root growth downstream of IDD10 in rice (*Oryza sativa*). *Ann. Bot.* **2019**, *124*, 947–960. [CrossRef] [PubMed]
16. Shabala, S.; Alnayef, M.; Bose, J.; Chen, Z.H.; Venkataraman, G.; Zhou, M.X.; Shabala, L.; Yu, M. Revealing the Role of the Calcineurin B-Like Protein-Interacting Protein Kinase 9 (CIPK9) in Rice Adaptive Responses to Salinity, Osmotic Stress, and K^+ Deficiency. *Plants* **2021**, *10*, 1513. [CrossRef] [PubMed]
17. Kurusu, T.; Hamada, J.; Nokajima, H.; Kitagawa, Y.; Kiyoduka, M.; Takahashi, A.; Hanamata, S.; Ohno, R.; Hayashi, T.; Okada, K.; et al. Regulation of Microbe-Associated Molecular Pattern-Induced Hypersensitive Cell Death, Phytoalexin Production, and Defense Gene Expression by Calcineurin B-Like Protein-Interacting Protein Kinases, OsCIPK14/15, in Rice Cultured Cells. *Plant Physiol.* **2010**, *153*, 678–692. [CrossRef]
18. Kanwar, P.; Sanyal, S.K.; Tokas, I.; Yadav, A.K.; Pandey, A.; Kapoor, S.; Pandey, G.K. Comprehensive structural, interaction and expression analysis of CBL and CIPK complement during abiotic stresses and development in rice. *Cell Calcium.* **2014**, *56*, 81–95. [CrossRef]
19. Xiang, Y.; Huang, Y.M.; Xiong, L.Z. Characterization of stress-responsive CIPK genes in rice for stress tolerance improvement. *Plant Physiol.* **2007**, *144*, 1416–1428. [CrossRef]
20. Kolukisaoglu, U.; Weinl, S.; Blazevic, D.; Batistic, O.; Kudla, J. Calcium sensors and their interacting protein kinases: Genomics of the Arabidopsis and rice CBL-CIPK signaling networks. *Plant Physiol.* **2004**, *134*, 43–58. [CrossRef]
21. Liu, C.; Mao, B.; Yuan, D.; Chu, C.; Duan, M. Salt tolerance in rice: Physiological responses and molecular mechanisms. *Crop J.* **2022**, *10*, 13–25. [CrossRef]
22. Li, J.; Long, Y.; Qi, G.N.; Li, J.; Xu, Z.J.; Wu, W.H.; Wang, Y. The Os-AKT1 Channel Is Critical for K^+ Uptake in Rice Roots and Is Modulated by the Rice CBL1-CIPK23 Complex. *Plant Cell* **2014**, *26*, 3387–3402. [CrossRef] [PubMed]
23. Lee, K.S.; Choi, W.Y.; Ko, J.C.; Kim, T.S.; Gregorio, G.B. Salinity tolerance of japonica and indica rice (*Oryza sativa* L.) at the seedling stage. *Planta* **2003**, *216*, 1043–1046. [CrossRef] [PubMed]

24. El Mahi, H.; Pérez-Hormaeche, J.; De Luca, A.; Villalta, I.; Espartero, J.; Gámez-Arjona, F.; Fernández, J.L.; Bundó, M.; Mendoza, I.; Mieulet, D.J.P.P. A critical role of sodium flux via the plasma membrane Na^+/H^+ exchanger SOS1 in the salt tolerance of rice. *Plant Physiol.* **2019**, *180*, 1046–1065. [CrossRef] [PubMed]
25. Ho, C.H.; Lin, S.H.; Hu, H.C.; Tsay, Y.F. CHL1 Functions as a Nitrate Sensor in Plants. *Cell* **2009**, *138*, 1184–1194. [CrossRef] [PubMed]
26. Straub, T.; Ludewig, U.; Neuhauser, B. The Kinase CIPK23 Inhibits Ammonium Transport in *Arabidopsis thaliana*. *Plant Cell* **2017**, *29*, 409–422. [CrossRef] [PubMed]
27. Martínez-Atienza, J.; Jiang, X.; Garciadeblas, B.; Mendoza, I.; Zhu, J.K.; Pardo, J.M.; Quintero, F.J. Conservation of the salt overly sensitive pathway in rice. *Plant Physiol.* **2007**, *143*, 1001–1012. [CrossRef]
28. Kumar Meena, M.; Kumar Vishwakarma, N.; Tripathi, V.; Chattopadhyay, D. CBL-interacting protein kinase 25 contributes to root meristem development. *J. Exp. Bot.* **2019**, *70*, 133–147. [CrossRef]
29. Gutaker, R.M.; Groen, S.C.; Bellis, E.S.; Choi, J.Y.; Pires, I.S.; Bocinsky, R.K.; Slayton, E.R.; Wilkins, O.; Castillo, C.C.; Negrao, S.; et al. Genomic history and ecology of the geographic spread of rice. *Nat. Plants* **2020**, *6*, 492–502. [CrossRef]
30. Mishra, R.; Joshi, R.K.; Zhao, K.J. Base editing in crops: Current advances, limitations and future implications. *Plant Biotechnol. J.* **2020**, *18*, 20–31. [CrossRef]
31. Zeng, D.C.; Liu, T.L.; Ma, X.L.; Wang, B.; Zheng, Z.Y.; Zhang, Y.L.; Xie, X.R.; Yang, B.W.; Zhao, Z.; Zhu, Q.L.; et al. Quantitative regulation of Waxy expression by CRISPR/Cas9-based promoter and 5′UTR-intron editing improves grain quality in rice. *Plant Biotechnol. J.* **2020**, *18*, 2385–2387. [CrossRef]
32. Wang, R.; Jing, W.; Xiao, L.Y.; Jin, Y.K.; Shen, L.; Zhang, W.H. The Rice High-Affinity Potassium Transporter1;1 Is Involved in Salt Tolerance and Regulated by an MYB-Type Transcription Factor. *Plant Physiol.* **2015**, *168*, 1076–1090. [CrossRef]
33. Chen, H.; Zou, Y.; Shang, Y.; Lin, H.; Wang, Y.; Cai, R.; Tang, X.; Zhou, J.M. Firefly luciferase complementation imaging assay for protein-protein interactions in plants. *Plant Physiol.* **2008**, *146*, 368–376. [CrossRef]
34. Leigh, J.W.; Bryant, D. POPART: Full-feature software for haplotype network construction. *Methods Ecol. Evol.* **2015**, *6*, 1110–1116. [CrossRef]

plants

MDPI

Article

Creation of Rice Doubled Haploids Resistant to Prolonged Flooding Using Anther Culture

Pavel Kostylev *, Nataliya Kalinina, Nataliya Vozhzhova, Valentina Golubova and Natalya Chertkova

FSBSI Agricultural Research Center "Donskoy", Nauchny Gorodok, 3, 347740 Zernograd, Rostov Region, Russia; kalinina74783@mail.ru (N.K.); nvozhzh@gmail.com (N.V.); valya_l7@rambler.ru (V.G.); tycik17082012@gmail.com (N.C.)

* Correspondence: p-kostylev@mail.ru

Abstract: Flood resistance in rice is very important in weed control, as weeds cannot overcome deep water. At present, there are no released varieties in Russia that would meet these requirements. The creation of such varieties will reduce production costs and pesticide load on the ecosystem. The object of the study was second-generation rice hybrids obtained by crossing the best varieties for economically valuable traits with samples carrying genes for resistance to prolonged flooding with water. To create double rice haploids resistant to prolonged flooding, the anther culture method was used, followed by molecular genetic evaluation of dihaploids for the presence of genes for resistance to prolonged flooding. An estimate of the growth energy under deep flooding was carried out according to our own method. As a result of the cultivation of anthers, 130 androgenic regenerated plants were obtained in 14 hybrid combinations. In terms of responsiveness to neoplasms, 60% of the panicles showed a positive result, while the rest 40% did not demonstrate callus formation. In total, 30 green regenerative lines were obtained from four rice hybrids, differing in visual morphological assessment. Large genotypic differences between the samples were revealed. These lines carry long-term flood resistance genes and can be used in rice breeding programs using dihaploids. As a result of the assessment of the growth energy in a number of obtained samples, the potential for rapid elongation of the first leaves, overcoming a large layer of water and accumulation of vegetative mass, was revealed.

Keywords: rice; grade; hybrid; lines; anther; induction; callus; regeneration; dihaploid; flood resistance

Citation: Kostylev, P.; Kalinina, N.; Vozhzhova, N.; Golubova, V.; Chertkova, N. Creation of Rice Doubled Haploids Resistant to Prolonged Flooding Using Anther Culture. *Plants* **2023**, *12*, 3681. https://doi.org/10.3390/plants12213681

Academic Editors: Weixun Wu, Yingxin Zhang, Guiai Jiao and Xiangjin Wei

Received: 13 September 2023
Revised: 6 October 2023
Accepted: 17 October 2023
Published: 25 October 2023

1. Introduction

Rice is one of the main food sources for more than half of the world's population and is the third largest grain producer. Breeders are constantly looking for biotic and abiotic stress resistance mechanisms to improve yields under stressful conditions. Great importance is attached to increasing flood resistance. If flooding persists for several days, then crops of unstable varieties die, and yields are lost [1,2].

During vegetative growth, rice uses various flood protection strategies. Genetic variability of plant response includes various schemes: (1) dormancy, which allows to withstand a long time under water, (2) a strategy of rapid stem elongation with changes in plant structure and metabolism [1].

The SUB1A locus belongs to the genes that control flood resistance, which manifests itself in waiting for the onset of an unfavorable factor. The dormant state is observed in rice varieties with the SUB1A locus, which limit starch mobilization and thus produce less ethanol and other fermentation products, and also affect the processes associated with aging [1,3].

The discovery of the main quantitative trait loci Sub1 (QTL) on rice chromosome 9, the principal contributor to flooding tolerance, was a significant step forward in flooding tolerance research [3].

Researchers have identified two additional loci associated with flood resistance: SNORKEL 1 and SNORKEL 2 (chromosome 12). Plants with these genes show rapid growth and coleoptile elongation during development under anaerobic conditions, which is called the "tube effect" [4].

This type of adaptation mechanism prevails in deep-sea lowland rice, where a significant elongation of shoots keeps the plant from completely submerging with a constant increase in flood water levels. SK1 and SK2, genes of ethylene sensitivity factors, which were found in the deep-sea rice variety C9285, play a vital role in lengthening the internodes of plant shoots in response to immersion. Both genes act in opposition to Sub1. The Sub1a gene allows the rice plant to withstand flash floods by limiting the elongation of shoots until flood waters recede. The slower lengthening of coleoptiles, the development of tolerant varieties adapted to the conditions of sudden floods, thus preserves the accumulated biomass, which is used to continue growth after leaving the water [4].

An important aspect of the use of rice flood resistance, most relevant for Russia, is the fight against weeds due to the deep water layer, which weeds cannot overcome and die. This leads to positive effects such as the absence of the need to use herbicides, a reduction in the cost of a unit of production, an increase in the quality of products that can be used in the production of products for children and diet food, and the absence of environmental damage [5,6].

Varieties with genes for this resistance have been found in Asia. It is necessary to transfer these genes into the germplasm of Russian varieties. To speed up this process, it is necessary to rationally combine classical breeding methods with biotechnological methods, which allows solving the tasks set in a shorter time [7]. Therefore, in addition to the classical methods of creating new rice varieties, androgenesis and haploidy are widely used [8]. Rice growing countries of the world have long used the cultivation of anthers on an artificial nutrient medium in breeding work [9–11]. The use of anther culture in breeding work allows for quickly obtaining homozygous plants that are resistant to various stressful environmental conditions, including drought, soil salinity, extreme temperatures and diseases.

Growing anthers on an artificial nutrient medium makes it possible to obtain rice haploids and homozygous dihaploids in 1–2 years. Extensive work on this technique is being carried out at the Federal State Budget Scientific Institution "Federal Science Center of Rice" [12–14] and the Primorsky Research Institute [15,16]. In order to obtain hundreds and thousands of lines in anther culture in vitro, it is necessary to ensure a large percentage of regeneration of green buds and plants in each hybrid combination [17–19]. Indicators of regenerative ability vary greatly depending on the hybrid and genotype, i.e., even within the same hybrid combination; different plants provide different rates of callus formation and regeneration of green plants from very low to very high. Thus, it is necessary to cover as many hybrid plant genotypes as possible for use in in vitro culture [20–22].

The aim of the study was to obtain rice dihaploids by cultivating anthers in vitro to accelerate selection for resistance to prolonged flooding.

2. Materials and Methods

Donor plants and growth conditions. The object of the study was rice hybrids of the second generation, obtained by crossing the best varieties for economically valuable traits with samples that do not have genes for resistance to prolonged flooding (Table 1).

The flood resistance locus SUB1A was transferred from donors Inbara-3, IR-64, Br-11, CR-1009, TDK-1. The donors of the SNORKEL 1, 2 genes were Khao Hlan On, Kharsu 80A, and Ma-Zhan Red. Local varieties Novator, Magnat, Contact and Kuboyar were used as a control.

Panicle selection and pre-treatment. The selection of panicles was carried out in the fields of the laboratory of rice breeding and seed production of the Agricultural Research Center "Donskoy" in Proletarsk, Rostov Region, in 2022 in the morning, in clear weather.

Table 1. Varieties—donors of resistance to prolonged flooding.

Variety	Gene	Country of Origin
Novator	–	Russia
Magnat	–	Russia
Contact	–	Russia
Kuboyar	–	Russia
Br-11	SUB1A	Bangladesh
CR-1009	SUB1A	India
Inbara-3	SUB1A	Indonesia
IR-64	SUB1A	Philippines
TDK-1	SUB1A	Laos
Khao Hlan On	SK1, SK2	Myanmar
Kharsu 80A	SK1, SK2	Pakistan
Ma-Zhan Red	SK1, SK2	China

A morphological trait suitable for the selection of shoots was the distance from the ear of the flag leaf to the ear of the next leaf, which should be from 5 to 10 cm [4,6,23].

Selected shoots were subjected to surface sterilization with 96% alcohol for 3 min. After that, they were placed in vessels with water, covered with a plastic bag, and exposed to low positive temperatures (5 °C) for 7–10 days in a refrigerator (Liebherr-Hausgeraete, Ochsenhausen, Germany).

Panicle sterilization. Before introduction into the culture, flag leaves were removed from the panicles, twigs with pollen grains in the stage of medium and late mononuclear microspores were selected according to morphological characteristics and the rest were removed. The twigs were wrapped in sterile gauze and loosely fastened with thread. Then, they were immersed in a sterilizing solution of 5% sodium hypochlorite (Sigma, St. Louis, MI, USA) for 10 min and washed three times in sterile distilled water.

Determination of the stage of development of microspores. From each panicle, from its middle part, 2–3 spikelets were isolated to determine the stage of development of microspores. Spikelets were placed on a glass slide and anthers were removed using dissecting needles. Then, the anther was cut across and the microspores were squeezed out, while two drops of acetocarmine (Merck Peruana S.A., Lima, Perú) were added, heated over an alcohol burner and left for staining (10–15 min). After that, the preparation was covered with a cover slip and examined under a microscope (ADF U300FL, Ningbo, China) (see Figure 1a). The most appropriate stage in the development of microspores is the transition from the late single-core to the early two-core stage. However, the middle uninuclear stage of microspore development has been identified as optimal for an effective androgenic response [1].

Preparation medium. Callus formation was induced from rice anthers on Blades nutrient medium containing 2.0 mg/L 2,4-D (Sigma, USA), 30 g/L sucrose (Panreac, Barcelona, Spain) and 8 g/L agar (MerckRGaA, Saint Louis, USA). For regeneration, morphogenic callus was transplanted onto Murashige and Skoog (MS) base medium with 1.0 mg/L NAA (Sigma, USA), 5.0 mg/L kinetin (Sigma, USA), 20 g/L sucrose, and 8 g/L agar (Table 2). The prepared nutrient medium was poured into test tubes with a diameter of 20 mm, autoclaved (Tuttnauer 3870 ELV, Beit Shemesh, Israel) for 15 min at a temperature of 121 °C and a pressure of 0.9–1.0 atm. Immediately after sterilization, tubes with hot medium were placed at an angle of about 30 °C to obtain an inclined agar and left to solidify [7,14].

Anther inoculation and callus induction. Anthers were inoculated under aseptic conditions in a laminar flow hood (BMB-II-Laminar-S, Miass, Russia), which was sterilized with ultraviolet lamps. The instruments used were also partially sterilized (SteriMaxsmart-WLD-TEC, Arenshausen, Germany). All items needed in the working process were treated with 96% ethyl alcohol and then burned in the flame of an alcohol burner. Sterile instruments were placed between sheets of thick wrapping paper previously sterilized by dry heat in an oven (BinderFD23, Tuttlingen, Germany) at 130 °C for 2 h (from the moment the set temperature was set).

Figure 1. Stages of obtaining rice regenerant plants in anther culture in vitro: (**a**)—microspores in the middle single-core stage; (**b**)—neoplasms on anthers; (**c**)—morphogenesis on calluses; (**d**)—cultivation of regenerating plants in the soil.

Table 2. Composition of media for the cultivation of rice anthers.

Media Components	**Induction Medium (Blaydes, 1966) (mg/L)**	**Regeneration Medium (Muraschige and Skoog, 1964) (mg/L)**
	Macro salts	
NH_4NO_3	1000	1650
KNO_3	1000	1900
$Ca(NO_3)_2 \times 4\ H_2O$	347	440
KH_2PO_4	300	170
$MgSO_4 \times 7\ H_2O$	35	370
KCl	65	–
	Micro salts	
$ZnSO_4 \times 7\ H_2O$	1.5	8.6
H_3BO_3	1.6	1.6
$MnSO_4 \times 4\ H^2O$	4.4	6.92
KI	0.8	0.83
$Na_2MoO_4 \times 2\ H_2O$	–	0.25
$CuSO_4 \times 5\ H_2O$	–	0.025
$CoCl_2 \times 6\ H_2O$	–	0.025
	Iron source	
$FeSO_4 \times 7\ H_2O$	27.8	27.8
Na_2EDTA	37.2	37.2

Table 2. *Cont.*

Media Components	Induction Medium (Blaydes, 1966) (mg/L)	Regeneration Medium (Muraschige and Skoog, 1964) (mg/L)
	Vitamins	
Nicotinic acid	0.5	0.5
Pyridoxine HCI	0.5	0.5
Thiamine HCI	0.5	0.5
	Other componentss	
myo-Inositol	100	100
Glycine	2.0	2.0
Agar	8.0	8.0
Sucrose	30.0	20.0
	Growth regulators	
2,4-D	2.0	–
NAA	–	1.0
Kinetin	–	5.0
	6.0	6.0

Anthers were separated from the spikelets and transferred to the medium in an amount of up to 30 pcs. into a test tube using a scalpel, tweezers and a dissecting needle, closed it with a foil cap and placed in a thermostat with a temperature of 28 ± 2 °C and a relative humidity of about 50% (Binder KV400, Tuttlingen, Germany) for 30–50 days (Figure 1b).

Planting calluses and plant regeneration. The morphology of calli is closely related to their ability to regenerate plants [14]. The morphological assessment of rice calluses can be described as follows:

(1) With meristematic foci, light shades, fine-grained, medium density (morphogenic);
(2) Spherical, white, light yellow, medium density (morphogenic);
(3) Dense, white, fine-grained (morphogenic);
(4) Loose, moist, with vascular cords (rhizogene);
(5) Granular, loose, brown, with large cells (very low ability to morphogenesis);
(6) Watery, dark brown, with large shapeless cells of different sizes (non-morphogenic).

Morphogenic calli 1 mm in size or more were placed on Murashige and Skoog regeneration medium poured into test tubes 20 mm in diameter (Figure 1c). Tubes with callus explants were incubated in an illuminated growth room with a temperature control of 25 ± 2 °C, illumination of 2000 Lux, and a photoperiod of 15 h/9 h. After 15–20 days, regenerated plants were formed in the light [6,7].

Acclimatization of regenerative plants and planting them in the soil. Regenerants with a developed root system and 4–5 leaves at least 8 cm long were planted in a pot culture (Figure 1d). The root system was thoroughly washed from the agar medium and left for a day in glasses with water. The soil was preliminarily sieved, sterilized (3 h in a dry oven at a temperature of +110 °C) and moistened. Containers with plants were placed in a bright room, without direct lighting, so that a sharp change in environmental conditions would not have a negative effect on the plants.

After that, the regenerator plants were transferred to the greenhouse, where their development continued until flowering, seed formation and maturation. The following microclimate parameters were favorable for rice plants: daytime temperature 25 °C, illumination more than 5000 lux; night temperature 20 °C, illumination 0 lux; humidity 70–80%; photoperiod 12 h [6,7].

One feeding session was carried out with a nutrient solution (1/2 MS). Plants were watered 3–4 times a week, depending on the drying of the soil in the growing vessels.

Diagnosis of plant ploidy. To restore fertility in haploid plants, it is necessary to induce chromosome duplication. However, rice haploids obtained by cultivating anthers usually undergo spontaneous diploidization. In this work, plants were not treated with polyploidizing agents.

The ploidy of the resulting regenerative plants at the stage of growth and development (before seed formation) was determined by morphological traits such as plant height and leaf size. Plants with normal morphological features were considered doubled haploids. Haploid plants [23] had a low plant height and narrow leaves. The tetraploids were very tall, with large leaves. Simultaneously, the content of nuclear DNA in different groups of regenerated plants was determined. The DNA content was determined using a flow cytometer (MuseCellAnalyzer, Austin, TX, USA). Green regenerative plants obtained from calli were evaluated. The plant material (leaves) was subjected to freeze-drying.

A leaf sample with an area of 1–2 cm^2 was crushed with a spatula in a Petri dish in 1 mL of cooled Tris-$MgCl_2$ buffer. The buffer contained 0.2 m Tris base (Molekula Ltd., Darlington, UK), 4 mM $MgCl_2 \times 6H_2O$ (Amresco, Solon, OH, USA) and 0.5% TritonX-100 (Sigma, USA) supplemented with 1 μL/mL β-mercaptoethanol (Sigma, St. Louis, MI, USA), 50 μg/mL propidium iodide (Sigma, USA) and 50 μg/mL RNase (Syntol, Moscow, Russia). The samples were filtered through a nylon membrane filter with a pore size of 50 μm. *Ficus benjamin* L. embryos isolated in a similar way with a known DNA content of 2C = 1.07 pg were used as an external standard [24,25].

Molecular genetic assessment of the presence of flood resistance genes.

All obtained regenerative rice lines were evaluated by two molecular markers (Table 3).

The genomic DNA of the isolated samples was extracted from freshly cut leaves by the CTAB method (Panreac, Barcelona, Spain) on a homogenizer (Bertin Precellys 24, Montigny le Bretonneux, France). Homogenization was carried out in two stages of 30 s, at a speed of 3000 rpm. The quantity and quality of the isolated DNA were assessed using a spectrophotometer (Implen Nanophotometr NP80, Munich, Germany). The isolated DNA was placed in an amplifier (Bio-RadT100, Richmond, CA, USA) for PCR. The samples were then placed in a gel electrophoresis chamber. The gel was prepared on the basis of a 0.5-fold TBE buffer 10 × (Evrogen, Moscow, Russia) (100 mL) and agarose (Amresco, Solon, OH, USA) (2 g). After PCR, 3–5 μL of 6× TriK dye (Biolabmix, Novosibirsk, Russia) was added to each tube with a sample after PCR. A molecular weight marker was also applied to the agarose gel. The duration of electrophoresis ranges from 30 min to 2 h, depending on the expected size of the amplificates. Then, the gel plates were placed for 20–30 min in a solution of ethidium bromide fluorescent dye (Sigma, USA) for staining. After that, the gel was removed, carefully washed in a cuvette, and photographed under ultraviolet light using a gel photodocumentation system (Bio-RadGelDocXR+, Richmond, VA, USA) [26].

Evaluation of resistance of rice samples to long-term flooding by the method of germination under stress conditions. To estimate the growth energy during prolonged flooding, our own method was developed. Glass test tubes (1.5 cm in diameter and 15 cm in height) were used for rice seed germination and plant growth. The seeds were placed in a test tube, filled with distilled water to a depth of 10 cm, and incubated in the light under room conditions at a temperature of 28 °C (Figure 2).

On the 5th, 7th, 9th, 11th days, the length of the sprout was measured to determine the growth dynamics.

Observations and statistical analysis. Mathematical and statistical data processing was carried out in MS Excel.

Figure 2. Germination of rice seeds in test tubes under stress.

Table 3. Molecular markers of quantitative trait loci associated with long-term flood resistance in rice.

Gene	Primer	Subsequence (5′–3′)	Size (bp)	Link
SUB1A	RM 7481 F RM 7481 R	CGA CCC AAT ATC TTT CTG CC ATT GGT CGT GCT CAA CAA G	95	Azarin et al., 2016 [27]
SNORKEL1	1F 1R	ATG TGC GGA GGT TGT CTC AT TCG TAG CGA CAG CCG TAC TG	743	Oe et al., 2021 [28]

3. Results and Discussion

Hybrids of Russian rice varieties with flood resistance donors were obtained in 2012–2013. The donors of the Sub1A gene were crossed with the varieties Novator, Contakt, Magnat, and the donors of the Snorkel and AG genes were crossed with the varieties Contakt and Kuboyar. The donors were extremely late-ripening and photosensitive; therefore, in subsequent years, mid-ripening recombinants capable of maturing in the Rostov region were selected. As a result of long-term selection of forms with a complex of economically valuable traits, a group of self-pollinated lines was formed, among which there were plants with resistance genes. To identify them, physiological and genetic studies were carried out, during which lines with genes of interest were selected. Subsequently, they were crossed with the best varieties and among themselves for the pyramiding of genes.

To speed up the selection process, it is necessary to obtain homozygous lines from second-generation hybrids using androgenesis.

The creation of double rice haploids resistant to prolonged flooding by the anther culture method is a two-stage process: the initial development of the callus and the subsequent regeneration of green plants from the embryogenic callus [13,25]. Features of androgenesis in hybrid combinations of rice were studied by cultivating anthers on Blades induction medium.

As a result of the experiment, in 2022, 12,604 anthers were extracted from 68 hybrid rice panicles (26 crossing combinations) and planted on an induction nutrient medium. The maximum number of anthers was planted in hybrid 5016/2—339 pieces, and the minimum was 4773/1—47 pieces.

When anthers of all genotypes were cultivated on an induction medium, neoplasms appeared from them—embryo-like structures (single embryoids and polyembryoids) and callus. When cultivated on a regenerative medium, as a rule, the callus and part of the embryonic structures remained unchanged. Some of the germ-like structures developed roots, while another part had solitary seedlings or clusters of seedlings. The appearance of callus and germ-like structures began on the 30–33rd day from the moment the anthers were planted on the nutrient medium. Dense or slightly transparent, well-defined neoplasms were classified as embryo-like structures. Callus formation continued for another four weeks; the anthers were on the medium for two months.

As follows from the above data, not all hybrid combinations of rice showed the ability to form callus and embryo-like structures. The amount of callus formation in the anther culture varied significantly both between hybrid combinations and in different plants from the same hybrid combination, which, apparently, is due to genotypic differences, as well as the influence of external factors (explant quality, growing conditions etc.). In total, 716 neoplasms were obtained, which is, on average, 10 pcs. per plant, taking into account non-susceptible ones (Table 4).

Table 4. Results of cultivation of rice anthers (2022).

№	Sample №	Plant №	Inoculated Anthers, pcs.	Number of Neoplasms, pcs.	Non-Morphogenic Callus, pcs.	Total Regenerating Plants, pcs.	Green Plants, pcs.	Albino Plants, pcs.
1	5022	1	243	4	4	0	0	0
		2	275	0	0	0	0	0
		3	92	0	0	0	0	0
2	5103	2	259	20	16	4	0	4
		4	245	2	2	0	0	0
		5	110	0	0	0	0	0
3	5007	1	214	0	0	0	0	0
		3	152	0	0	0	0	0
		4	114	0	0	0	0	0
4	5005	1	299	37	34	3	0	3
		2	86	0	0	0	0	0
		3	225	1	1	0	0	0
5	5029	3	270	0	0	0	0	0
		5	132	1	1	0	0	0
		8	304	1	1	0	0	0
		10	284	0	0	0	0	0
6	5006	1	277	0	0	0	0	0
		2	189	0	0	0	0	0
		5	120	0	0	0	0	0
7	5093	1	194	1	1	0	0	0
		3	82	0	0	0	0	0
		4	272	0	0	0	0	0
8	5019	1	251	8	5	3	0	3
		2	126	3	2	1	0	1
		3	289	1	1	0	0	0
9	5003	1	306	0	0	0	0	0
		3	258	1	1	0	0	0
10	5009	1	212	0	0	0	0	0
		2	119	84	67	17	5	12
		4	278	12	12	0	0	0
11	5010	1	183	0	0	0	0	0
		2	277	94	87	7	5	2
12	5011	1	271	0	0	0	0	0
		3	186	0	0	0	0	0
13	5008	1	86	3	3	0	0	0
		2	184	0	0	0	0	0
		3	279	21	21	0	0	0
14	5020	1	243	26	15	11	0	11
		2	47	0	0	0	0	0
		3	210	13	13	0	0	0
15	5018	1	132	1	1	0	0	0
		2	298	5	4	1	0	1
		3	297	23	20	3	0	3

Table 4. *Cont.*

№	Sample №	Plant №	Inoculated Anthers, pcs.	Number of Neoplasms, pcs.	Non-Morphogenic Callus, pcs.	Total Regenerating Plants, pcs.	Green Plants, pcs.	Albino Plants, pcs.
16	4565	2 3 5	82 195 245	3 85 46	3 82 43	0 3 3	0 2 0	0 1 3
17	4773	1 2 3	47 114 59	4 0 1	4 0 1	0 0 0	0 0 0	0 0 0
18	5016	2 3 4	339 120 216	1 0 0	1 0 0	0 0 0	0 0 0	0 0 0
19	4758	1	209	4	2	2	0	2
20	5021	1 2 3	248 112 140	62 3 0	37 1 0	25 2 0	0 0 0	25 2 0
21	4641	1 2	193 194	5 69	4 48	1 21	0 18	1 3
22	5017	1 2 3	51 255 195	4 2 42	2 1 32	2 1 10	0 0 0	2 1 10
23	4526	1	80	10	3	7	0	7
24	4688	1 2 3	85 117 57	12 0 0	9 0 0	3 0 0	0 0 0	3 0 0
25	4617	1	83	1	1	0	0	0
26	4585	1 2	104 94	0 0	0 0	0 0	0 0	0 0
Sum		69	12,604	716	586	130	30	100
Average		2.5	185.35	10.53	8.6	1.91	0.44	1.47
Minimum		1	47	0	0	0	0	0
Maximum		4	339	94	87	25	18	25

According to the ability to neoplasms, 60% of the volume of the studied material (39 panicles) showed a positive result, while 40% (29 pcs.) did not demonstrate callus formation. The most sensitive to callus formation were the following four hybrid combinations: 5009/2—84 pcs., 5010/2—94 pcs., 4565/3—85 pcs., 4641/2—69 pcs. (Table 2). The same samples showed the ability to carry out morphogenesis, while the remaining combinations formed a non-morphogenic callus (in some cases up to 100%). Samples 5007, 5006, 5011 and 4585 showed no response to in vitro culture.

It was found that not only hybrid combinations differed among themselves in the their callus formation abilities, but also individual plants within the same combination. For example, in sample 5103, anthers were planted from three plants on a nutrient medium with 259, 245, and 110 pieces. Of these, 20, 2 and 0 calli were formed, respectively, but only 4 regenerated plants appeared from the first one. From sample 4641, an equal number of anthers was taken from two plants, including 193 and 194 pieces. However, in the first one, 5 calli were found, and in the second, 69 were found, that is, 14 times more. The former subsequently developed only 1 regenerating plant, which was albino, while the latter had 18 green plants and 3 without chlorophyll. This indicates the presence of genetic factors that significantly affect the ability of cells to callus formation and regeneration.

Green seedlings and albinos developed from the formed neoplasms during cultivation on a regenerative medium; structures developing by the type of root formation and structures without development were also noted.

The ability of calli to carry out morphogenesis was assessed by plant regeneration. The studied rice samples formed 130 regenerated plants from 14 hybrid combinations. Of these, only 25 plants were green. In total, four samples were isolated that formed regenerants without chlorophyll defects in the leaves—5009/2 [(Inbara 3 × Novator) × Contact] – 3 pcs., 5010/2 [(Inbara 3 × Novator) × Contact] – 6 pcs., 4565/3 (IR64 × Magnate) – 2 pcs., 4641/2 [(Inbara-3 × Contact) × (Khao Hlan On × Kuboyar)] – 14 pcs. Albino plants died at the early stages of development, as they were not capable of photosynthesis and, accordingly, of the autotrophic type of nutrition. The largest number of albinos was formed in the sample 5021/1—25 pcs.

As a result of evaluating the efficiency of anther cultivation in rice hybrids, it was found that the largest number of neoplasms per 100 cultivated anthers (76.9) was obtained in sample 5009/2, which significantly exceeded the average value in the experiment (Table 5).

Table 5. Results of assessment of the effectiveness of anther culture of isolated rice samples (2022).

Sample №	Plant №	Number of Neoplasms/ per 100 Anthers	The Number of All Regenerants/ per 100 Anthers	Number of Green Regenerants/ per 100 Anthers	The Number of All Regenerants/ per 100 Neoplasms
5009	2	70.6 *	14.3 *	4.2	20.2
5010	2	33.9	2.5	1.8	7.5
4565	3	43.6	1.5	1.0	3.5
4641	2	35.6	10.8	9.3 *	30.4 *
Average value		45.9	7.3	4.1	15.4
Standard deviation		17.0	6.3	3.7	12.3

Note: * is a significant difference from the average value. 5009 (Inbara-3 × Novator) × Contact (gene Sub1A). 5010 (Inbara-3 × Novator) × Contact (gene Sub1A). 4565 IR-64 × Magnate (gene Sub1A). 4641 (Inbara-3 × Contact) × (Khao Hlan On × Kubojar) (genes Sub1A and SK1).

Morphogenesis and the ability of neoplasms to form seedlings, including green and albinos, were assessed on the basis of the indicator "the number of all regenerants per 100 planted anthers." More plants were regenerated based on the hybrid combination 5009/2, which was significantly higher than the average value. In general, more regenerants were formed in samples 5009/2 and 4641/2: 14.3 and 10.8, respectively.

Green plants are of practical interest; therefore, the most important indicator of anther culture is the sign "number of green regenerants per 100 isolated anthers". According to this characteristic, sample 4641/2 had a higher value (9.3). The indicators of other genotypes were at the level of the average value.

To assess the ability of neoplasms to regenerate seedlings, the number of all regenerants per 100 neoplasms was determined. The maximum value of this characteristic was in the hybrid combination 4641/2 (30.4). On average, new growths of sample 5009/2 (20.2) also regenerated plants well. The remaining combinations formed fewer plants, but within the average of the experiment.

Accession 4641 was obtained as a result of stepwise hybridization of the Inbara-3 variety (Sub1A gene), which is resistant to deep flooding, with the Russian early maturing variety Contact, and then the fifth-generation line was used when crossing with a vigorously growing accession (Khao Hlan On × Kubojar) carrying the Sk1 gene. These genes allow seeds to germinate quickly and plants to vigorously overcome the water layer. Apparently, this ability also affected the regenerative ability of callus cells, dramatically increasing the number of regenerative plants.

It was established that regenerated plants were obtained from combinations with long-term flood resistance genes: (Inbara-3 × Novator) × Contact (5009/2; 5010/2), IR-64 × Magnat (4565/3)—Sub1A gene, from the combination (Inbara-3 × Contact) × (Khao Hlan On x Kubojar) (4641)—Sk1 gene.

Following from the results obtained in this work, there are strong differences in the manifestation of signs of androgenesis in the anther culture between second-generation rice hybrids carrying the target Sub1A, Snorkel 1, 2 genes. At the same time, the low ability for androgenesis in these hybrids could depend on many factors. It has been established that the composition of the nutrient medium for growing rice anthers most closely matches the type of nutrition of this crop. There are studies showing that the rate and frequency of callus induction can be increased up to 27.9%, and even higher in some types of genes [6,29]. These data are consistent with the studies conducted in our work, namely, the inclusion of 2,4-D in the induction medium of 2.0 mg/L made it possible to obtain up to 70% of neoplasms per 100 cultivated anthers.

In the course of the work, the quantity and quality of calli were taken into account for each hybrid combination. Morphogenic calli were light, opaque, compact and had green areas containing chlorophyll, which were zones of morphogenesis. This is confirmed in the works of other authors [30].

The activation of morphogenesis processes and plant regeneration is stimulated by the use of 1.0 mg/L of alpha-naphthylacetic acid (NAA) in combination with 5.0 mg/L of kinetin. For example, in some works, samples with a low regenerative capacity (0.31–0.72%) were isolated, in addition to medium (6.07–7.61%) and high (21.65–27.9%) [7,28]. The results are also confirmed in our studies. Among the selected hybrid combinations of rice, the number of all regenerants per 100 neoplasms reached a maximum value of 30%, including green ones—from 1 to 9%, i.e., samples showed low and medium regenerative capacities.

In most rice plants obtained by cultivating anthers, spontaneous duplication of chromosomes occurs [26]. This is confirmed in our studies (Table 6). The obtained green regenerated plants were divided into three groups depending on the content of nuclear DNA: haploids, diploids and tetraploids.

Table 6. Content of nuclear DNA in the population of rice regenerants obtained from second-generation hybrids in anther culture in vitro.

Index	Haploids	Diploids	Tetraploids
Number of plants, pcs.	9	11	5
DNA content, pg:			
M	0.901	1.880	3.762
±SEM	0.012	0.023	0.048
min	0.790	1.654	3.590
max	1.112	2.015	3.960
Cv, %	8.3	9.6	10.0

All plants within their group were characterized by insignificant variability in the content of DNA in cell nuclei. Our data are comparable with the results of other authors, according to which the content of nuclear DNA in the main set of chromosomes in *O. sativa* rice ranges from 0.91 to 1.00 pg [24,26]. The ratio of the average values of the content of nuclear DNA in dihaploids and haploids was not a multiple of two. This may indirectly indicate the loss of some parts of the chromosomes in haploids during cultivation, which leads to changes in the morphotype of the regenerants.

Assessing the level of ploidy in regenerants is the most important key step in applying androgenesis to a breeding program. In our study, morphological evaluation was found to be sufficiently reliable to distinguish diploids from other ploidy levels and was performed quickly and easily. Although flow cytometry is an attractive approach for assessing the ploidy level of regenerated plants, its use is still limited in many laboratories due to the high cost of equipment and higher analysis costs [29,31,32]. The flow cytometry method in combination with morphological assessment can be used in rice breeding to identify the ploidy of regenerants, as well as to cull haploids in order to exclude the stage of growing unpromising forms under ex vitro conditions.

Regenerative anther-derived rice lines surviving in pot culture under greenhouse conditions were assessed using selected molecular markers identifying alleles of the Sub1A, Sk1, 2, and AG genes.

The Sub1 locus is considered to be the main gene controlling flood resistance. As a result of molecular genetic analyzes conducted in 2023, the presence of Sub1A long-term flood resistance genes was found in rice regenerants. Electrophoregrams of sample identification for the Sub1A gene are shown in Figure 3.

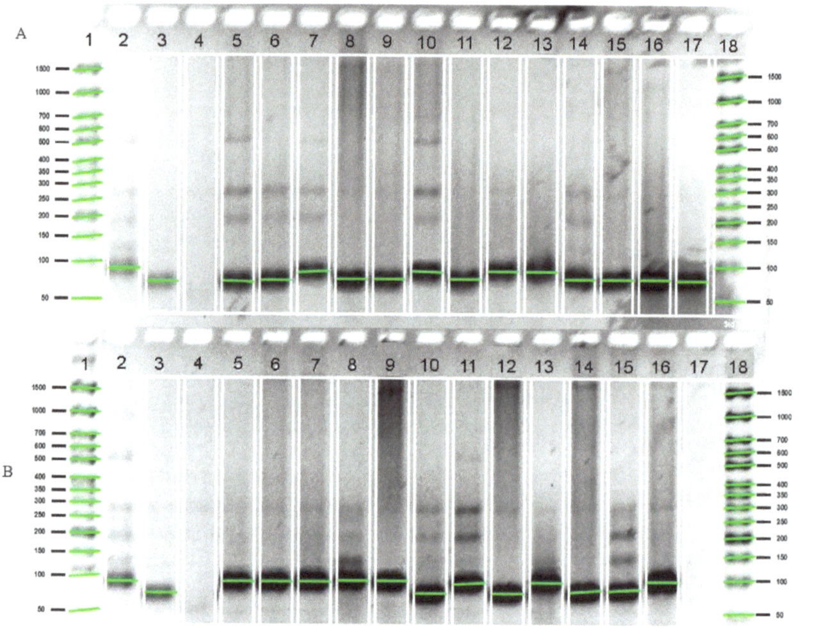

Figure 3. Electrophoregram of screening rice regenerative plants for the presence of the SUB1A gene with a molecular marker RM7481. (**A**) Line 1—molecular weight marker Biolabmix Step 50+ (50–1500 bp), 2—Inbara 3 (Sub 1A donor), 3—Novator (recessive), 4—deionized water (internal experience control), 5—4641/1, 6—5010/4, 7—5010/3, 8—4641/2, 9—4641/14, 10—4641/13, 11—4565/1, 12—4565/2, 13—4641/7, 14—5009/1, 15—4641/3, 16—5009/3, 17—5010/6, 18—Molecular weight marker Biolabmix Step 50+ (50–1500 bp). (**B**) Line 1—molecular weight marker Biolabmix Step 50+ (50–1500 bp), 2—Inbara 3 (Sub 1 donor), 3—Novator (recessive), 4—deionized H_2O (internal control experience), 5—4641/10, 6—4641/12, 7—4641/8, 8—4641/5, 9—5010/2, 10—4641/9, 11—5010/5, 12—4641/6, 13—5010/1, 14—4641/11, 15—4641/4, 16—5009/2, 17—deionized water, 18—molecular weight marker Biolabmix Step 50+ (50–1500 bp).

In Figure 3A, samples 5010/3, 4641/13, 4565/2, 4641/7 (lines 7, 10, 12, 13) corresponds to the Inbara 3 control (line 2), while in Figure 3B, samples 4641/10, 4641/12, 4641/8, 4641/5, 5010/2, 5010/1, 5009/2 (lines 5–9, 11, 13 and 16) correspond to the Inbara 3 control (line 2). These accessions carry the Sub1A long-term flood resistance gene. The Sub1A locus belongs to the genes that control resistance to flooding, which manifests itself in waiting for the impact of an unfavorable factor.

When a submerged coleoptile reaches the surface of the water, the hollow structure of the coleoptile allows O_2 to escape from the surface to the submerged parts of the plant. Thus, elongation of coleoptiles is a strategy of the rice plant to avoid stress during prolonged flooding, and is called the "tube effect" [33]. Plants with Snorkel 1, 2 (Sk 1, 2) genes show rapid growth and coleoptile elongation during development under anaerobic conditions.

In Figure 4A, the samples (line 6, 8–11, 13, 14, 18) correspond to the control Khao Hlan On (line 3) and samples (lines 5, 7, 12, 15–17) correspond to the Bakhus control (line 2). In Figure 4B, the samples (lines 6, 8–11, 13, 16, 17) correspond to the control Khao Hlan On (line 3). The isolated accessions carry the Sk1 long-term flood resistance gene.

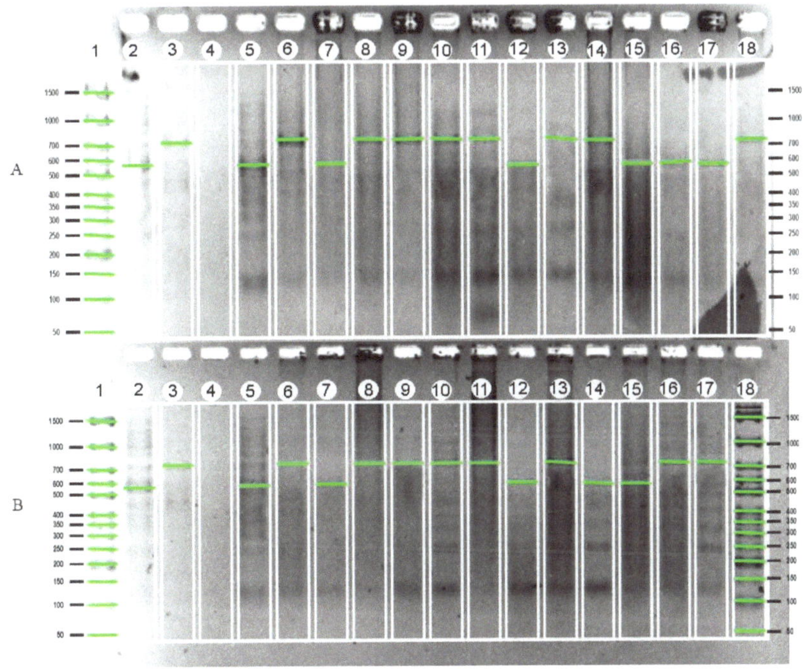

Figure 4. Electrophoregram of screening rice regenerative plants for the presence of the gene Sk1 with molecular marker F1R1. (**A**) Line 1—Molecular weight marker Biolambix Step 50+ (50–1500 bp), 2—Bahus (vigorously growing), 3—Khao Hlan On (vigorously growing), 4—deionized H_2O (internal control experience), 5—Boyarin (slow growing), 6—4641/14, 7—4641/2, 8—4641/1, 9—4641/13, 10—5009/2, 11—4641/5, 12—4641/7, 13—5009/1, 14—4641/8, 15—5009/3, 16—5010/6, 17—4641/10, 18—5010/2. (**B**) Line 1—Molecular weight marker Biolambix Step 50+ (50–1500 bp), 2—Bahus (vigorously growing), 3—Khao Hlan On (vigorously growing), 4—deionized H_2O (internal control experience), 5—Boyarin (slow growing), 6—4641/3, 7—4565/2, 8—4641/12, 9—4641/9, 10—5010/5, 11—4641/6, 12—5010/1, 13—4641/11, 14—4641/4, 15—4565/1, 16—5010/3, 17—5010/4, 18—Marker (Biolabmix Step 50+).

The results of the studies showed significant differences between plants that are decsendants of regenerants in terms of their growth rate under water in test tubes (Figure 5). After 5 days from the beginning of germination, the average length of sprouts varied from 0 to 25 mm. On the 7th day, the differences between the samples became more significant, from 8 to 60 mm. On the 11th day of growth, plant height varied from 24 to 119 mm (Figure 5).

By the 11th day, the maximum length of plants was more than 108–118 mm in samples with a locus in the genotype Sk1: 1—4641/12, 2—4641/3 and 3—4641/1. They rose above the layer of water and the top edge of the tube. Samples with the gene Sub1A: 12—4641/7; 13—4641/5; 14—5009/2 with a stem length 24–30 mm were noted.

Thus, as a result of the physiological evaluation of rice samples, vigorously growing homozygous lines with the greatest potential for growth and development were identified. Of interest are also samples that inhibit their growth to save nutrients, which is due to the Sub1A gene.

Regenerant plants were planted in a greenhouse in lysimeters, where they grew to adulthood (Figure 6). All of them have expanded, forming from 3 to 20 panicles per plant. The height of the plants was slightly lower than in the field, and varied from 26 to 74 cm. The length of the panicle ranged from 4 to 17 cm. The number of spikelets on the panicle ranged from 40 to 128 pieces. At the same time, high sterility was observed. The number of grains on the panicles of diploid forms ranged from 18 to 104 pieces. Tetraploid plants formed from two to four grains (Table 7). Haploid forms were completely sterile and did not form grain.

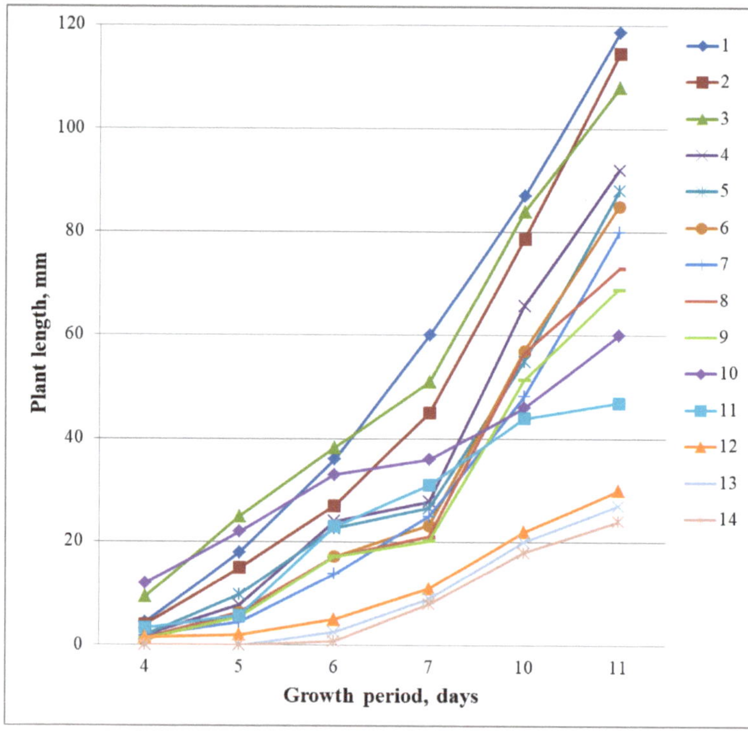

Figure 5. Growth dynamics of rice plants in test tubes under water. Regenerated lines: 1—4641/12; 2—4641/3; 3—4641/1; 4—4641/11; 5—4641/8; 6—4641/10; 7—4641/2; 8—4641/4; 9—4641/14; 10—5010/6; 11—4641/9; 12—4641/7; 13—4641/5; 14—5009/2.

Figure 6. Appearance of rice regenerating plants, 2023 Zernograd.

Table 7. Morphological features of rice regenerating plants.

Hybrid Number *	Plant No.	Ploidy	Sk1 Gene	Sub1A Gene	Plant Height, cm	Panicle Length, cm	Number of Panicles on a Plant, pcs	Number of Spikelets on a Panicle, pcs	Number of Grains on a Panicle, pcs	Weight of 1000 Grains, g
4565	1	1			51	11	20	128	0	0.0
4565	2	1		Sub1A	41	14	12	75	0	0.0
4641	1	2	Sk1		57	14.5	15	96	39	25.0
4641	2	2			60	13.5	9	116	52	22.1
4641	3	2	Sk1		47	12.5	10	109	60	24.8
4641	4	4			58	15.5	8	55	4	38.3
4641	5	2	Sk1	Sub1A	57	14	10	114	89	30.1
4641	6	4	Sk1		57	18	6	40	2	44.8
4641	7	2		Sub1A	65	16.5	8	126	104	28.7
4641	8	2	Sk1	Sub1A	58	15.5	8	50	32	24.2
4641	9	4	Sk1		48	13	11	36	4	47.2
4641	10	2		Sub1A	50	14	11	101	63	25.6
4641	11	2	Sk1		74	17	12	127	98	30.1
4641	12	2	Sk1	Sub1A	51	14	12	43	25	27.6
4641	13	4	Sk1	Sub1A	51	15.5	7	49	2	35.5
4641	14	4	Sk1		55	15.5	3	64	4	51.7
5009	1	1	Sk1		53	14.5	10	98	0	0.0
5009	2	2	Sk1	Sub1A	33	10	7	51	18	28.9
5009	3	1			62	13.5	40	40	0	0.0
5010	1	1		Sub1A	26	5.5	3	96	0	0.0
5010	2	1	Sk1	Sub1A	31	7.5	3	82	0	0.0
5010	3	1	Sk1	Sub1A	39	5	4	48	0	0.0
5010	4	1	Sk1		30	4	5	44	0	0.0
5010	5	1	Sk1	Sub1A	44	13	10	117	0	0.0
5010	6	2			57	14	7	101	3	20.0

* Note: Пa: 4565 IR-64 × Magnate, 4641 (Inbara-3 × Contact) × (Khao Hlan On × Kubojar), 5009 (Inbara-3 × Novator) × Contact, 5010 (Inbara-3 × Novator) × Contact.

The mass of 1000 grains in diploid forms ranged from 22.1 to 30.1 g. In tetraploid forms, the mass of 1000 grains ranged from 35.5 to 51.7 g. Plants of different levels of ploidy differed significantly in the number and size of spikelets (Figure 7).

Figure 7. Morphological features of panicles and spikelets of rice regenerating plants: (**a**)—tetraploid, (**b**)—diploid, (**c**)—haploid, (**d**)—spikelets in order: tetraploid, diploid, haploid.

The average number of spikelets on the panicle in tetraploids was 48.8 pcs., in diploids, it was 94.0 and in haploids, it was 80.9. The spikelet length in tetraploids reached 9.0 mm, in diploids, it was 7.3 and in haploids, it was 5.0.

Morphological differences between plants did not depend on the presence of flood-resistance genes (Sub1A, Sk1), since plants did not experience this stress in the greenhouse.

The largest number of regenerating plants (14) was formed in the hybrid 4641, of which 10 were diploid and formed normal seeds.

Seeds from rice regenerant plants were sown in the field. The plants that grew out of them were subjected to prolonged flooding with water to combat the hedgehog. Under these conditions, the sample 4641/1 (Inbara-3 × Contact) × (Khao Hlan On × Kubojar) carrying the Sk1 gene performed well. In the field, it was marked under the number 1477 (Figure 8).

Figure 8. Sample 1477 from seeds of the regenerate plant 4641/1 (Inbara-3 × Contact) × (Khao Hlan On × Kubojar).

4. Conclusions

The selected lines, which are characterized by good responsiveness to anther cultivation in vitro and carry the genes for resistance to prolonged flooding (Sub1A, Sk1), are used in rice breeding programs using biotechnology. On the basis of the obtained androgenic plants, dihaploid lines were formed, which will subsequently be included in breeding work to create rice varieties with economically valuable and adaptive traits.

Author Contributions: N.K., N.V., V.G., N.C.—preparation of experiments, laboratory experiments, data collection, data analysis, and interpretation, preparation of the manuscript, final revision of the text; P.K.—conceptualisation of research, preparation of the manuscript, critical analysis of the text. All authors have read and agreed to the published version of the manuscript.

Funding: The work was supported by grant No. 22-26-00246 of the Russian Science Foundation (RSF).

Institutional Review Board Statement: Not applicable.

Data Availability Statement: Not applicable.

Conflicts of Interest: The authors declare no conflict of interest.

References

1. Mishra, R.; Rao, G. In-vitro Androgenesis in Rice: Advantages, Constraints and Future Prospects. *Rice Sci.* **2016**, *23*, 57–68. [CrossRef]
2. Oladosu, Y.; Rafii, M.Y.; Arolu, F.; Chukwu, S.C.; Muhammad, I.; Kareem, I.; AdekunleSalisu, M.; Arolu, I.W. Submergence Tolerance in Rice: Review of Mechanism, Breeding and Future. *Sustainability* **2020**, *12*, 1632. [CrossRef]
3. Haque, M.A.; Rafii, M.Y.; Yusoff, M.M.; Ali, N.S.; Yusuff, O.; Arolu, F.; Anisuzzaman, M. Flooding tolerance in Rice: Adaptive mechanism and marker-assisted selection breeding approaches. *Mol. Biol. Rep.* **2023**, *50*, 2795–2812. [CrossRef]
4. Bailey-Serres, J.; Fukao, T.; Ronald, P.; Ismail, A.; Heuer, S.; Mackill, D. Submergence Tolerant Rice: SUB1's Journey from Landrace to Modern Cultivar. *Rice* **2010**, *3*, 138–147. [CrossRef]
5. Dubina, E.; Shilovsky, V.; Kostylev, P.; Garkusha, S.; Kovalev, V.; Yesaulova, L.; Balyasny, I.; Straholysova, M.; Dinh, T.; Le, L. Sub1A Gene in Rice Breeding for Tolerance to Flooding as a Factor in Weed Control. *Rice Farming* **2017**, *2*, 20–26.
6. Goncharova, Y. *The Use of the Anther Culture Method in rice Breeding*; Research Institute of Rice: Krasnodar, Russia, 2012; p. 91.
7. Savenko, E.; Korotenko, T.; Glazyrina, V.; Shundrina, L. Accelerated production of genetically stable (homozygous) forms of rice based on selectively valuable samples with target genes of resistance to pyriculariosis by anther culture in vitro method. *Proc. Kuban State Agrar. Univ.* **2020**, *85*, 213–219. [CrossRef]
8. Goncharova, Y.; Kharitonov, E.; Malyuchenko, E.; Sheleg, V. Rice adaptability to abiotic stresses analysis of breeding efficiency. In Proceedings of the Achievements and Prospects for the Development of Rice Breeding and Cultivation in Temperate Countries, International scientific Internet Conference, Krasnodar, Russia, 26–27 November 2015; pp. 40–46.
9. Ilyushko, M.; Romashova, M. The effect of intensity and quality of illumination on the regenerative ability of *Oryza sativa* L. rice callus obtained in in vitro androgenesis. *Russ. Agric. Sci.* **2021**, *3*, 41–45. [CrossRef]
10. Bishnoi, U.; Jain, R.; Gupta, K.R.; Chowdhury, V.K. High frequency androgenesis in indica Basmati rice hybrids using liquid culture media. *Plant Cell Tissue Organ Cult.* **1995**, *61*, 153–159.
11. Sarao, N.; Gosal, S. In vitro Androgenesis for Accelerated Breeding in Rice. *Biotechnol. Crop Improv.* **2018**, *1*, 407–435. [CrossRef]
12. Malysheva, N.; Savenko, E.; Glazyrina, V.; Shundrina, L. Obtaining, evaluation and selection of dihaploid rice lines with economically valuable traits. *Rice Farming* **2012**, *21*, 14–18.
13. Bushman, N.; Vereshchagina, S. The effectiveness of nutrient media for the induction of callus formation in rice hybrids. *Rice Farming* **2013**, *1*, 13–16.
14. Savenko, E.; Mukhina, Z.; Glazyrina, V.; Shundrina, L. Optimization of cellular technologies in vitro for accelerated generation of rice *Oryza sativa* L. *Bull. State Nikitsk. Bot. Gard.* **2022**, *144*, 114–121.
15. Guchenko, S. Comparative characteristics and selection of dihaploid rice lines by economically valuable traits. *Far East. Agrar. Bull.* **2016**, *1*, 10–15. Available online: https://vestnik.dalgau.ru/arkhiv_nomerov/?PAGEN_1=3 (accessed on 13 September 2023).
16. Ilyushko, M.; Romashova, M. Variability of rice haploids obtained in anther culture in vitro. *Russ. Agric. Sci.* **2019**, *2*, 11–14. [CrossRef]
17. Goncharova, Y.; Malyuchenko, E. The culture of anthers as a method of creating material for the study of various areas of breeding work. *Proc. Kuban State Agrar. Univ.* **2017**, *66*, 70–74. [CrossRef]
18. Savenko, E.; Glazyrina, V.; Shundrina, L. Variability of traits in populations of DH rice lines. *Rice Farming* **2022**, *2*, 6–10. [CrossRef]
19. Ilyushko, M.; Romashova, M.; Zhang, J.; Deng, L.; Liu, D.-J.; Zhang, R.; Guchenko, S. Intracallus variability of doubled rice haploids obtained in in vitro androgenesis. *Agric. Biol.* **2020**, *55*, 533–543. [CrossRef]
20. Goncharova, Y.; Kharitonov, E.; Bushman, N.; Vereshchagina, S. Comparative analysis of the effectiveness of nutrient media for the induction of callus formation in rice hybrids. *Rep. Russ. Acad. Agric. Sci.* **2013**, *6*, 6–9.
21. Romashova, M.; Ilyushko, M.; Companietz, S. Method of Obtaining Rice Regenerants in Anther Culture In Vitro. R.U. Patent 2681339 C1, 6 March 2019.
22. Pattnaik, S.S.; Dash, B.; Bhuyan, S.S.; Katara, J.L.; Parameswaran, C.; Verma, R.; Ramesh, N.; Samantaray, S. Anther culture efficiency in quality hybrid rice: A comparison between hybrid rice and its ratooned plants. *Plants* **2020**, *9*, 1306. [CrossRef]
23. Ilyushko, M.; Skaptsov, M.; Romashova, M. Nuclear DNA content in rice (*Oryza sativa* L.) regenerants obtained in anther culture in vitro. *Agric. Biol.* **2018**, *53*, 531–538. [CrossRef]

24. Savenko, E.; Glazyrina, V.; Shundrina, L. Practical application of biotechnological methods for breeding rice (*Oryza sativa* L.), from bioproducts to bioeconomy. In Proceedings of the IV Interregional Scientific and Practical Conference (with International Participation), Barnaul, Russia, 23–24 September 2021; pp. 128–132.
25. Rout, P.; Naik, N.; Ngangkham, U.; Verma, R.L.; Katara, J.L.; Singh, O.N.; Samantaray, S. Doubled Haploids generated through anther culture from an elite long duration rice hybrid, CRHR32: Method optimization and molecular characterization. *Plant Biotechnol.* **2016**, *33*, 177–186. [CrossRef]
26. Ahmadi, B.; Ebrahimzadeh, H. In vitro androgenesis: Spontaneous vs. artificial genome doubling and characterization of regenerants. *Plant Cell Rep.* **2020**, *39*, 299–316. [CrossRef]
27. Azarin, K.V.; Usatov, A.V.; Alabushev, A.V.; Kostylev, P.I.; Makarenko, M.S.; Kovalevich, A.A. Validation of SSR markers associated with Submergence Tolerance in Rice (*Oryza sativa* L.). *Am. J. Agric. Biol. Sci.* **2016**, *11*, 142–147. [CrossRef]
28. Oe, S.; Sasayama, D.; Luo, Q.; Fukayama, H.; Hatanaka, T.; Azuma, T. Growth responses of seedlings under complete submergence in rice cultivars carrying both the submergence-tolerance gene sub1a-1 and the floating genes snorkels. *Plant Prod. Sci.* **2021**, *25*, 70–77. [CrossRef]
29. Mayakaduwa, D.; Silva, T.D. A cytological indicator allows rapid assessment of microspore maturity, leading to improved in vitro anther response in Indica rice (*Oryza sativa* L.). *In Vitro Cell. Dev. Biol. Plant* **2017**, *53*, 591. [CrossRef]
30. Sahoo, S.A.; Jha, Z.; Verulkar, S.B.; Srivastava, A.K.; Suprasanna, P. High-throughput cell analysis based protocol for ploidy determination in anther-derived rice callus. *Plant Cell Tissue Organ Cult.* **2019**, *137*, 187. [CrossRef]
31. Palanisamy, D.; Marappan, S.; Ponnuswamy, R.D.; Mahalingam, P.S.; Bohar, R.; Vaidyanathanl, S. Accelerating hybrid rice breeding through the adoption of doubled haploid technology for R-line development. *Biologia* **2019**, *74*, 1259. [CrossRef]
32. Fukao, T.; Xu, K.; Ronald, P.; Bailey-Serres, J. A variable cluster of ethylene responsive-like factors regulates metabolic and developmental acclimation responses to submergence in rice. *Plant Cell* **2006**, *18*, 2021–2034.
33. Singh, N.; Dang, T.T.M.; Vergara, G.V.; Pandey, D.M.; Sanchez, D.; Neeraja, C.N.; Septiningsih, E.M.; Mendioro, M.; Tecson-Mendoza, E.M.; Ismail, A.M.; et al. Molecular marker survey and expression analyses of the rice submergence-tolerance gene SUB1A. *Theor. Appl. Genet.* **2010**, *121*, 1441–1453. [CrossRef]

plants

Article

Identification of Genetic Loci for Rice Seedling Mesocotyl Elongation in Both Natural and Artificial Segregating Populations

Fangjun Feng [1,2,†], Xiaosong Ma [1,2,3,†], Ming Yan [1,2], Hong Zhang [4], Daoliang Mei [4], Peiqing Fan [1], Xiaoyan Xu [1], Chunlong Wei [1], Qiaojun Lou [1,2], Tianfei Li [1,2], Hongyan Liu [1,2,*], Lijun Luo [1,2,3,*] and Hanwei Mei [1,2,*]

1 Shanghai Agrobiological Gene Center, Shanghai 201106, China; ffj@sagc.org.cn (F.F.); mxs09@sagc.org.cn (X.M.)
2 Key Laboratory of Grain Crop Genetic Resources Evaluation and Utilization, Ministry of Agriculture and Rural Affairs, Shanghai 201106, China
3 National Key Laboratory of Crop Genetic Improvement, Huazhong Agricultural University, Wuhan 430070, China
4 Anji Administrative Station of Water and Soil Conservation, Huzhou 313300, China
* Correspondence: lhy@sagc.org.cn (H.L.); lijun@sagc.org.cn (L.L.); hmei@sagc.org.cn (H.M.)
† These authors contributed equally to this work.

Abstract: Mesocotyl elongation of rice seedlings is a key trait for deep sowing tolerance and well seedling establishment in dry direct sowing rice (DDSR) production. Subsets of the Rice Diversity Panel 1 (RDP1, 294 accessions) and Hanyou 73 (HY73) recombinant inbred line (RIL) population (312 lines) were screened for mesocotyl length (ML) via dark germination. Six RDP1 accessions (Phudugey, Kasalath, CA902B21, Surjamkuhi, Djimoron, and Goria) had an ML longer than 10 cm, with the other 19 accessions being over 4 cm. A GWAS in RDP1 detected 118 associated SNPs on all 12 chromosomes using a threshold of FDR-adjusted $p < 0.05$, including 11 SNPs on chromosomes 1, 4, 5, 7, 10, and 12 declared by $-\log_{10}(P) > 5.868$ as the Bonferroni-corrected threshold. Using phenotypic data of three successive trials and a high-density bin map from resequencing genotypic data, four to six QTLs were detected on chromosomes 1, 2, 5, 6, and 10, including three loci repeatedly mapped for ML from two or three replicated trials. Candidate genes were predicted from the chromosomal regions covered by the associated LD blocks and the confidence intervals (CIs) of QTLs and partially validated by the dynamic RNA-seq data in the mesocotyl along different periods of light exposure. Potential strategies of donor parent selection for seedling establishment in DDSR breeding were discussed.

Keywords: *Oryza sativa* L.; mesocotyl elongation; GWAS; linkage mapping; dry direct sowing rice (DDSR)

Citation: Feng, F.; Ma, X.; Yan, M.; Zhang, H.; Mei, D.; Fan, P.; Xu, X.; Wei, C.; Lou, Q.; Li, T.; et al. Identification of Genetic Loci for Rice Seedling Mesocotyl Elongation in Both Natural and Artificial Segregating Populations. *Plants* **2023**, *12*, 2743. https://doi.org/10.3390/plants12142743

Academic Editors: Xiangjin Wei, Yingxin Zhang, Weixun Wu, Guiai Jiao and Mikeal L. Roose

Received: 23 April 2023
Revised: 6 July 2023
Accepted: 7 July 2023
Published: 24 July 2023

1. Introduction

Rice production is facing critical challenges due to shortages of irrigation water and labor resources; rapid increases in resource and labor costs and relatively stable grain prices have led to poor profits and deficits in the industry. Seedling transplanting in puddled fields is still the predominant cultivation system in major rice production countries, especially in high-yielding acreages. This traditional system, more precisely termed as transplanted–flooded rice by Liu et al. (2015), is water- and labor-consuming [1]. Field puddling and seedling transplanting consume up to 30% of the total water for the whole cropping season [2], which is not biologically necessary for the growth and development of rice plants. In rainfed rice cultivation, field preparation and transplanting may fail or be severely delayed when rainfall is inadequate or postponed in the window of crop rotation [3,4]. Although mechanized transplanting and wet direct sowing have been widely used in rice production, dry direct sowing rice (DDSR), especially when mechanized, has been suggested as an optimal rice cultivation system to save natural and labor resources and to consequently increase farmers' incomes [1,5]. Based on a positive review of previous reports and their own experimental results, Liu et al. (2015) nominated DDSR as a sustainable and feasible alternative to transplanted rice as it can produce a similar (or even higher) grain yield, largely save water and labor resources, and enhance water and nitrogen use efficiency [1].

The adaptation of rice varieties to dry direct sowing and the proper management of soil conditions, sowing time, and depth are key issues to assure rapid and full seedling establishment, which is beneficial to weed competition, rice plant growth and development, and the final grain yield. Rice varieties with less sensitivity to sowing depth, i.e., tolerance to deep sowing, are beneficial for allowing the seeds to germinate rapidly using the higher moisture in deep soil layers, for which mesocotyl elongation is a key underlying characteristic [6,7]. Screening several sets of rice germplasm collections has not only identified accessions with a largely elongated mesocotyl, but also showed the low proportion of such accessions in those populations. It was also found that there was a relatively higher percentage of accessions with elongated mesocotyls in landraces than in improved semidwarf varieties and in upland rice than in paddy rice [8–12].

Linkage analysis and genome-wide association studies (GWASs) have been widely adopted to identify loci significantly associated with important and complex agronomic traits in rice. Dozens of QTLs have been identified for mesocotyl elongation in different populations. Five QTLs have been mapped for mesocotyl elongation on chromosomes 1, 3, 5, and 7 using an F_3 population developed from a cross between a *japonica* cultivar, Labelle, and an *indica* cultivar, Black Gora [9]. Thirteen loci for mesocotyl elongation on chromosomes 1, 3, 4, 5, 6, and 9 were identified using a GWAS in our previous study [12]. Eight QTLs on chromosomes 1, 3, 6, 7, 8, and 12 were identified for mesocotyl elongation using a doubled-haploid population from a cross between IR64 and Azucena [13]. Eleven QTLs were identified for mesocotyl elongation on chromosomes 1, 3, 4, 5, 6, 9, and 11 using an RIL population [14]. Lee et al. (2012) detected five QTLs for mesocotyl elongation on chromosomes 1, 3, 7, 9, and 12 [15]. Liu et al. (2020) reported 16 unique loci for mesocotyl elongation through a GWAS [16]. Wang et al. (2021) identified 14 QTLs for mesocotyl elongation via QTL sequencing using 12 F_2 populations, and five unique QTLs with a GWAS using two diverse panels [17]. Gothe et al. (2022) also identified eight QTLs for mesocotyl elongation using a GWAS [18]. Several genes were characterized that regulated mesocotyl elongation via different molecular mechanisms, including examples like *GY1* [19], *OsGSK2* [20], *OsPAO5* [21], and *OsSMAX1* [22].

A large collection of global diverse rice germplasm accessions and its genotypic dataset based on high-density SNP markers, such as the open-access resources of the Rice Diversity Panels (RDP1 and RDP2) and genotypic data generated with a high-density rice array, provide great opportunities for the international rice research community to investigate the genetic basis of various phenotypic traits [23–28]. The screening of mesocotyl elongation in a subset of RDP1 showed rich genetic variation, supporting GWAS for this characteristic.

Seed vigor is usually defined as the ability of genotypes or seed samples to obtain strong seedlings via quick germination and fast growth at the early stage. Drought-resistant rice cultivars have been shown to have higher seed vigor and better seedling establishment after dry direct sowing than popular irrigated rice cultivars, especially under stress conditions like drought and phosphorus deficiency [29]. Early studies found that mesocotyl elongation is the key characteristic contributing to seedling vigor under aerobic soil conditions, even though elongations of both the mesocotyl and the coleoptile are anatomically beneficial to seedling emergence [7,30]. On the other hand, the seedling vigor of a hybrid (F_1) is usually higher than the average performance of two parental lines, even higher than the better parent, showing heterosis at the early seedling stage. Among reciprocal hybrids and their parental lines (three *japonica* cultivars and the *indica* cultivar IR50), significant correlation was observed between the emergence index of the second leaf (LEI) and germination index (GI): LEI = 1.561 + 0.793 × GI (r = 0.890 ***). There was high-level midparent heterosis (MPH) for characteristics related to seedling growth in most cases of hybrid combinations, especially for subspecific hybrids. For example, the leaf area index had an MPH of 2.4–27.6% among those hybrids at 23 days after sowing (DAS) [30]. As a commercial hybrid variety of water-saving and drought-resistant rice (WDR) [31,32], Hanyou 73 (HY73, Huhan7A × Hanhui 3) has been widely cultivated in drought-prone areas in China, showing promising potential to save labor and water resources in DDSR production. Superior seedling emergence and moderate mesocotyl elongation of HY73 were also observed in the field trial of DDSR with a sowing depth as high as 8–10 cm. An interesting question is whether the genetic basis of mesocotyl elongation or the heterosis of the hybrid determines the superior seedling emergence under dry sowing conditions. A set of recombinant inbred lines were developed from the hybrid of HY73. Linkage mapping of mesocotyl elongation was conducted using a high-density bin map from genotyping via the resequencing approach.

In this study, the associated loci and QTL in two different populations, together with candidate genes, are jointly analyzed to identify important genomic regions for mesocotyl elongation. Based on the seedling morphological and agronomical features of the germplasm accessions with the highest mesocotyl lengths, strategies for selecting the favored donor parental lines in DDSR breeding are also discussed.

2. Results

2.1. Phenotypic Variation in Mesocotyl Elongation in RDP1 and HY73 RILs

In the subset of RDP1, a wide range of mesocotyl lengths were observed, while a large proportion of the accessions had low mesocotyl elongation ability in darkness, showing a very high peak of frequency at the left end (ML < 1 cm) together with a long tail at the right end (Figure 1A). The ANOVA results showed highly significant variation in mesocotyl elongation among the accessions (Table 1). Six accessions from five different countries had MLs higher than 10cm, including Phudugey (19.0 cm, Bhutan), Kasalath (14.1 cm, India), CA902B21 (13.6 cm, Chad), Surjamkuhi (12.3 cm, India), Djimoron (12.3 cm, Guinea), and Goria (10.6 cm, Bangladesh). There were another 19 accessions with MLs longer than 4 cm, including Karkati 87 (Bangladesh), MTU 9 (India), PTB 30 (India), Arias (Indonesia), Azerbaidjanica (Azerbaijan), P373 (Pakistan), Jhona 349 (India), N12 (India), NSF-TV13 (unknown), Sinampaga selection (Philippines), Basmati 217 (India), WC 6 (China), Basmati (Pakistan), T26 (India), Rondo (China), DM56 (Bangladesh), Lemont (USA), Sadri Belyi (Azerbaijan), and 9524 (India).

For the RIL population from the hybrid HY73, the seedlings of two parental lines, Huhan7A and Hanhui 3, had mesocotyl lengths of about 2.5 cm and 4.0 cm, respectively. The subset of the RIL population had nearly normal distribution and ranges of around 1 cm to 9 cm, suggesting noticeable transgressive segregation in mesocotyl lengths, especially on the side of high phenotypic values (Figure 1B). The ANOVA results showed highly significant variation among the lines (Table 1). There were also significant variations among the trials and line–trial interactions, suggesting a high sensitivity in mesocotyl elongation phenotyping to environments.

As measured in this experiment and another sand culture trial in 2019 with a sowing depth of 8 cm, the seedlings of the hybrid (HY73, F_1) had an average elongated mesocotyl length of about 2.5 cm, while the sterile maternal parent (Huhan7A) and paternal restorer line (Hanhui 3) had mesocotyl lengths of 2–2.5 cm and 3.5–4 cm, respectively (Figure 2). This result supports the recessive (or at least semirecessive) inheritance of mesocotyl elongation in rice.

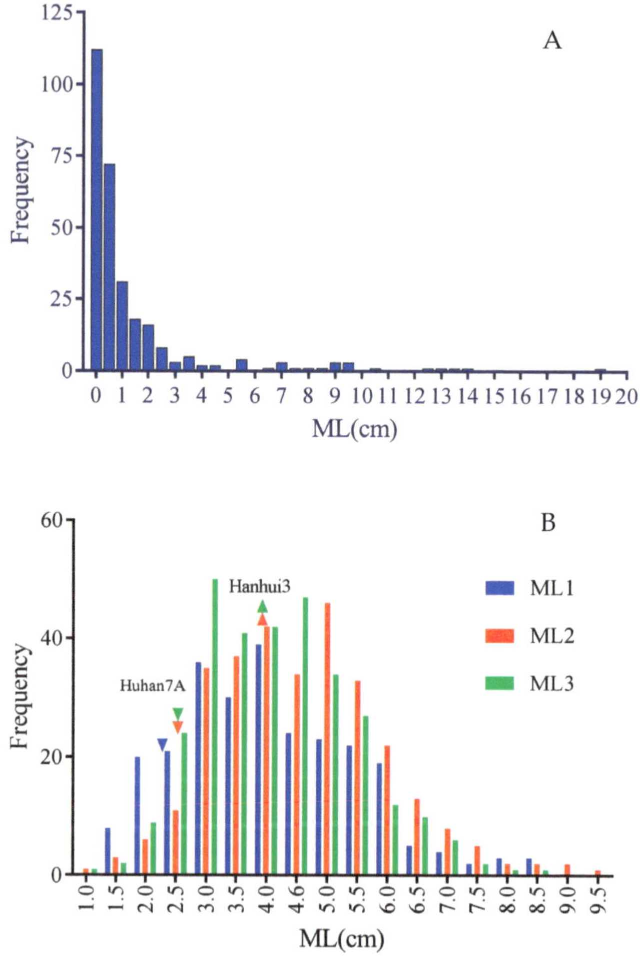

Figure 1. Frequency distribution of mesocotyl lengths measured in subsets of RDP1 (**A**) and HY73 RIL (**B**) populations. Downward and upward triangles represent the average MLs of Huhan 7A and Hanhui 3 (data missed in trial 1), respectively. ML1, ML2, and ML3 represent mesocotyl length measured in three successive trials.

Table 1. ANOVA of mesocotyl elongation measured in subsets of RDP1 and HY73 RIL populations.

Population	Source of Variation	SS (Type III)	*df*	MS	*F* Value	*p* Value
RDP1	Accessions	1,578,722.41	293	5369.80	335.02	0.000
	Residual	26,494.84	1653	16.03		
HY73 RILs	Lines	10,363.36	311	33.32	25.74	0.000
	Trials	442.50	2	221.25	170.92	0.000
	Lines × trials	1913.13	555	3.45	2.66	0.000
	Residual	7750.12	5987	1.29		

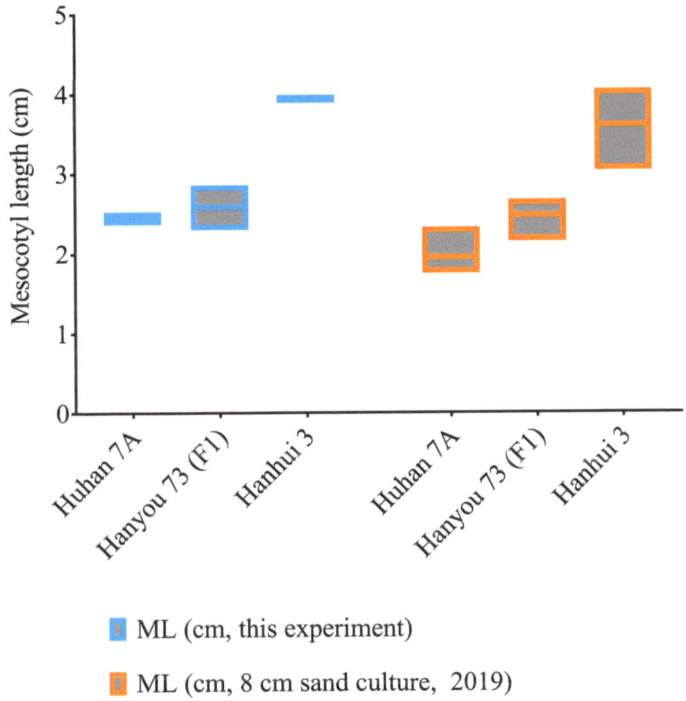

Figure 2. Mesocotyl lengths of the hybrid HY73 and two parental lines, Huhan 7A and Hanhui 3.

2.2. Genome-Wide Association Study in RDP1

A total of 118 associated SNPs were declared on all 12 rice chromosomes using the threshold of FDR-adjusted $p < 0.05$. Among them, 11 SNPs reached the stringent threshold of $-\log_{10}$ (0.05/36901) = 5.868, according to the Bonferroni correction for multiple tests (Table S1). There were at least 17 peaks of $-\log_{10}$ (p) values showing association signals along the rice genome, including three peaks on each of chromosomes 1 and 7; two on each of chromosomes 6, 11, and 12; and one on each of chromosomes 2–5 and 10. Eleven SNP markers with high $-\log_{10}$ (p) values fell into the six peaks of association signals on rice chromosomes 1, 4, 5, 7, 10, and 12 (Figure 3).

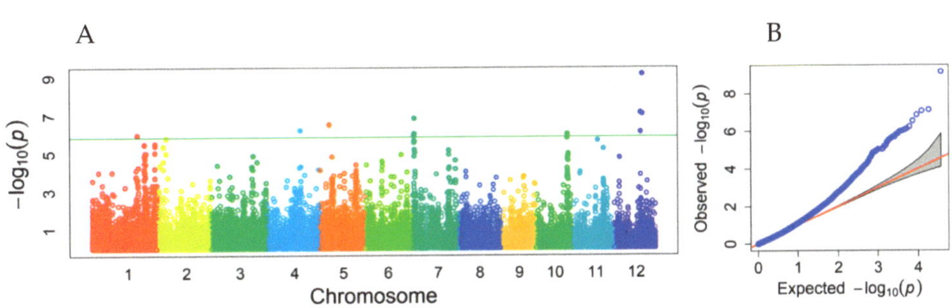

Figure 3. Manhattan plot (**A**) and QQ plot (**B**) showing associated loci for mesocotyl elongation identified in a subset of the RDP1 population. Varied colors represent different chromosomes. The horizontal line represents the threshold of −log10 (0.05/36901) = 5.868.

2.3. Linkage Mapping in HY73 RILs

Using the phenotypic values measured in three successive trials (ML1, ML2, and ML3), four, five, and six QTLs were detected for the elongated mesocotyl length, respectively (Table 2), including three on chromosome 1 and one on chromosome 9 for ML1; three on chromosome 1 and one on each of chromosomes 5 and 6 for ML2; and three on chromosome 1 and one on each of chromosomes 2, 6, and 10 for ML3. Among them, two identical intervals, Chr1_bin348–Chr1_bin350 and Chr6_bin6431–Chr6_bin6433, were repeatedly mapped for ML1 and ML3 and ML2 and ML3, respectively. The superior allele of the QTL interval Chr1_bin348–Chr1_bin350 was derived from Hanhui 3, and the superior allele of the QTL interval Chr6_bin6431–Chr6_bin6433 was derived from Huhan 7A. The significant interval on chromosome 1 Chr1_bin147–Chr1_bin149 for ML2 was close to Chr1_bin140–Chr1_bin146 for ML3, and superior alleles were both derived from Hanhui 3. Three significant intervals, Chr1_bin1342–Chr1_bin1348 for ML1, Chr1_bin1322–Chr1_bin1327 for ML2, and Chr1_bin1396–Chr1_bin1398 for ML3, were located adjacent to one another on the end of chromosome 1 (Figure 4), of which the superior alleles were derived from Hanhui 3. The abovementioned colocated or adjacent QTLs had the same direction of additive effects (Table 2), suggesting one QTL for ML at each location.

Table 2. QTLs for mesocotyl elongation mapped in HY73 RIL population.

Trait	Chr.	Left Marker	Right Marker	LOD	PVE (%)	Add (cm)
ML1	1	Chr1_bin348	Chr1_bin350	10.68	14.12	0.53
ML1	1	Chr1_bin899	Chr1_bin902	3.58	4.46	0.30
ML1	1	Chr1_bin1342	Chr1_bin1348	3.98	4.92	0.31
ML1	9	Chr9_bin8476	Chr9_bin8481	4.50	5.64	−0.33
ML2	1	Chr1_bin147	Chr1_bin149	7.67	7.69	0.40
ML2	1	Chr1_bin612	Chr1_bin620	12.02	11.96	0.49
ML2	1	Chr1_bin1322	Chr1_bin1327	4.54	4.28	0.29
ML2	5	Chr5_bin5536	Chr5_bin5537	3.24	3.03	0.24
ML2	6	Chr6_bin6431	Chr6_bin6433	7.49	7.24	−0.38
ML3	1	Chr1_bin140	Chr1_bin146	5.78	5.70	0.29
ML3	1	Chr1_bin348	Chr1_bin350	8.51	8.53	0.35
ML3	1	Chr1_bin1396	Chr1_bin1398	5.79	5.74	0.28
ML3	2	Chr2_bin1887	Chr2_bin1895	4.09	3.99	−0.24
ML3	6	Chr6_bin6431	Chr6_bin6433	3.94	3.82	−0.23
ML3	10	Chr10_bin8945	Chr10_bin8950	4.98	4.92	0.29

Chr, chromosome; LOD, logarithm of the odds; PVE, phenotypic variance explained; Add, additive effect. ML1, ML2, and ML3 represent the mesocotyl lengths measured in three trials.

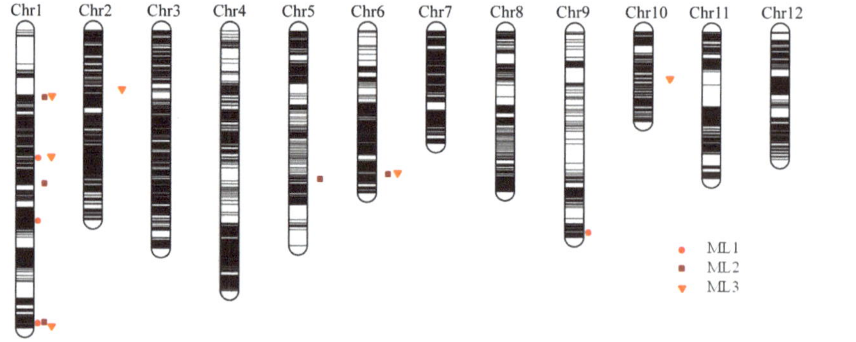

Figure 4. The high-density bin map constructed in a subset of HY73 RIL population and QTLs for elongated mesocotyl lengths measured in three successive trials. ML1, ML2, and ML3 represent the mesocotyl lengths measured in three trials.

2.4. Candidate Genes Predicted from Associated or Linked Genomic Regions

A primary candidate gene pool was developed by including 1954 annotated genes from either the LD blocks anchored by the associated SNPs in the RDP1 population or the CIs of QTLs detected in the HY73 RIL population (Table S2). Among them, 64 candidate genes were validated from their dynamic expression patterns during mesocotyl growth blocking by light, i.e., DEGs in the mesocotyl tissues of the seedlings exposed to light for 20 min, 60 min, or 360 min vs. the seedlings in darkness (Table S3) [33]. Three candidate genes were differentially expressed in all three light exposure periods, including *LOC_Os01g09100*, *LOC_Os07g03120*, and *LOC_Os09g37600*, annotated as OsWRKY10—a superfamily of TFs with WRKY and zinc finger domains, unknown expressed protein, and lysM domain-containing GPI-anchored protein precursor, respectively. Another 12 genes showed shifted expressions after light exposure for two time points, including 11 genes at 60 min light exposure together with the shorter (20 min) or longer (360 min) light exposure, and only one gene at both 20 min and 360 min light exposure (Table S3). It is noticeable that the expression levels of all three genes were induced by light exposure for three time periods, while those DEGs detected at two time points also kept the same direction, i.e., upregulated or downregulated in both light treatments.

3. Discussion

3.1. Genetic Variations in Mesocotyl Elongation Have Been Investigated in Different Rice Germplasm Populations

A total of 3677 rice accessions, or 3971 accessions if including the 294 accessions in this study, have been screened for mesocotyl elongation in 11 phenotypic experiments, as publicly reported since 1996 (Table S4). These natural variation populations cover almost all taxonomic branches of the germplasm collections in *Oryza sativa* L., e.g., landraces or varieties, upland or lowland rice, traditional or modern semidwarf varieties, together with weedy rice and *O. nivara* accessions. Benefitting greatly from powerful technical solutions, several rice germplasm collections have been developed to enrich the genetic variation, i.e., to contain a high percentage of the total genetic variations in the whole gene pool in subsets with smaller population sizes, e.g., the mini-core collection of Chinese rice germplasm [34], RDP1 [23,24], and 3010 rice accessions in 3K-RGP [35]. Using the diverse rice germplasm collections presented in this study and previous reports [12,36], together with special collections like upland rice and weedy rice, we can be highly confident that phenotypic screening results up to now have covered the major proportion of the genetic variation for mesocotyl elongation in the gene pool of rice germplasm collections.

However, varied protocols have been employed in different phenotypic screening experiments, roughly following two main strategies. A few methods have been modified from the standard protocol of a seed germination test. Seedlings are cultured on moist filter papers, gel media, or within filter paper rolls in plastic boxes, bottles, or other containers with more space than Petri dishes; then, they are easily separated for ML measurement. Other methods try to cover the seeds with moist soil, sand, or humus layers, simulating direct sowing in the field. Absolute darkness should be maintained throughout the whole procedure as light exposure at the early stage of germination might inhibit mesocotyl growth. Additionally, as submergence or extremely high moisture promote coleoptile growth but inhibit mesocotyl elongation, soil or sand water content needs to be uniform and well controlled among the samples. The method of humus soil culture with a 6 cm sowing depth is used in ML phenotyping and has been suggested as an optimal protocol [16]. However, exposing seedling tips to light after emergence from a 6 cm depth will block the further growth of the mesocotyl, thus causing an underestimation of mesocotyl elongation ability above 4–5 cm, similar to the results measured from the 5 cm sand culture [12]. If the whole culture procedure is conducted in darkness, precisely controlling the covering depth seems unnecessary. To evaluate the mesocotyl elongation potential of germplasm accessions, dark germination should still be first choice, being the cheapest and quickest method. But soil or sand culture with a specific sowing depth, close to or slightly higher than the largest sowing depth required in rice production, would be a sufficient protocol for screening germplasm accessions or segregating populations in DDSR breeding programs. Allowing for sowing depth ranges that actually occur in the field, a sand culture with 8 cm depth was employed in the screening trials of seed bulks of inbred or BC progenies by the authors' team and was quite effective in developing new lines with deep sowing tolerance. Using nylon net sheets to keep the seeds at a relatively equal depth and hold the seedlings after "digging out" at the end of the screening trials, this protocol is suitable for selection in large populations as large amounts of seeds can be screened rigorously in the limited space of the seedling nurseries.

Significant genetic variations in mesocotyl elongation, together with bias distribution in rice germplasm populations, were observed in those screening experiments [6,8,10–12,16–18,36,37]. Largely different ranges were reported, mainly at the side of high phenotypic values (Table S4). In addition to the difference caused by screening methods, varied genetic diversity among those populations should be the major reason, implying higher efficiency by using diverse germplasm collections. A higher proportion of germplasm accessions with moderate MLs (with approximately half having an ML > 1 cm) than several other screening results was observed and explained to be caused by different genotypes [16].

3.2. Novel QTLs and Candidate Genes Were Identified through GWAS and Linkage Analysis

Mesocotyl length is a quantitative trait and displays a substantial amount of variation among genotypes. Several previous studies have reported the associated QTLs and candidate genes for rice mesocotyl elongation through linkage analysis or GWAS [12,15–18,20,38]. Some known QTL intervals and novel QTLs could be detected through linkage analysis and GWAS in this study (Table S2). Among them, few overlapping cases were observed between the associated LD blocks in RDP1 and the CIs of QTLs in the HY73 RIL population. One LD block on rice chromosome 1, i.e., block 605 (marker 5506–5509, physical distance 41,826,155–41,836,778), and two sole SNPs (physical positions 41,956,669 and 41,957,405) were located within the CI of the QTL mapped for ML1, ML2, and ML3 on the end of rice chromosome 1 (physical distance 41,114,455–42,917,379). The LD block 362 (markers 2919–2921, 2923, and 2925–2931; physical distance 27,626,617–27,726,591) was located quite close to the CI of the QTL on chromosome 6 for ML2 and ML3 (Figures 3 and 4, Table S2). Comparing the candidate genes from our previous GWAS in the mini-core collection of Chinese rice germplasm plus drought-resistant parental lines [12] and the RDP1 population in this study, only one candidate gene (*LOC_Os01g71410*, glycosyl hydrolase family 17) was commonly predicted. This gene did not show differential expression in the light exposure experiment. However, based on the strategy of candidate genes within the LD block, instead of being hit by the top SNP [12], more genes could be predicted as candidates from this associated locus. Thus, more putative candidate genes from the adjacent chromosomal regions, e.g., *LOC_Os01g71340* (glycosyl hydrolase family 17), *LOC_Os01g71350* (glycosyl hydrolases family 17), *LOC_Os01g71420* (Ser/Thr protein phosphatase family protein), *LOC_Os01g71790* (NAM), *LOC_Os01g71820* (glycosyl hydrolase family 17), *LOC_Os01g71830* (glycosyl hydrolase family 17), *LOC_Os01g71860* (glycosyl hydrolase family 17), and *LOC_Os01g71970* (GRAS family transcription factor containing protein), could be included in the candidate gene list to be further investigated as they showed upregulated expression in the mesocotyls of rice seedlings after light exposure, but *LOC_Os01g71420* encoding the Ser/Thr protein phosphatase family protein was downregulated (Table S3). This chromosomal region also hosted the QTL *qMel-1* that was primarily mapped in the Kasalath/Nipponbare backcross inbred line population and was finely mapped in the derived population from the chromosomal segment substitution line [15]. The importance of this chromosomal region was further supported by more repeatedly mapped QTLs or associated SNPs for mesocotyl length and by being adjacent to the genes of *gy1* and *sd1* [13,19,37].

Within the closely located regions of associated SNPs (LD block 362) and QTLs for ML2 and ML3 on chromosome 6, one candidate gene, *OsFtsH2* (*LOC_Os06g45820*) encoding FtsH protease (homologue of *AtFtsH2/H8*), was upregulated after 60 min light exposure in the mesocotyl of rice seedlings (Table S3). These QTLs and candidate genes identified in this study will help to better understand the genetic mechanism controlling rice mesocotyl elongation.

3.3. Varied Strategies Could Be Considered to Select Donor Parents for DDSR Breeding Programs Concerning More Environmental Issues

The dry direct sowing rice production system faces more unfavorable environmental issues if compared with transplanting rice or wet direct sowing rice cropping. Inadequate rainfall, together with limited irrigation capacity or decreases in irrigation, thus drought proneness, is the inherent scenario of a DDSR production system. Focusing on the seed germination and seedling establishment stage, the soil moisture and temperature are two major environmental factors. A water deficit in the top soil layer (e.g., 0–5 cm) is the most frequent limiting factor delaying seed sowing or causing failed, or partially bad seedling establishment. An earlier study showed an association between mesocotyl elongation and seedling vigor [39]. Rice germplasm collections or segregating populations like RILs were also screened for traits related to seedling vigor [9,40–43]. As those screening trials were conducted under light and shallow sowing conditions, non-ML-driven seedling vigor traits were detected. Germplasm accessions with mesocotyl elongation, drought resistance, and other seedling vigor traits could be nominated as the donor parents for deep sowing tolerance in DDSR breeding.

4. Materials and Methods

4.1. Phenotyping Mesocotyl Elongation in the Rice Diversity Panel 1 (RDP1) and HY73 Recombinant Inbred Line (RIL) Population

A subset of 294 accessions with sufficient seeds from 423 accessions in Rice Diversity Panel 1 (RDP1) [23] was phenotypic for mesocotyl elongation. For each accession, 15 brown rice grains were sterilized in 2.5% sodium hypochlorite solution for 20 min, rinsed with sterilized water three times, and then embedded on MS media. The mesocotyl lengths of eight seedlings were measured after germination and growth at 30 °C in darkness for 7 days.

A subset of 312 lines, randomly selected from 1320 lines in the RIL population derived from the commercial hybrid rice Hanyou 73 (HY73, F_1 of Huhan7A × Hanhui 3), was phenotyped for mesocotyl elongation in three replicated trials (conducted successively). For each line in one trial, 12 brown rice grains were embedded into 12 wells along the long side of the 96-well plate with perforated well bottoms, inserting a black plastic straw in each well to keep the seedling growing straight. The mesocotyl length of seedlings was measured after germination and growth at 30 °C in darkness for 14 days, abbreviated as ML1, ML2, and ML3 for the three trials. All seedlings were exposed to light to prevent subsequent mesocotyl elongation during the period of measurement.

4.2. Genotypic Data of RDP1 and HY73 RIL Population

The RDP1 genotypic data of 36,901 SNPs were generated from the Affymetrix SNP array containing 44,100 SNP variants, which is publicly available via the NCBI dbSNP Database (accession codes 469281739 to 469324700) [26].

The whole set of HY73 RIL populations had 1320 lines at the F_9 generation. Genotyping via the resequencing approach was used to generate a genotypic dataset containing more than 5M SNPs, with the Shuhui 498 genome assembly as a reference (http://www.mbkbase.org/R498/ (accessed on 22 November 2017)) [44]. Following the previously reported pipeline [45], a set of 316,089 high-quality SNPs was used to construct a bin map containing 10,574 bins (Ma XS et al., unpublished data). For the subset of 312 RILs used in this study, a compressed version of the bin map containing 5,571 bins, after merging the adjacent redundant bins, was used for linkage mapping.

4.3. GWAS and Linkage Mapping

Using the mean mesocotyl length values of the accessions in the RDP1 subset as the phenotypic data, together with the genotypic data of 36,901 SNPs, a GWAS was conducted using the CMLM algorithm implemented in the software package GAPIT v2.25, using the parameter "PCA.total = 2" [46]. A threshold of FDR-adjusted $p < 0.05$ was used to declare significantly associated markers, including those declared by the stringent threshold of $-\log_{10}(0.05/36901) = 5.868$ following Bonferroni correction for multiple tests.

The mean values of mesocotyl lengths of HY73 RILs in each trial (ML1, ML2, and ML3) were used as phenotypic data in linkage mapping. The inclusive composite interval mapping for the additive and dominant QTL (ICIM-ADD) in the software package of QTL IciMapping v4.2 was used to locate the significant QTLs [47]. A threshold of LOD values was determined using permutation tests of 1000 times with $p < 0.05$. The confidence intervals (CIs) were defined using the one-LOD-value drop strategy, where the physical positions on the Shuhui 498 genome were flanked by the first and last bin markers of each CI.

4.4. Candidate Gene Prediction

The genotypic data files of RDP1 (sativas413.ped, sativas413.map, and sativas413.fam) were retrieved from the zip file RiceDiversity.44K.MSU6.Genotypes_PLINK.zip downloaded from the website of RiceDiversity.org. Plink was used to transfer the data file to the format compatible for Haploview (e.g., sativas413.chr1.ped, sativas413.chr1.info, and sativas413.chr1.nosex). Haploview was used to estimate the linkage disequilibrium (LD) blocks for each chromosome, using a max LD comparison distance = 100 Kb and maximum amount of missing data allowed per individual = 0.5 [48]. The annotated genes within the LD blocks containing associated SNPs, or hit by sole associated SNPs, were retrieved from the rice genome annotation database v6.1 (http://rice.uga.edu (accessed on 31 May 2021)) to serve as the primary pool of putative candidate genes. The annotated genes within the CIs of QTLs in the HY73 RIL population were also included in this candidate gene pool (Table S3). As the Nipponbare or Shuhui 498 genome assembly was used as the reference in the genotyping pipelines of the RDP1 and HY73 populations, respectively, the physical positions of the CIs detected in the HY73 population were transferred to the physical positions on the Nipponbare reference genome by blastall 2.2.17 (-p blastn).

Information about overlapping with other gene/QTL mapping results and/or repeated identification in gene transcriptional, functional, or regulating studies was used to predict more promising candidate genes. For instance, our previous transcriptome sequencing experiment measured the differentially expressed genes (DEGs) among the samples of mesocotyl tissue of Zhaxima seedlings in darkness and light exposure conditions for 20 min, 60 min, and 360 min [33]. Zhaxima has an extreme phenotype for ML, so its ML gene expression patterns should be particularly informative about candidate loci. Several publications also reported candidate gene prediction according to their GWAS results for mesocotyl elongation [12,16,36,37]. Wide involvement of multiple plant hormones in regulating mesocotyl elongation, e.g., gibberellin, cytokinin, ethylene, jasmonic acid, strigolactone, and brassinosteroid, was reported in several studies in rice [19,49–54]. Similar to the strategy used in previous studies [16,33], those genes from repeatedly detected genomic regions and/or related to phytohormone metabolism or signal transduction would be regarded as candidate genes with higher reliability.

Supplementary Materials: The following supporting information can be downloaded at https://www.mdpi.com/article/10.3390/plants12142743/s1: Table S1. List of 118 significantly associated SNPs declared based on the threshold of the FDR-adjusted $p < 0.05$, including 11 associated SNPs with $-\log_{10}(P) > 5.868$ as the conserve threshold after Bonferroni adjustment. Table S2. List of the 1954 primarily predicted candidate genes within the LD blocks containing associated markers in RDP1 or within the CIs of QTL in HY73 population. Table S3. List of 64 candidate genes screened from the primary pool supported by dynamically differential expressions in mesocotyl along the procedure of light exposure. Table S4. Result summary of phenotypic screening for mesocotyl elongation in germplasm and RIL populations reported in 11 publications and this study.

Author Contributions: Conceptualization, H.M. and L.L.; investigation, F.F., M.Y., H.Z., D.M., P.F., X.X., Q.L., T.L. and C.W.; resources, X.M. and H.L.; data curation, F.F. and X.M.; original draft preparation, H.M.; writing—review and editing, H.M., X.M. and F.F. All authors have read and agreed to the published version of the manuscript.

Funding: This work was supported by grants from the National Natural Science Foundation of China (31930080, 31671672), Shanghai Natural Science Foundation (18ZR1433300, 19ZR1446900, and 20ZR1449100), and Shanghai Municipal Commission of Agriculture (2014-7-1-4).

Data Availability Statement: The data presented in this study are available on request from the corresponding authors.

Acknowledgments: The authors are grateful to Sheng Teng for sharing the RDP1 seed samples, and to Susan McCouch and partners who originally constructed and publicly released the RDP1 and genotypic data.

Conflicts of Interest: The authors declare no conflict of interest.

References

1. Liu, H.; Hussain, S.; Zheng, M.; Peng, S.; Huang, J.; Cui, K.; Nie, L. Dry Direct-Seeded Rice as an Alternative to Transplanted-Flooded Rice in Central China. *Agron. Sustain. Dev.* **2015**, *35*, 285–294. [CrossRef]
2. Chauhan, B.S.; Opeña, J. Effect of Tillage Systems and Herbicides on Weed Emergence, Weed Growth, and Grain Yield in Dry-Seeded Rice Systems. *Field Crop. Res.* **2012**, *137*, 56–69. [CrossRef]
3. Saleh, A.F.M.; Bhuiyan, S.I. Crop and Rain Water Management Strategies for Increasing Productivity of Rainfed Lowland Rice Systems. *Agric. Syst.* **1995**, *49*, 259–276. [CrossRef]
4. Mazid, M.; Bhuiyan, S.I.; Mannan, M.; Wade, L. Dry-Seeded Rice for Enhancing Productivity of Rainfed Drought-Prone Lands: Lessons from Bangladesh and the Philippines. In *Direct Seeding: Research Issues and Opportunities. Proceedings of the International Workshop on Direct Seeding in Asian Rice Systems: Strategic Research Issues and Opportunities, Bangkok, Thailand, 25–28 January 2000*; Pandey, S., Ed.; IRRI: Los Banos, Philippines, 2002; pp. 185–200.
5. Kumar, V.; Ladha, J.K. Direct Seeding of Rice. Recent Developments and Future Research Needs. *Adv. Agron.* **2011**, *111*, 297–413.
6. Alibu, S. Relationship between Coleoptile and Mesocotyl Elongation of Upland Rice (*Oryza Sativa* L.) Seedlings under Submergence and Soil-Sand Culture. *African J. Agric. Res.* **2011**, *6*, 6463–6472.
7. Mgonja, M.A.; Dilday, R.H.; Skinner, S.L.; Collins, F.C. Association of Mesocotyl Elongation with Seedling Vigor in Rice. *Proceeding Ark. Acad. Sci.* **1988**, *42*, 52–55.
8. Chung, N.-J. Elongation Habit of Mesocotyls and Coleoptiles in Weedy Rice with High Emergence Ability in Direct-Seeding on Dry Paddy Fields. *Crop Pasture Sci.* **2010**, *61*, 911. [CrossRef]
9. Redoña, E.D.; Mackill, D.J. Mapping Quantitative Trait Loci for Seedling Vigor in Rice Using RFLPs. *Theor. Appl. Genet.* **1996**, *92*, 395–402. [CrossRef]
10. Wu, M.; Zhang, G.; Lin, J.; Cheng, S. Screening for Rice Germplasms with Specially-Elongated Mesocotyl. *Rice Sci.* **2005**, *12*, 226–228.
11. Luo, J.; Tang, S.Q.; Pei-Song, H.U.; Louis, A.; Jiao, G.A.; Jian, T. Analysis on Factors Affecting Seedling Establishment in Rice. *Rice Sci.* **2007**, *14*, 27–32. [CrossRef]
12. Wu, J.; Feng, F.; Lian, X.; Teng, X.; Wei, H.; Yu, H.; Xie, W.; Yan, M.; Fan, P.; Li, Y. Genome-Wide Association Study (GWAS) of Mesocotyl Elongation Based on Re-Sequencing Approach in Rice. *BMC Plant Biol.* **2015**, *15*, 218. [CrossRef]
13. Cao, L.; Zhu, J.; Ren, L.; Zhao, S.; Yan, Q. Mapping QTLs and Epistasis for Seedling Vigor in Rice (*Oryza sativa* L.). *Acta Agron. Sin.* **2002**, *28*, 809–815.
14. Cai, H.; Morishima, H. QTL Clusters Reflect Character Associations in Wild and Cultivated Rice. *Theor. Appl. Genet.* **2002**, *104*, 1217–1228. [CrossRef] [PubMed]
15. Lee, H.S.; Sasaki, K.; Higashitani, A.; Ahn, S.N.; Sato, T. Mapping and Characterization of Quantitative Trait Loci for Mesocotyl Elongation in Rice (*Oryza sativa* L.). *Rice* **2012**, *5*, 13. [CrossRef] [PubMed]
16. Liu, H.; Zhan, J.; Li, J.; Lu, X.; Liu, J.; Wang, Y.; Zhao, Q.; Ye, G. Genome-Wide Association Study (GWAS) for Mesocotyl Elongation in Rice (*Oryza sativa* L.) under Multiple Culture Conditions. *Genes* **2020**, *11*, 49. [CrossRef]
17. Wang, Y.; Liu, J.; Meng, Y.; Liu, H.; Ye, G. Rapid Identification of QTL for Mesocotyl Length in Rice Through Combining QTL-Seq and Genome-Wide Association Analysis. *Front. Genet.* **2021**, *12*, 713446. [CrossRef] [PubMed]
18. Gothe, R.M.; Bhatia, D.; Kamboj, A.; Sandhu, N.; Dhillon, B.S. Genetic Variation for Anaerobic Germination and Emergence from Deeper Soil Depth in *Oryza nivara* Accessions. *Rice Sci.* **2022**, *29*, 304–308. [CrossRef]

19. Xiong, Q.; Ma, B.; Lu, X.; Huang, Y.; He, S.; Yang, C.; Yin, C.; Zhao, H.; Zhou, Y.; Zhang, W.; et al. Ethylene-Inhibited Jasmonic Acid Biosynthesis Promotes Mesocotyl/Coleoptile Elongation of Etiolated Rice Seedlings. *Plant Cell* **2017**, *29*, 1053–1072. [CrossRef] [PubMed]
20. Sun, S.; Wang, T.; Wang, L.; Li, X.; Jia, Y.; Liu, C.; Huang, X.; Xie, W.; Wang, X. Natural Selection of a GSK3 Determines Rice Mesocotyl Domestication by Coordinating Strigolactone and Brassinosteroid Signaling. *Nat. Commun.* **2018**, *9*, 2523. [CrossRef] [PubMed]
21. Lv, Y.; Shao, G.; Jiao, G.; Sheng, Z.; Xie, L.; Hu, S.; Tang, S.; Wei, X.; Hu, P. Targeted Mutagenesis of *POLYAMINE OXIDASE 5* That Negatively Regulates Mesocotyl Elongation Enables the Generation of Direct-Seeding Rice with Improved Grain Yield. *Mol. Plant* **2021**, *14*, 344–351. [CrossRef]
22. Zheng, J.; Hong, K.; Zeng, L.; Wang, L.; Kang, S.; Qu, M.; Dai, J.; Zou, L.; Zhu, L.; Tang, Z. Karrikin Signaling Acts Parallel to and Additively with Strigolactone Signaling to Regulate Rice Mesocotyl Elongation in Darkness. *Plant Cell* **2020**, *32*, 2780–2805. [CrossRef] [PubMed]
23. Eizenga, G.C.; Ali, M.L.; Bryant, R.J.; Yeater, K.M.; Mcclung, A.M.; Mccouch, S.R. Registration of the Rice Diversity Panel 1 for Genomewide Association Studies. *J. Plant Regist.* **2014**, *8*, 109–116. [CrossRef]
24. McCouch, S.R.; Wright, M.H.; Tung, C.W.; Maron, L.G.; Mcnally, K.L.; Fitzgerald, M.; Singh, N.; Declerck, G.; Agosto-Perez, F.; Korniliev, P. Open Access Resources for Genome-Wide Association Mapping in Rice. *Nat. Commun.* **2016**, *7*, 10532. [CrossRef] [PubMed]
25. Norton, G.J.; Douglas, A.; Lahner, B.; Yakubova, E.; Guerinot, M.L.; Pinson, S.R.M.; Tarpley, L.; Eizenga, G.C.; McGrath, S.P.; Zhao, F.J.; et al. Genome Wide Association Mapping of Grain Arsenic, Copper, Molybdenum and Zinc in Rice (*Oryza sativa* L.) Grown at Four International Field Sites. *PLoS ONE* **2014**, *9*, e89685. [CrossRef]
26. Zhao, K.; Tung, C.W.; Eizenga, G.C.; Wright, M.H.; Ali, M.L.; Price, A.H.; Norton, G.J.; Islam, M.R.; Reynolds, A.; Mezey, J.; et al. Genome-Wide Association Mapping Reveals a Rich Genetic Architecture of Complex Traits in Oryza Sativa. *Nat. Commun.* **2011**, *2*, 467. [CrossRef]
27. Davidson, H.; Shrestha, R.; Cornulier, T.; Douglas, A.; Travis, T.; Johnson, D.; Price, A.H. Spatial Effects and GWA Mapping of Root Colonization Assessed in the Interaction Between the Rice Diversity Panel 1 and an Arbuscular Mycorrhizal Fungus. *Front. Plant Sci.* **2019**, *10*, 633. [CrossRef]
28. Yamane, K.; Garcia, R.; Imayoshi, K.; Mabesa-Telosa, R.C.; Banayo, N.P.M.C.; Vergara, G.; Yamauchi, A.; Cruz, P.S.; Kato, Y. Seed Vigour Contributes to Yield Improvement in Dry Direct-Seeded Rainfed Lowland Rice. *Ann. Appl. Biol.* **2018**, *172*, 100–110. [CrossRef]
29. Takahashi, N. Adaptive Importance of Mesocotyl and Coleoptile Growth in Rice Under Different Moisture Regimes. *Funct. Plant Biol.* **1978**, *5*, 511–517. [CrossRef]
30. de Leon, J.C.; Abe, T.; Sasahara, T. Variations in Morpho-Physiological Traits Relating to Seedling Vigor and Heterosis in Reciprocal Crosses of Rice. *Breed. Sci.* **2001**, *51*, 57–61. [CrossRef]
31. Luo, L.; Mei, H.; Yu, X.; Xia, H.; Chen, L.; Liu, H.; Zhang, A.; Xu, K.; Wei, H.; Liu, G.; et al. Water-Saving and Drought-Resistance Rice: From the Concept to Practice and Theory. *Mol. Breed.* **2019**, *29*, 145. [CrossRef]
32. Luo, L.J. Breeding for Water-Saving and Drought-Resistance Rice (WDR) in China. *J. Exp. Bot.* **2010**, *61*, 3509–3517. [CrossRef] [PubMed]
33. Feng, F.; Mei, H.; Fan, P.; Li, Y.; Xu, X.; Wei, H.; Yan, M.; Luo, L. Dynamic Transcriptome and Phytohormone Profiling along the Time of Light Exposure in the Mesocotyl of Rice Seedling. *Sci. Rep.* **2017**, *7*, 11961. [CrossRef]
34. Zhang, H.; Zhang, D.; Wang, M.; Sun, J.; Qi, Y.; Li, J.; Wei, X.; Han, L.; Qiu, Z.; Tang, S. A Core Collection and Mini Core Collection of *Oryza sativa* L. in China. *Theor. Appl. Genet.* **2011**, *122*, 49–61. [CrossRef]
35. Wang, W.; Mauleon, R.; Hu, Z.; Chebotarov, D.; Tai, S.; Wu, Z.; Li, M.; Zheng, T.; Fuentes, R.R.; Zhang, F.; et al. Genomic Variation in 3,010 Diverse Accessions of Asian Cultivated Rice. *Nature* **2018**, *557*, 43–49. [CrossRef] [PubMed]
36. Zhao, Y.; Zhao, W.; Jiang, C.; Wang, X.; Xiong, H.; Todorovska, E.G.; Yin, Z.; Chen, Y.; Wang, X.; Xie, J.; et al. Genetic Architecture and Candidate Genes for Deep-Sowing Tolerance in Rice Revealed by Non-Syn GWAS. *Front. Plant Sci.* **2018**, *9*, 332. [CrossRef] [PubMed]
37. Lu, Q.; Zhang, M.; Niu, X.; Wang, C.; Xu, Q.; Yue, F.; Shan, W.; Yuan, X.; Yu, H.; Wang, Y. Uncovering Novel Loci for Mesocotyl Elongation and Shoot Length in Indica Rice through Genome-Wide Association Mapping. *Planta* **2015**, *243*, 645–657. [CrossRef]
38. Cao, L.; Zhu, J.; Yan, Q.; Libin, H.E.; Wei, X.; Cheng, S. Mapping QTLs with Epistasis for Mesocotyl Length in a DH Population from Indica-Japonica Cross of Rice (*Oryza sativa*). *Chin. J. Rice Sci.* **2002**, *16*, 221–224.
39. Mgonja, M.A.; Ladeinde, T.A.O.; Aken'Ova, M.E. Genetic Analysis of Mesocotyl Length and Its Relationship with Other Agronomic Characters in Rice (*Oryza sativa* L.). *Euphytica* **1993**, *72*, 189–195. [CrossRef]
40. Cui, K.H.; Peng, S.B.; Xing, Y.Z.; Xu, C.G.; Yu, S.B.; Zhang, Q. Molecular Dissection of Seedling-Vigor and Associated Physiological Traits in Rice. *Theor. Appl. Genet.* **2002**, *105*, 745–753. [CrossRef]
41. Guo, T.; Yang, J.; Li, D.; Sun, K.; Luo, L.; Xiao, W.; Wang, J.; Liu, Y.; Wang, S.; Wang, H.; et al. Integrating GWAS, QTL, Mapping and RNA-Seq to Identify Candidate Genes for Seed Vigor in Rice (*Oryza sativa* L.). *Mol. Breed.* **2019**, *39*, 1–16. [CrossRef]
42. Dang, X.; Thi, T.G.T.; Dong, G.; Wang, H.; Edzesi, W.M.; Hong, D. Genetic Diversity and Association Mapping of Seed Vigor in Rice (*Oryza Sativa* L.). *Planta* **2014**, *239*, 1309–1319. [CrossRef] [PubMed]

43. Anandan, A.; Anumalla, M.; Pradhan, S.K.; Ali, J. Population Structure, Diversity and Trait Association Analysis in Rice (*Oryza sativa* L.) Germplasm for Early Seedling Vigor (ESV) Using Trait Linked SSR Markers. *PLoS ONE* **2016**, *11*, e0152406. [CrossRef] [PubMed]
44. Du, H.; Yu, Y.; Ma, Y.; Gao, Q.; Cao, Y.; Chen, Z.; Ma, B.; Qi, M.; Li, Y.; Zhao, X.; et al. Sequencing and *de novo* Assembly of a near Complete Indica Rice Genome. *Nat. Commun.* **2017**, *8*, 15324. [CrossRef] [PubMed]
45. Huang, X.; Feng, Q.; Qian, Q.; Zhao, Q.; Wang, L.; Wang, A.; Guan, J.; Fan, D.; Weng, Q.; Huang, T.; et al. High-Throughput Genotyping by Whole-Genome Resequencing. *Genome Res.* **2009**, *19*, 1068–1076. [CrossRef]
46. Lipka, A.E.; Tian, F.; Wang, Q.; Peiffer, J.; Li, M.; Bradbury, P.J.; Gore, M.A.; Buckler, E.S.; Zhang, Z. GAPIT: Genome Association and Prediction Integrated Tool. *Bioinformatics* **2012**, *28*, 2397–2399. [CrossRef]
47. Meng, L.; Li, H.; Zhang, L.; Wang, J. QTL IciMapping: Integrated Software for Genetic Linkage Map Construction and Quantitative Trait Locus Mapping in Biparental Populations. *Crop J.* **2015**, *3*, 269–283. [CrossRef]
48. Barrett, J.C.; Fry, B.; Maller, J.; Daly, M.J. Haploview: Analysis and Visualization of LD and Haplotype Maps. *Bioinformatics* **2005**, *21*, 263–265. [CrossRef]
49. Chen, F.; Jiang, L.; Zheng, J.; Huang, R.; Wang, H.; Hong, Z.; Huang, Y. Identification of Differentially Expressed Proteins and Phosphorylated Proteins in Rice Seedlings in Response to Strigolactone Treatment. *PLoS ONE* **2014**, *9*, e93947. [CrossRef]
50. Hu, Z.; Yamauchi, T.; Yang, J.; Jikumaru, Y.; Tsuchida-Mayama, T.; Ichikawa, H.; Takamure, I.; Nagamura, Y.; Tsutsumi, N.; Yamaguchi, S.; et al. Strigolactone and Cytokinin Act Antagonistically in Regulating Rice Mesocotyl Elongation in Darkness. *Plant Cell Physiol.* **2014**, *55*, 30–41. [CrossRef]
51. Tamiru, M.; Abe, A.; Utsushi, H.; Yoshida, K.; Takagi, H.; Fujisaki, K.; Undan, J.R.; Rakshit, S.; Takaichi, S.; Jikumaru, Y.; et al. The Tillering Phenotype of the Rice Plastid Terminal Oxidase (PTOX) Loss-of-function Mutant Is Associated with Strigolactone Deficiency. *New Phytol.* **2014**, *202*, 116–131. [CrossRef]
52. Liang, Q.; Wang, C.; Ma, D.; Li, L.; Cui, Z.; Wang, X.; Qian, Q.; Cai, B.; Feng, Y.; Chen, W. Cortical Microtubule Disorganized Related to an Endogenous Gibberellin Increase Plays an Important Role in Rice Mesocotyl Elongation. *Plant Biotechnol.* **2016**, *33*, 59–69. [CrossRef]
53. Kameoka, H.; Kyozuka, J. Downregulation of Rice DWARF 14 LIKE Suppress Mesocotyl Elongation via a Strigolactone Independent Pathway in the Dark. *J. Genet. Genom.* **2015**, *42*, 119–124. [CrossRef] [PubMed]
54. Riemann, M.; Müller, A.; Korte, A.; Furuya, M.; Weiler, E.W.; Nick, P. Impaired Induction of the Jasmonate Pathway in the Rice Mutant Hebiba. *Plant Physiol.* **2003**, *133*, 1820–1830. [CrossRef] [PubMed]

plants

Article

Genetic Diversity and Association Mapping of Grain-Size Traits in Rice Landraces from the Honghe Hani Rice Terraces System in Yunnan Province

Mengli Ma [1,2], En Lei [2], Tiantao Wang [1], Hengling Meng [1], Wei Zhang [2] and Bingyue Lu [1,2,*]

1 Key Laboratory for Research and Utilization of Characteristic Biological Resources in Southern Yunnan, Honghe University, Mengzi 661199, China
2 College of Biological and Agricultural Sciences, Honghe University, Mengzi 661199, China
* Correspondence: lby202@126.com

Abstract: The Honghe Hani Rice Terraces System (HHRTS) of Yunnan Province is an important agricultural and cultural heritage landscape. Until now, a large number of local rice landraces have been planted. Mining excellent genes contained in these landraces provides a reference for variety improvement and new variety breeding. In this study, 96 rice landraces collected from the Hani terraces were planted in Honghe Mengzi, Yunnan Province, in 2013, 2014, 2015, and 2021, and five major grain traits were measured and analyzed. The genomic variation of 96 rice landraces was scanned by 201 simple sequence repeat (SSR) markers. The genetic diversity, population structure, and genetic relationships of the natural population were analyzed. The mixed linear model (MLM) method of the TASSEL software was used to analyze the associations between markers and traits. A total of 936 alleles were amplified by 201 pairs of SSR primers. The average number of observed alleles (Na), the effective number of alleles (Ne), Shannon's information index (I), heterozygosity (H), and the polymorphism information content (PIC) per marker were 4.66, 2.71, 1.08, 0.15, and 0.55, respectively. Ninety-six landraces were divided into two groups by population structure, clustering, and principal component analysis, and *indica* rice was the main group. The coefficients of variation of the five traits ranged from 6.80 to 15.24%, and their broad heritabilities were more than 70%. In addition, there were positive correlations among the same grain traits between different years. Through MLM analysis, 2, 36, 7, 7, and 4 SSR markers were significantly associated with grain length (GL), grain width (GW), grain thickness (GT), grain length–width ratio (LWR), and thousand-grain weight (TGW), respectively. The explanation rates of phenotypic variation were 16.31 (RM449, Chr. 1)—23.51% (RM316, Chr. 9), 10.84 (RM523, Chr. 3; RM161/RM305, Chr. 5)—43.01% (RM5496, Chr. 1), 11.98 (RM161/RM305, Chr. 5)—24.72% (RM275, Chr. 6), 12.68 (RM126, Chr. 8)—36.96% (RM5496, Chr. 1), and 17.65 (RM4499, Chr. 2)—26.32% (RM25, Chr. 8), respectively. The associated markers were distributed on 12 chromosomes of the genome.

Keywords: Honghe Hani Rice Terraces; rice landraces; grain-size traits; genetic diversity; association mapping

Citation: Ma, M.; Lei, E.; Wang, T.; Meng, H.; Zhang, W.; Lu, B. Genetic Diversity and Association Mapping of Grain-Size Traits in Rice Landraces from the Honghe Hani Rice Terraces System in Yunnan Province. *Plants* **2023**, *12*, 1678. https://doi.org/10.3390/plants12081678

Academic Editors: Guangcun He, Xiangjin Wei, Yingxin Zhang, Weixun Wu and Guiai Jiao

Received: 6 March 2023
Revised: 31 March 2023
Accepted: 15 April 2023
Published: 17 April 2023

1. Introduction

Rice (*Oryza sativa* L.), one of the most important food crops in the world, is a staple food for more than half of the world's population [1]. China is the largest rice producer in the world, with its rice-sown area reaching 30 million hectares (National Bureau of Statistics, 2022). Yunnan Province is recognized as one of the original centers of cultivated rice. It is also the largest center of genetic and ecological diversity of rice germplasm resources in China [2]. The Honghe Hani Rice Terraces System (HHRTS) has a long history as a world heritage cultural landscape and still grows a large number of local rice landraces, especially red rice resources [3]. From ancient times, the Hani people believed that traditional red rice had a higher nutritional value and that consumption of the same made the body healthier.

In 2014, Yuanyang terraced red rice was listed as one of the "six famous rices" in Yunnan Province (https://nync.yn.gov.cn/ (accessed on 15 October 2022)). Hani Terrace is located in Southern Yunnan, with its core area being located in Yuanyang County (east longitude 102°27′ to 103°13′, north latitude 22°49′ to 23°19′). The annual average sunshine in this area is 1670 h, with the average temperature being 15.4 °C. Environmental conditions there are suitable for rice cultivation. The Hani people live on a hillside at an altitude of 1400 to 2000 m above sea level. They have a history of rice farming of more than 1500 years with varied rice landraces [4].

Rice grain-size traits usually include grain length (GL), grain width (GW), grain thickness (GT), grain length–width ratio (LWR), and thousand-grain weight (TGW). Rice grain traits are important indicators of both rice quality and yield. Therefore, the study of the inheritance and molecular mechanisms of grain-size traits plays an important role in improving rice yield and quality. Grain size is a typical quantitative trait controlled by polygenes [5]. In recent years, researchers have mapped a large number of quantitative trait loci (QTLs) related to grain type using different genetic populations, including F_2 populations, backcross populations (BCs), doubling haploid populations (DHs), recombinant inbred lines (RILs), and chromosome segment replacement lines (CSSLs) [6]. Based on incomplete statistics, more than 500 QTLs related to grain size were mapped on 12 chromosomes of rice (http://www.ricedata.cn/index.htm (accessed on 6 March 2022)). Using natural variation among cultivated varieties, researchers cloned several major QTLs related to grain size, including *GW2* [7], *GS2* [8], *TGW2* [9], *OsLG3* [10], *GS3* [11], *qGL3* [12], *qTGW3* [13], *GL4* [14], *GS5* [15], *GW5* [16], *GS6* [17], *GW6* [18], *TGW6* [19], *GL6* [20], *GLW7* [21], *GW7* [22], *GW8* [23], *GS9* [24], and *GW10* [25]. The cloning of these genes updated and improved the grain-size regulatory network, laying a key theoretical foundation and providing genetic resources for high-yield and high-quality rice breeding. Though several major QTLs were identified using a family-based mapping approach, a family population can only identify an allelic variation between two parents, with the optimal allelic variation among species not being found. Association mapping, which utilizes allelic variation in natural populations, can detect many natural allelic variations from germplasm accessions with a much higher resolution compared to linkage mapping. In recent years, excellent allelic variation in grain size in rice was discovered by association analysis [26].

In the present study, a rice association panel containing 96 rice landraces collected from HHRTS in Yunnan was developed to unravel the genetic basis of grain-size traits using 201 simple sequence repeat (SSR) markers. The objectives of the study were to: (i) determine the genetic diversity and population structure of the association panel, (ii) investigate phenotypic traits and their variations in different environments, and (iii) identify significant marker–trait associations for grain-size traits.

2. Results

2.1. Genetic Diversity Analysis

In total, 936 polymorphic bands were detected from 96 rice landraces by 201 pairs of SSR primers. The observed number of alleles ranged from 2 to 13, with an average of 4.66. The average value of I was 1.08, the value for the primer RM80 being the highest (I = 2.05) and that for the primer RM596 the lowest (I = 0.31). Values of H ranged from 0.00 (RM5994) to 1.00 (RM161, RM1385, RM598, RM479, RM484, RM147, RM319, RM512, RM317, and RM269), with an average of 0.15 (Tables 1 and S1). As an important index for species genetic diversity evaluation, the PIC values of 201 primer pairs ranged from 0.20 (RM596) to 0.86 (RM206), with an average of 0.55, indicating that these rice landraces from HHRTS had high genetic diversity (Tables 1 and S1).

Table 1. Genetic diversity parameters of rice landraces in the Honghe Hani terraces based on SSR markers.

Parameter	Mean ± SE	Minimum	Maximum
Na	4.66 ± 0.15	2.00	13.00
Ne	2.71 ± 0.09	1.19	6.73
I	1.08 ± 0.03	0.31	2.05
H	0.15 ± 0.02	0.00	1.00
PIC	0.55 ± 0.01	0.20	0.86

Na: observed number of alleles; Ne: effective number of alleles; I: Shannon's information index; H: heterozygosity; PIC: polymorphism information content.

2.2. Population Structure and Genetic Relationships

When the population structure of 96 rice landraces in the Hani terraces was analyzed using Structure v.2.3.4, a peak appeared along with an obvious inflection point when K = 2 (Figure 1A). Hence, the 96 rice materials were divided into two groups (Figure 1B), namely, an *indica* rice group (Q1) and a *japonica* rice group (Q2); Q1 contained 81 materials and Q2 contained 15 materials, suggesting that *indica* rice was the main variety in the Hani terraces. To further verify the results of the population structure analysis, we performed clustering and PCA and also divided the rice landraces into two categories (Figure 1C,D). The results were consistent with the population structure analysis based on the STRUCTURE model, indicating that the population structure of this association mapping population was relatively simple and could effectively reduce the influence of population structure on association analysis.

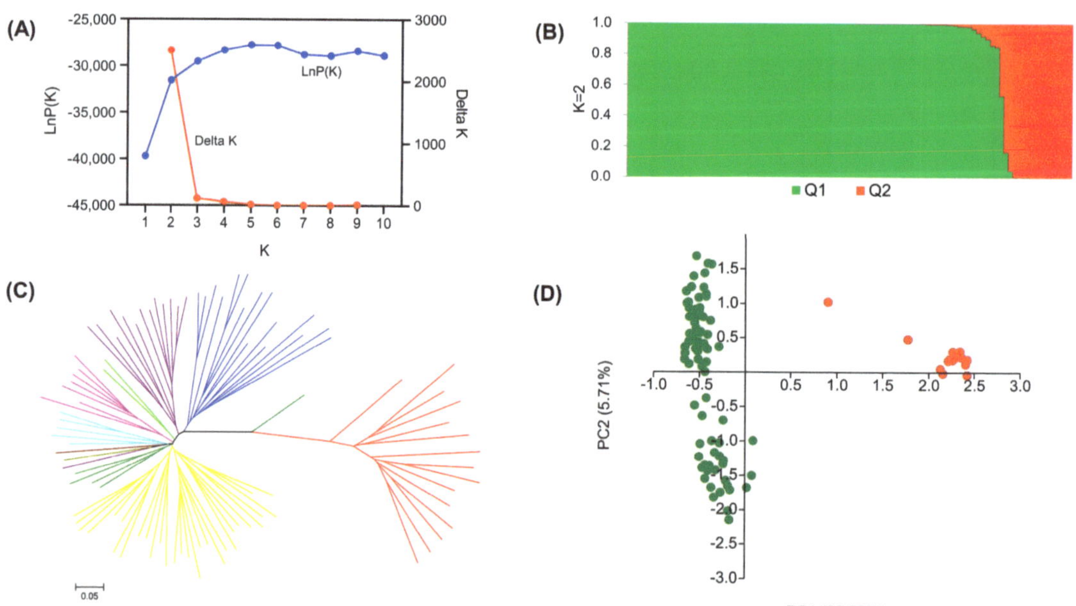

Figure 1. Population structure analysis using Bayesian clustering (**A**,**B**), neighbor joining ((**C**), the red branches represent *japonica* accessions, and the other color branches represent *indica* accessions), and the principal component (**D**) method.

2.3. Phenotypic Distribution of Grain-Size Traits

Five grain traits were evaluated in 96 rice landraces in 2013, 2014, 2015, and 2021 (Table 2). The coefficients of variation (CVs) of the five traits ranged from 6.80 to 15.24%.

Among the five traits, TGW showed maximum variation, ranging from 11.42 to 15.24%, with abundant variation among different materials. The minimum CV was found in grain thickness, ranging from 6.80 to 7.73%, where the range of variation was relatively narrow. The absolute values for skewness and kurtosis of the five traits in the four environments were less than 1 and 2 (Table 2), respectively, and their distributions were continuous (Figure 2A–E), belonging to typical quantitative traits. The broad-sense heritability (H_B^2) of GL, GW, GT, LWR, and TGW reached 92.63, 87.85, 92.64, 72.39, and 86.11%, respectively, indicating that the grain-size traits were mainly controlled by genetic effects and less by environmental factors.

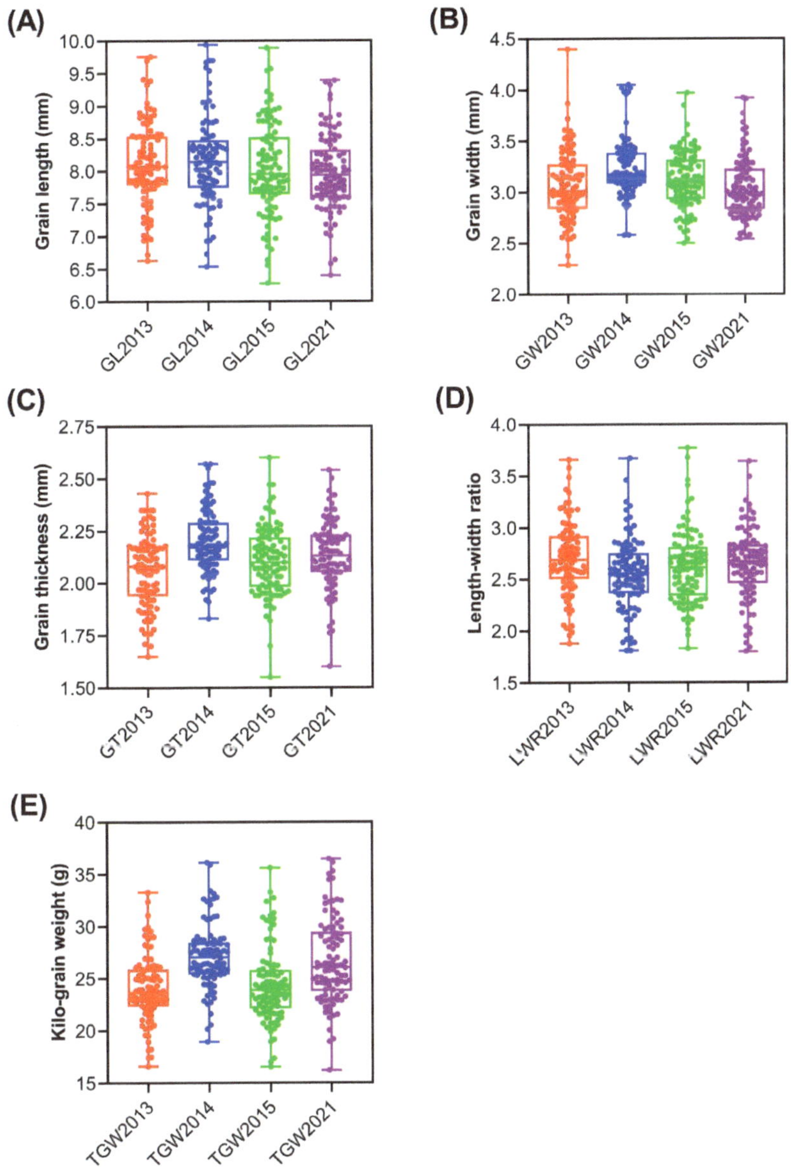

Figure 2. Boxplots (**A–E**) of five grain traits in 2013, 2014, 2015, and 2021.

Table 2. Descriptive statistics and heritability in a broad sense for grain-size traits in rice.

Traits	Year	Mean ± SE	Minimum	Maximum	SD	Skewness	Kurtosis	CV (%)	H^2_B (%)
GL (mm)	2013	8.12 ± 0.06	6.63	9.76	0.63	0.11	0.13	7.80	92.63
	2014	8.17 ± 0.07	6.54	9.94	0.67	0.34	0.42	8.20	
	2015	8.03 ± 0.07	6.28	9.89	0.70	0.10	−0.04	8.74	
	2021	7.97 ± 0.06	6.40	9.39	0.59	0.14	0.31	7.39	
GW (mm)	2013	3.06 ± 0.03	2.29	4.40	0.33	0.69	1.84	10.91	87.85
	2014	3.24 ± 0.03	2.58	4.05	0.28	0.95	1.94	8.51	
	2015	3.11 ± 0.03	2.50	3.97	0.28	0.25	0.32	8.89	
	2021	3.04 ± 0.03	2.54	3.92	0.30	0.85	0.58	9.72	
GT (mm)	2013	2.06 ± 0.02	1.65	2.43	0.17	−0.23	−0.49	8.21	92.64
	2014	2.20 ± 0.02	1.83	2.57	0.15	0.30	0.07	6.80	
	2015	2.11 ± 0.02	1.55	2.60	0.17	−0.01	1.11	7.88	
	2021	2.13 ± 0.02	1.60	2.54	0.17	−0.28	0.72	7.73	
LWR	2013	2.71 ± 0.04	1.88	3.66	0.36	0.08	0.12	13.38	72.39
	2014	2.56 ± 0.04	1.81	3.67	0.35	0.22	0.85	13.51	
	2015	2.62 ± 0.04	1.83	3.77	0.36	0.63	0.93	13.65	
	2021	2.65 ± 0.03	1.80	3.64	0.34	−0.14	0.63	12.90	
TGW (g)	2013	24.06 ± 0.33	16.62	33.28	3.19	0.33	0.53	13.25	86.11
	2014	27.16 ± 0.32	18.96	36.11	3.10	0.38	0.93	11.42	
	2015	24.42 ± 0.37	16.56	35.61	3.60	0.72	0.76	14.75	
	2021	26.64 ± 0.41	16.22	36.45	4.06	0.34	−0.11	15.24	

The phenotypic data for each trait showed positive correlations in both the *indica* and *japonica* groups between different years (Figure 3 and Table S2). For the *indica* group, the correlation coefficients of GL, GW, GT, LWR, and TGW were 0.57–0.73 ($p < 0.001$), 0.36–0.60 ($p < 0.001$), 0.23–0.46 ($p < 0.05$), 0.52–0.73 ($p < 0.001$), and 0.63–0.76 ($p < 0.001$), respectively. For the *japonica* group, the correlation coefficients of GL, GW, GT, LWR, and TGW were 0.74–0.87 ($p < 0.01$), 0.31–0.80, 0.18–0.61, 0.81–0.88 ($p < 0.001$), and 0.33–0.78, respectively. This indicated relatively high genetic stability in these traits across the years (Table S2). In addition, there were significant correlations among most of the grain-size traits, especially in the *indica* group (Table S2), indicating that these grain traits were interrelated. The largest correlation coefficient among the five traits was −0.80 (LWR and GW, 2021) in the *japonica* group and the smallest was 0.01 (TGW and LWR, 2014) in the *indica* group (Table S2).

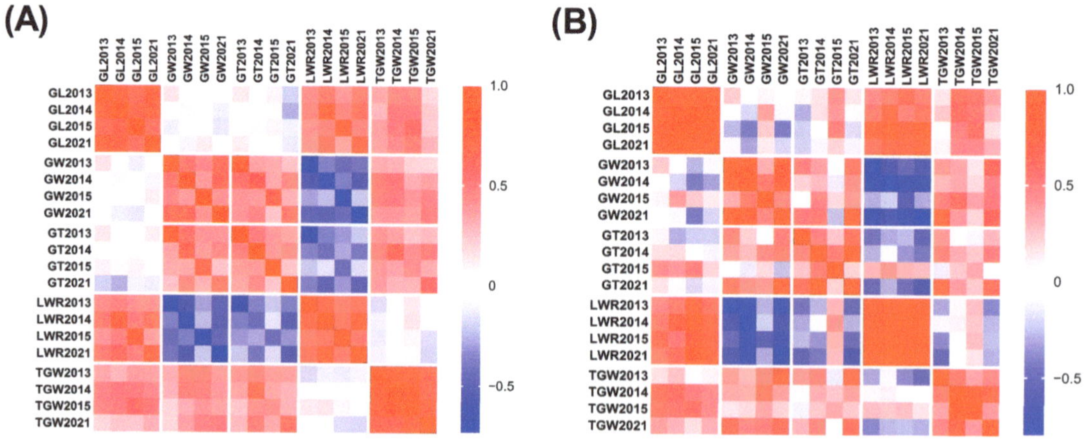

Figure 3. Correlation heatmaps for the *indica* group (**A**) and the *japonica* group (**B**) for five grain traits in 2013, 2014, 2015, and 2021.

2.4. Marker–Trait Associations for Grain-Size Traits

Using the MLM model in Tassel v.3.0, marker–trait association analysis of five grain traits was carried out (Table 3 and Figure 4). In the 2013 and 2021 environments, RM449

(Chr. 1) and RM316 (Chr. 9) were found to be significantly associated with GL traits, explaining 16.31% and 23.51% of the phenotypic variations, respectively (Table 3). For GW, 5, 17, 15, and 3 SSR markers were found to be significantly associated in 2013, 2014, 2015, and 2021, respectively. These loci could explain 10.84 (RM425)—43.01% (RM5496) of the phenotypic variations (Table 3). Of these associated markers, RM6092 (Chr. 1), RM452/RM550 (Chr. 2), RM229 (Chr. 11), and RM519 (Chr. 12) were repeatedly identified in different environments (Figure 5). For GT, 1, 5, and 1 marker-GT association pairs were detected in 2013, 2014, and 2021 respectively, with the marker R^2 ranging from 11.98 (RM161/RM305)—24.72% (RM275) (Table 3 and Figure 5). Seven markers distributed on chromosomes 1, 4, 8, 11, and 12 were found to be associated with LWR, with their R^2 values ranging from 12.68 (RM126) to 36.96% (RM5496) (Table 3 and Figure 5). RM25 (2013), RM316 (2014), RM4499 and RM316 (2015), and RM1901 (2021) were significantly associated with TGW, with their R^2 values ranging from 17.65 to 26.32% (Table 3). It is worth noting that some markers were also associated with more than one phenotype. For instance, RM5496 (Chr. 1), RM202 (Chr. 11), and RM519 (Chr. 12) were found to be associated with GW and LWR, RM449 (Chr. 1) with GL and LWR, RM4499 (Chr. 2) and RM190 (Chr. 6) with GT and TGW, RM161/RM305 (Chr. 5) and RM432 (Chr. 7) with GW and GT, and RM316 (Chr. 9) was found to be significantly associated with GL, GW, and TGW (Figure 5).

Table 3. SSR markers significantly associated with five grain traits in four years.

Traits	Year	Locus	Chr.	*p*-Value	R^2 (%)
GL	2013	RM449	1	1.11×10^{-3}	16.31
	2021	RM316	9	2.95×10^{-3}	23.51
GW	2013	RM6092	1	3.74×10^{-3}	26.79
	2013	RM1	1	7.10×10^{-3}	37.26
	2013	RM452/RM550	2	7.77×10^{-4}	19.04
	2013	RM229	11	3.96×10^{-3}	25.03
	2013	RM519	12	6.11×10^{-3}	26.57
	2014	RM414	1	7.85×10^{-3}	29.24
	2014	RM425	2	3.26×10^{-3}	12.86
	2014	RM523	3	7.57×10^{-3}	10.84
	2014	RM7097	3	4.24×10^{-3}	25.92
	2014	RM570	3	5.60×10^{-3}	28.92
	2014	RM169	5	1.46×10^{-3}	28.83
	2014	RM161/RM305	5	2.72×10^{-3}	13.31
	2014	RM6313	5	7.32×10^{-3}	10.92
	2014	RM253	6	4.74×10^{-3}	28.10
	2014	RM3827	6	3.04×10^{-3}	18.19
	2014	RM4085	8	9.98×10^{-3}	19.55
	2014	RM596	10	1.01×10^{-3}	16.31
	2014	RM228	10	4.54×10^{-3}	36.56
	2014	RM6296	12	8.51×10^{-3}	11.21
	2014	RM277	12	1.45×10^{-3}	15.88
	2014	RM519	12	5.59×10^{-3}	25.90
	2014	RM235	12	8.83×10^{-3}	42.32
	2015	RM84	1	1.39×10^{-3}	40.03
	2015	RM5496	1	2.38×10^{-3}	43.01
	2015	RM246/RM237	1	4.95×10^{-3}	25.54
	2015	RM472	1	5.82×10^{-3}	21.38
	2015	RM563	3	3.70×10^{-3}	18.1
	2015	RM261	4	1.97×10^{-3}	19.57
	2015	RM252	4	6.47×10^{-3}	26.12
	2015	RM540	6	5.90×10^{-3}	27.18
	2015	RM432	7	2.86×10^{-3}	21.85
	2015	RM346	7	9.28×10^{-3}	23.7

Table 3. *Cont.*

Traits	Year	Locus	Chr.	*p*-Value	R^2 (%)
GW	2015	RM316	9	3.15×10^{-3}	22.8
	2015	RM7557	11	2.22×10^{-3}	19.43
	2015	RM202	11	5.85×10^{-3}	18.81
	2015	RM144	11	3.92×10^{-3}	30.78
	2015	RM19	12	8.22×10^{-3}	20.66
	2021	RM6092	1	4.67×10^{-3}	27.28
	2021	RM452/RM550	2	4.26×10^{-3}	15.57
	2021	RM229	11	8.79×10^{-4}	29.01
GT	2013	RM138	2	9.95×10^{-3}	12.73
	2014	RM4499	2	6.19×10^{-4}	23.47
	2014	RM161/RM305	5	4.79×10^{-3}	11.98
	2014	RM190	6	5.26×10^{-3}	18.81
	2014	RM432	7	6.65×10^{-3}	19.21
	2014	RM331	8	8.42×10^{-3}	23.81
	2021	RM275	6	4.10×10^{-3}	24.72
LWR	2013	RM449	1	3.81×10^{-3}	13.18
	2014	RM202	11	9.19×10^{-3}	19.02
	2014	RM519	12	7.16×10^{-4}	32.94
	2021	RM5496	1	7.80×10^{-3}	36.96
	2021	RM471	4	5.82×10^{-3}	24.54
	2021	RM337	8	3.32×10^{-3}	20.04
	2021	RM126	8	3.33×10^{-3}	12.68
TGW	2013	RM25	8	7.09×10^{-3}	26.32
	2014	RM316	9	6.57×10^{-3}	20.52
	2015	RM4499	2	4.56×10^{-3}	17.65
	2015	RM316	9	6.63×10^{-3}	20.72
	2021	RM190	6	7.62×10^{-3}	17.88

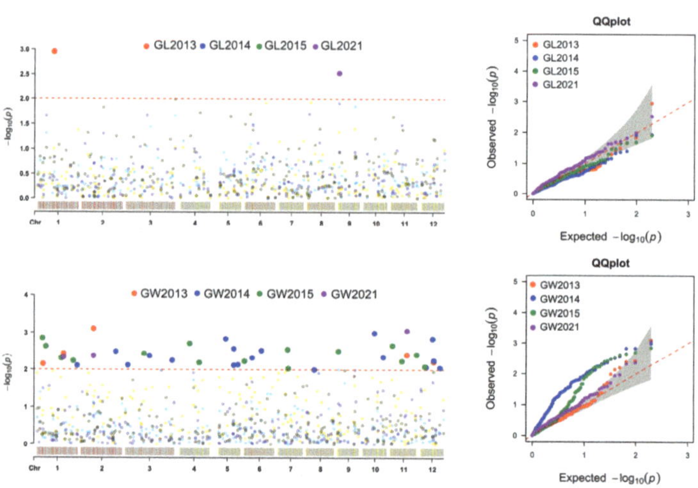

Figure 4. *Cont.*

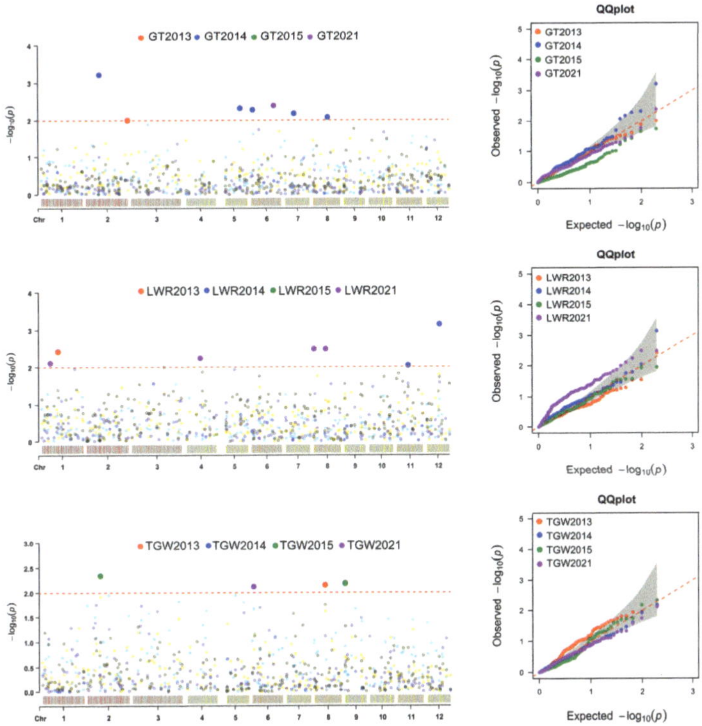

Figure 4. Manhattan plots and quantile–quantile (QQ) plots for grain-size traits based on the MLM model. For the Manhattan plot, the x-axis presents the 12 rice chromosomes and the y-axis the LOD (−log(*p*-value)) values. For the QQ plot, the x-axis presents LOD (−log(*p*-value)) values and the y-axis the expected LOD (−log(*p*-value)) values.

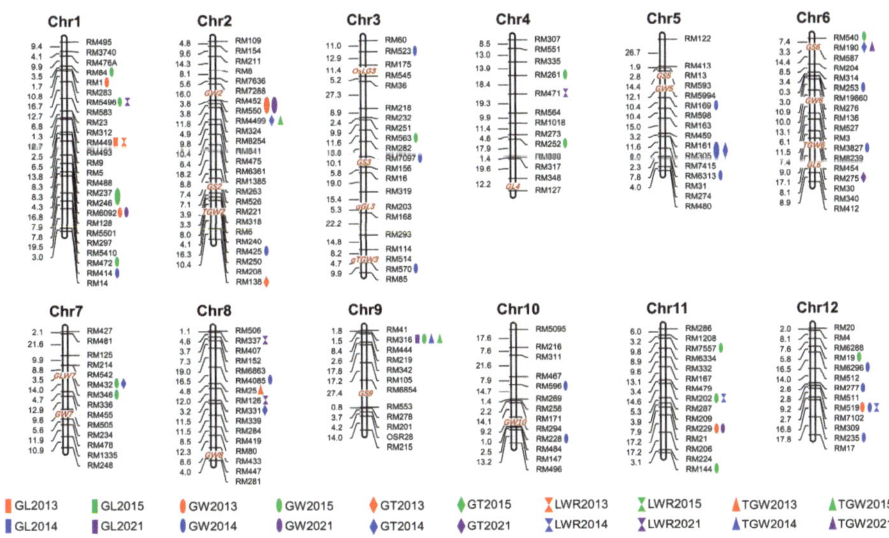

Figure 5. Distribution of significant marker associations for five grain traits. Red, green, blue, and pink shapes indicate 2013, 2014, 2015, and 2021, respectively.

3. Discussion

3.1. Molecular Genetic Diversity

The molecular marker technique can effectively evaluate the genetic diversity and genetic relationships among rice varieties. This is important for the effective protection and utilization of rice germplasm resources [27]. Rice in Yuanyang Hani terraces has been planted for thousands of years. These landraces continued in the long-term cultivation process without being eliminated, indicating that these varieties had strong adaptability and rich genetic variation [28]. However, in recent years, the large-scale popularization of modern bred varieties led to a gradual decrease in the planting area for traditional rice landraces. The analysis of the genetic diversity of rice landraces in the Hani terraces can provide a theoretical basis for the protection and utilization of these rice resources. In this study, a total of 936 polymorphic bands were amplified by 201 SSR markers, with an average of 4.66 bands per primer pair, which was higher compared to a report by Liu et al. (Na = 2.161) [29] but slightly lower compared to a report of Xu et al. (Na = 5.065~5.313) [30]. The high band amplification number showed that rice from the Hani terraces had rich genetic diversity. The average PIC value for SSR markers in this study was similar to that reported in a previous study on 48 pairs of SSR markers (average PIC = 0.652~0.660) [30], which was significantly higher compared to the results of Liu et al. (0.256) [29]. This indicated that the SSR markers used in this study effectively reflected the genetic diversity of rice landraces in the Hani terraces. The I value of 201 SSR markers was higher than 1.000, which further confirmed the rich genetic diversity of the rice landraces in the Hani terraces. Lower heterozygosity (H = 0.15) was consistent with the characteristics of self-pollination and high homozygosity of rice genotypes.

3.2. Population Structure and Genetic Relationships

Population structure is an important factor that affects the results of association analysis. Mixing of subpopulations enhances the linkage disequilibrium (LD) intensity of the whole population, resulting in pseudo-association. Hence, the analysis of population genetic structure is the premise of association mapping. In this study, 96 rice landraces in the Hani terraces were analyzed by three different methods: hierarchical Bayesian analysis, NJ cluster analysis, and principal component analysis. Ninety-six landraces were divided into two sub-populations, namely, Q1 and Q2. Q1 was the main group and included 81 landraces (84.38%), while Q2 comprised 15 landraces (15.62%). This was consistent with the results of previous studies on rice landraces in the Hani terraces [3,28,29]. The uncomplicated population structure was suitable for association analysis.

3.3. Phenotypic Evaluation

In this study, the heritability of each trait was more than 70%. Higher heritability indicated that the inheritance of grain-size traits was more stable and less affected by environment, which was consistent with the results of Edzesi et al. [31], Feng et al. [32], and Dang et al. [33]. The correlation analysis among characters showed that there was a positive correlation between TGW and GL, GW, and GT in the *indica* group as the main cultivation type of the Hani terraces. However, there was no significant correlation between LWR and TGW in the *indica* or *japonica* group, which was consistent with previous research results [33–36]. An increase in GL, GW, and GT contributed to an increase in grain weight, with little contribution from LWR. In addition, there was a significant negative correlation between LWR and GT (r = 0.39–0.64, $p < 0.001$) in the *indica* group, which indicated that the long-grain landraces were less thick compared to the round-grain landraces in the Hani terraces.

3.4. Marker–Trait Associations

Rice grain size is a typical quantitative trait controlled by multiple genes, which are important indicators of rice yield and appearance. Bi-parental linkage mapping proved effective in mining rice grain-size QTLs. However, the limited number of parents limits the opportunities to recombine the offspring. QTLs cannot be detected when there is no difference in alleles between the parents. Compared to linkage mapping, association analysis uses natural populations as research materials. With natural populations crossing naturally for many generations, recombination is sufficient, the mapping accuracy is higher, and multiple alleles at the same locus can be detected at the same time. In this study, using SSR-trait association analysis, 2, 36, 7, 7, and 4 QTLs controlling GL, GW, GT, LWR, and TGW were detected on 12 rice chromosomes (Figure 5). Compared to previous studies, a few QTLs were similar to known QTLs or genes. RM449 (GL, LWR), RM414 (GW), RM169 (GW), and RM161/RM305 (GW, GT) within the marker intervals RM449-RM237 (*qGL1-2*), RM431-PSM370 (*qGW1-3*), and RM413-RM161 (*qGW5-1*) were detected by Lin et al. [37] using single segment substitution lines (GLU-SSSLs). RM169 on chromosome 5 was found close to the cloned GW5 gene [16], and multiple grain-size QTLs were identified near RM169 [1,26,38–40], indicating that this locus played a role in the regulation of grain size in many rice varieties. Zhang et al. [41] used 274 SSR markers to analyze 12 agronomic traits, including GL, GW, LWR, and TGW. The detected loci RM81A (GL), RM144 (GL), RM277 (GL, LWR), RM237 (GW, LWR), RM19 (GW), and RM252 (TGW) were consistent with GL (RM449) and GW (RM144, RM277, RM237, RM19, and RM252) QTLs detected in this study. Zhao et al. [42] identified 53 QTLs related to grain size in two years. The marker intervals of *qKGW1.1*, *qGW2.5*, *qLWR3.2*, *qKGW3.1*, *qGL3.4*, *qLWR3.3*, *qGL6.2*, *qGW6.3*, *qGT9*, *qKGW10*, *qGL11.1*, *qLWR11*, and *qGL11.2* were similar or overlapped with the markers RM6092 (GW, Chr. 1), RM425 (GW, Chr. 2), RM563 (GW, Chr. 3), RM7097 (GW, Chr. 3), RM570 (GW, Chr. 3), RM3827 (GW, Chr. 6), RM316 (GL, GW, TGW, Chr. 6), RM228 (GW, Chr. 10), RM202 (GW, LWR, Chr. 11), and RM229 (GW, Chr. 12) in the present study. Of these, RM452 (GW, Chr. 2), RM425 (GW, Chr. 2), RM570 (GW, Chr. 3), RM169 (GW, Chr. 5), RM190 (GT and TGW, Chr. 6), RM275 (GT, Chr. 6), RM432 (GW and GT, Chr. 7), and RM228 (GW, Chr. 10) were close to the cloned *GW2* [7], *TGW2* [9], *qTGW3* [13], *GW5* [16], *GS6* [17], *TGW6* [19], *GLW7* [21], and *GW10* [25] (Figure 5), respectively. Recently, Zhu et al. [43] finely mapped a TGW QTL, qTGW10-20.8, in the RM228-RM18A region of chromosome 10. However, further verification was required to determine whether the locus detected in this study was related to the cloned gene/QTL allele.

Previous studies revealed that grain-size traits were controlled by a set of QTLs, some of which were QTL clusters. Several SSRs identified in this study overlapped with previous QTL clusters, including RM563 (Chr. 3) and RM169 (Chr. 5), located in QTL clusters RM517-RM411 (Chr. 3) and RM413-RM598 (Chr. 5), studied by Lu et al. [1]. RM7097 (GW) on chromosome 3 was located in the RM411-RM7097 interval controlling seven grain traits, as revealed by Yin et al. [44]. Shi et al. [45] detected *qGL2*, *qGW2*, *qLWR2*, *qGT2*, and *qTGW2a* in the RM322-YP9506 region of chromosome 2, which was consistent with the position of RM452-RM4499, controlling GW, GT, and TGW, detected in this study. Zhang et al. [46] identified a QTL cluster controlling GL, GW, and TGW at the end of the long arm of chromosome 2, which was consistent with the RM138 (GW) locus detected in this study. We detected the RM190 locus controlling GT and TGW on chromosome 6, which was previously identified as being linked to grain size in many populations [47–49], including the cloned *GS6* gene, a member of the GRAS gene family, which plays a negative role in regulating the grain size of rice [17]. RM190 is closely linked to the *Wx* gene that controls amylose synthesis, and a few studies showed that rice grain shape was closely related to amylose content [50]. RM5496 (GW, LWR), detected on chromosome 1, explained 43.01% (GW) and 36.96% (LWR) of phenotypic variations. With the p-value set to 0.05, this locus was significantly associated with GL in three environments, with R^2 ranging from 31.12 to 32.62%. This indicated that this locus made a major contribution to grain size. RM316, controlling GL (2021), GW (2015), and TGW (2014, 2015), was detected in the short arm of chromosome 9, with R^2 values being above 20%. When $p < 0.05$, RM316 was significantly associated with GL (2014, 2015) and GW (2021). This showed that the RM316 locus played an important role in regulating grain-size traits of rice landraces in the Hani terraces. As no genes controlling grain traits were cloned at the RM316 locus, it is of great significance to further mine the potential genes at this locus.

4. Materials and Methods

4.1. Rice Material and Phenotyping

Ninety-six rice landraces were collected from Yuanyang County, Honghe Hani, and Yi Autonomous Prefecture of Yunnan Province. Detailed sample information is shown in Table S3. All experimental materials were planted during the rice-growing seasons of 2013, 2014, 2015, and 2021 in the experimental farm of Honghe University, Mengzi, China. A random block design was adopted, and each rice landrace was planted in two rows. Ten plants were planted in each row, with a row spacing of 15 cm × 20 cm between plants and two repetitions. Protective rows were set up around the test field. Field management, including soil fertility and irrigation in the experimental field, was the same in different years. Random samples were harvested, and traits were inspected after natural air-drying. GL, GW, and GT were measured using electronic digital calipers (Guilin Measuring and Cutting Tool Co., Ltd., Guilin, China). Ten grains were measured for each variety, and measurements were repeated five times. Five hundred fully filled grains were randomly selected to measure grain weight, which was repeated twice. With a difference of not more than 5%, grain weight was then converted into TGW.

4.2. DNA Extraction and Genotyping

Fresh young leaves of rice were collected at the tillering stage for genomic DNA extraction. DNA was extracted using the cetyltrimethylamine bromide (CTAB) method. DNA concentration was detected by spectrophotometry, and its quality was detected by 1% agarose gel electrophoresis. Finally, DNA was diluted in a working solution of 20 ng/μL. A total of 201 pairs of SSR primers uniformly distributed on rice chromosomes were obtained from the Gramene database (www.gramene.org (accessed on 25 March 2013)). The primers were synthesized by Sangon Biotech Co., Ltd., Shangai. The chromosomes of SSR primers are shown in Table S4. The 10 μL PCR reaction volume contained 10 ng template DNA, 0.2 μM primers, 2.5 mM dNTP, 1.2 μL 10 × PCR buffer, 25 mM $MgCl_2$, and 0.5 U rTaq DNA polymerase. The PCR amplification procedure was pre-denatured at 95 °C for 5 min, followed by 30 cycles at 95 °C for 30 s, 55~58 °C for 30 s, 72 °C for 30 s, and extended at 72 °C for 8 min at the end of the cycle. The amplified products were separated by electrophoresis with 8% non-denaturing polyacrylamide gel and stained with 1% silver nitrate [51].

4.3. Phenotypic Data Analysis

Excel 2016 was used for data processing and organization. Past v.3.0 software (https://www.nhm.uio.no/english/research/resources/past/ (accessed on 22 November 2022)) was used for phenotypic data descriptive statistics and trait correlation analysis. GraphPad Prism v.8.0 software (GraphPad Prism Software Inc., San Diego, CA, USA) was used to draw a phenotypic data box diagram and correlation heatmap. The following formula was used to estimate the generalized heritability of five traits: $H_B^2 = \sigma_g^2/(\sigma_g^2 + \sigma_e^2/n)$, where σ_g^2 is the genetic variance, σ_e^2 is the error variance, and n is the number of replications [52].

4.4. Genetic Diversity, Phylogenetic Analysis, and Population Structure

Polymorphic bands of SSR electrophoresis were recorded and organized based on a previous study [53]. The observed number of alleles (Na), the effective number of alleles (Ne), and Shannon's information index (I) for each pair of primers were calculated using POPGENE v.1.32 [54]. The polymorphism information content (PIC) and the heterozygosity (H) of each pair of primers were calculated by PowerMarker v.3.25 [55]. The structure of the total population of varieties was analyzed by STRUCTURE v.2.3.4 [56]. The results of the operation repeated five times under different K values were uploaded to Structure

Harvester v.0.6.94 (https://taylor0.biology.ucla.edu/structureHarvester/ (accessed on 8 December 2022)) [57] and the optimal number of groups was determined by LnP (K) and Delta (K). Based on the allelic genotype frequency of each locus, the Nei genetic distance, calculated by PowerMarker v.3.25, was used to construct the neighbor-joining (NJ) cluster tree, which was observed using MEGA v.5.0 (http://www.megasoftware.net (accessed on 10 December 2022)). Principle component analysis (PCA) was undertaken using PAST v.3.0.

4.5. Association Mapping

The mixed linear model (MLM) in TASSEL v.3.0 was used to analyze the correlation between SSR markers and GL, GW, GT, LWR, and TGW, considering the factors of population structure (Q) and kinship (K) of the materials. The Q value for population structure was calculated using Structure v.2.3.4 and the K value among individuals was calculated using TASSEL v.3.0. It was considered that there was a significant association of the target trait with the marker at a level of $p < 0.01$. The results of the association analysis were visualized by CMplot (https://cran.r-project.org/web/packages/CMplot/ (accessed on 15 December 2022)).

5. Conclusions

The genetic diversity, genetic relationships, and population structure of 96 rice landraces in the Hani terraces were analyzed using 201 SSR markers widely distributed on 12 chromosomes. The results showed that rice landraces in the Hani terraces harbored rich genetic diversity. All landraces were divided into two groups, with *indica* rice being the main group. Further, the MLM model of the TASSEL software was used for the SSR marker–trait association analysis of GL, GW, GT, LWR, and TGW in four environments. The results showed that 2, 36, 7, 7, and 4 SSR markers were significantly associated with GL, GW, GT, LWR, and TGW, respectively, with individual phenotypic variations ranging from 10.84 to 43.01%. RM6092 (Chr. 1), RM452/RM550 (Chr. 2), RM316 (Chr. 9), RM229 (Chr. 11), and RM519 (Chr. 12) were detected repeatedly in different environments. Nine SSR markers (RM5496, RM449, RM4499, RM161/RM305, RM190, RM432, RM316, RM202, and RM519) were significantly associated with different grain-size traits. These mapping results provide a theoretical foundation for further fine mapping, cloning, and molecular-assisted breeding of related genes.

Supplementary Materials: The following supporting information can be downloaded at: https://www.mdpi.com/article/10.3390/plants12081678/s1, Table S1: Genetic diversity parameters of 96 rice landraces collected from the Hani terraces based on 201 SSR markers; Table S2: Correlation analysis of five grain-size traits of Hani terrace rice landraces in four environments; Table S3: Codes, names, and collection places of 96 rice landraces from the Honghe Hani Rice Terraces System; Table S4: Chromosome information for 201 pairs of SSR markers.

Author Contributions: Conceptualization, B.L.; methodology, M.M. and E.L.; software, M.M.; validation, M.M., T.W. and W.Z.; formal analysis, H.M.; investigation, T.W. and E.L.; resources, M.M.; data curation, M.M.; writing—original draft preparation, M.M.; writing—review and editing, B.L.; funding acquisition, B.L. All authors have read and agreed to the published version of the manuscript.

Funding: This research was funded by Youth Top-notch Talent Support Program of Yunnan Province (no. YNWR–QNBJ–2020–207), the Program of Applied Basic Research of Yunnan Province (no. 2013FZ124), and the Scientific Research Fund of the Yunnan Provincial Education Department (no. 2020J0671).

Data Availability Statement: The datasets supporting the conclusions of this article are included within the article.

Conflicts of Interest: The authors declare no conflict of interest.

References

1. Lu, B.Y.; Yang, C.Y.; Xie, K.; Zhang, L.; Wu, T.; Li, L.F.; Liu, X.; Jiang, L.; Wan, J.M. Quantitative trait loci for grain-quality traits across a rice F_2 population and backcross inbred lines. *Euphytica* **2013**, *192*, 25–35. [CrossRef]
2. Zeng, Y.W.; Zhang, H.L.; Li, Z.C.; Shen, S.Q.; Sun, J.N.; Wang, M.X.; Liao, D.Q.; Liu, X.; Wang, X.K.; Xiao, F.H.; et al. Evaluation of genetic diversity of rice landraces (*Oryza sativa* L.) in Yunnan, China. *Breed. Sci.* **2007**, *57*, 91–99. [CrossRef]
3. Ma, M.L.; Zhou, X.M.; Zheng, Y.; Zhang, T.T.; Zhang, X.Q.; Lu, B.Y. Genetic diversity and population structure analysis of red rice from Hani terraced fields based on SRAP markers. *Mol. Plant Breed.* **2019**, *17*, 2231–2237.
4. Xu, F.R.; Tang, C.F.; Yu, T.Q.; Dai, L.Y.; Zhang, H.S. Diversity of paddy rice varieties from Yuanyang Hani's terraced fields in Yunnan, China. *Acta Ecol. Sin.* **2010**, *30*, 3346–3357.
5. Huang, R.Y.; Jiang, L.R.; Zheng, J.S.; Wang, T.S.; Wang, H.C.; Huang, Y.M.; Hong, Z.L. Genetic bases of rice grain shape: So many genes, so little known. *Trends Plant Sci.* **2013**, *18*, 218–226. [CrossRef]
6. Kang, Y.W.; Chen, Y.Y.; Zhang, Y.X. Research progress and breeding prospects of grain size associated genes in rice. *Chin. J. Rice Sci.* **2020**, *34*, 479–490.
7. Song, X.J.; Huang, W.; Shi, M.; Zhu, M.Z.; Lin, H.X. A QTL for rice grain width and weight encodes a previously unknown RING-type E3 ubiquitin ligase. *Nat. Genet.* **2007**, *39*, 623–630. [CrossRef]
8. Hu, J.; Wang, Y.X.; Fang, Y.X.; Zeng, L.J.; Xu, J.; Yu, H.P.; Shi, Z.Y.; Pan, J.J.; Zhang, D.; Kang, S.J.; et al. A rare allele of *GS2* enhances grain size and grain yield in rice. *Mol. Plant* **2015**, *8*, 1455–1465. [CrossRef] [PubMed]
9. Ruan, B.; Shang, L.; Zhang, B.; Hu, J.; Wang, Y.; Lin, H.; Zhang, A.; Liu, C.; Peng, Y.; Zhu, L.; et al. Natural variation in the promoter of *TGW2* determines grain width and weight in rice. *New Phytol.* **2020**, *227*, 629–640. [CrossRef]
10. Yu, J.P.; Xiong, H.Y.; Zhu, X.Y.; Zhang, H.L.; Li, H.H.; Miao, J.L.; Wang, W.S.; Tang, Z.S.; Zhang, Z.Y.; Yao, G.X. *OsLG3* contributing to rice grain length and yield was mined by Ho-LAMap. *BMC Biol.* **2017**, *15*, 28. [CrossRef]
11. Mao, H.L.; Sun, S.Y.; Yao, J.L.; Wang, C.R.; Yu, S.B.; Xu, C.G.; Li, X.H.; Zhang, Q.F. Linking differential domain functions of the *GS3* protein to natural variation of grain size in rice. *Proc. Nat. Acad. Sci. USA* **2010**, *107*, 19579–19584. [CrossRef] [PubMed]
12. Zhang, X.J.; Wang, J.F.; Huang, J.; Lan, H.X.; Wang, C.L.; Yin, C.F.; Wu, Y.Y.; Tang, H.J.; Qian, Q.; Li, J.Y.; et al. Rare allele of *OsPPKL1* associated with grain length causes extra-large grain and a significant yield increase in rice. *Proc. Nat. Acad. Sci. USA* **2012**, *109*, 21534–21539. [CrossRef] [PubMed]
13. Hu, Z.J.; Lu, S.J.; Wang, M.J.; He, H.H.; Sun, L.; Wang, H.R.; Liu, X.H.; Jiang, L.; Sun, J.L.; Xin, X.Y.; et al. A novel QTL *qTGW3* encodes the GSK3/SHAGGY-like kinase OsGSK5/OsSK41 that interacts with OsARF4 to negatively regulate grain size and weight in rice. *Mol. Plant* **2018**, *11*, 736–749. [CrossRef]
14. Wu, W.G.; Liu, X.Y.; Wang, M.H.; Meyer, R.S.; Luo, X.J.; Ndjiondjop, M.; Tan, L.B.; Zhang, J.W.; Wu, J.Z.; Cai, H.W.; et al. A single-nucleotide polymorphism causes smaller grain size and loss of seed shattering during African rice domestication. *Nat. Plants* **2017**, *3*, 17064. [CrossRef]
15. Li, Y.; Fan, C.; Xing, Y.; Jiang, Y.; Luo, L.; Sun, L.; Shao, D.; Xu, C.; Li, X.; Xiao, J. Natural variation in *GS5* plays an important role in regulating grain size and yield in rice. *Nat. Genet.* **2011**, *43*, 1266–1269. [CrossRef] [PubMed]
16. Weng, J.F.; Gu, S.H.; Wan, X.Y.; Gao, H.; Guo, T.; Su, N.; Lei, C.L.; Zhang, X.; Cheng, Z.J.; Guo, X.P.; et al. Isolation and initial characterization of *GW5*, a major QTL associated with rice grain width and weight. *Cell Res.* **2008**, *18*, 1199–1209. [CrossRef] [PubMed]
17. Sun, L.; Li, X.; Fu, Y.; Zhu, Z.; Tan, L.; Liu, F.; Sun, X.; Sun, X.; Sun, C. GS6, a member of the GRAS gene family, negatively regulates grain size in rice. *J. Integr. Plant Biol.* **2013**, *55*, 938–949. [CrossRef] [PubMed]
18. Shi, C.L.; Dong, N.Q.; Guo, T.; Ye, W.W.; Shan, J.X.; Lin, H.X. A quantitative trait locus *GW6* controls rice grain size and yield through the gibberellin pathway. *Plant J.* **2020**, *103*, 1174–1188. [CrossRef] [PubMed]
19. Ishimaru, K.; Hirotsu, N.; Madoka, Y.; Murakami, N.; Hara, N.; Onodera, H.; Kashiwagi, T.; Ujiie, K.; Shimizu, B.; Onishi, A. Loss of function of the IAA-glucose hydrolase gene *TGW6* enhances rice grain weight and increases yield. *Nat. Genet.* **2013**, *45*, 707–711. [CrossRef]
20. Wang, A.; Hou, Q.; Si, L.; Huang, X.; Luo, J.; Lu, D.; Zhu, J.; Shangguan, Y.; Miao, J.; Xie, Y. The PLATZ transcription factor *GL6* affects grain length and number in rice. *Plant Physiol.* **2019**, *180*, 2077–2090. [CrossRef]
21. Si, L.; Chen, J.; Huang, X.; Gong, H.; Luo, J.; Hou, Q.; Zhou, T.; Lu, T.; Zhu, J.; Shangguan, Y.; et al. *OsSPL13* controls grain size in cultivated rice. *Nat. Genet.* **2016**, *48*, 447–456. [CrossRef]
22. Wang, S.K.; Li, S.; Liu, Q.; Wu, K.; Zhang, J.Q.; Wang, S.S.; Wang, Y.; Chen, X.B.; Zhang, Y.; Gao, C.X.; et al. The *OsSPL16-GW7* regulatory module determines grain shape and simultaneously improves rice yield and grain quality. *Nat. Genet.* **2015**, *47*, 949–954. [CrossRef] [PubMed]
23. Wang, S.; Wu, K.; Yuan, Q.; Liu, X.; Liu, Z.; Lin, X.; Zeng, R.; Zhu, H.; Dong, G.; Qian, Q. Control of grain size, shape and quality by *OsSPL16* in rice. *Nat. Genet.* **2012**, *44*, 950–954. [CrossRef]
24. Zhao, D.S.; Li, Q.F.; Zhang, C.Q.; Zhang, C.; Yang, Q.Q.; Pan, L.X.; Ren, X.Y.; Lu, J.; Gu, M.H.; Liu, Q.Q. *GS9* acts as a transcriptional activator to regulate rice grain shape and appearance quality. *Nat. Commun.* **2018**, *9*, 1240. [CrossRef]
25. Zhan, P.; Wei, X.; Xiao, Z.; Wang, X.; Ma, S.; Lin, S.; Li, F.; Bu, S.; Liu, Z.; Zhu, H. *GW10*, a member of P450 subfamily regulates grain size and grain number in rice. *Theor. Appl. Genet.* **2021**, *134*, 3941–3950. [CrossRef] [PubMed]
26. Xu, F.F.; Liang, J.; Huang, Y.; Chuan, T.; CHEN, Y.L.; BAO, J.S. Association mapping of quantitative trait loci for yield-related agronomic traits in rice (*Oryza sativa* L.). *J. Integr. Agr.* **2016**, *15*, 2192–2202. [CrossRef]

27. Qun, X.; Hong, C.; Wang, C.H.; Yu, H.H.; Yuan, X.P.; Wang, Y.P.; Yue, F.; Tang, S.X.; Wei, X.H. Genetic diversity and structure of new inbred rice cultivars in China. *J. Integr. Agr.* **2012**, *11*, 1567–1573.
28. Ma, M.L.; Zheng, Y.; Zhou, X.M.; Zhang, T.T.; Zhang, X.Q.; Lu, B.Y. Genetic Diversity Analysis of Red Rice from Hani's Terraced Fields in Yunnan Province. *Crops* **2018**, *5*, 21–26.
29. Liu, C.C.; Zhao, F.W.; Wu, X.X.; Zhang, C.Q.; Zhu, K.Z.; Xue, D.Y.; Wu, J.Y.; Huang, S.W.; Xu, X.Y.; Jin, Y.G.; et al. Genetic diversity and population structure analysis of currently cultivated rice landraces from Hani's terraced fields in Yunnan Province. *Chin. J. Rice Sci.* **2015**, *29*, 28–34.
30. Xu, F.; Dong, C.; Yang, W.; Tang, C.; Zhang, E.; Yang, Y.; Zhang, F.; Zhang, H. Comparison of genetic diversity of rice landraces planted in two periods in Hani's terraced fields in Yuanyang county, Yunnan Province, China using microsatellite markers. *Chin. J. Rice Sci.* **2011**, *25*, 381–386.
31. Edzesi, W.M.; Dang, X.; Liang, L.; Liu, E.; Zaid, I.U.; Hong, D. Genetic diversity and elite allele mining for grain traits in rice (*Oryza sativa* L.) by association mapping. *Front. Plant Sci.* **2016**, *7*, 787. [CrossRef]
32. Feng, Y.; Lu, Q.; Zhai, R.; Zhang, M.; Xu, Q.; Yang, Y.; Wang, S.; Yuan, X.; Yu, H.; Wang, Y. Genome wide association mapping for grain shape traits in indica rice. *Planta* **2016**, *244*, 819–830. [CrossRef]
33. Dang, X.; Giang Tran Thi, T.; Mawuli Edzesi, W.; Liang, L.; Liu, Q.; Liu, E.; Wang, Y.; Qiang, S.; Liu, L.; Hong, D. Population genetic structure of *Oryza sativa* in East and Southeast Asia and the discovery of elite alleles for grain traits. *Sci. Rep.* **2015**, *5*, 1–13. [CrossRef] [PubMed]
34. Jiang, G.H.; Hong, X.Y.; Xu, C.G.; Li, X.H.; He, Y.Q. Identification of quantitative trait loci for grain appearance and milling quality using a doubled-haploid rice population. *J. Integr. Plant Biol.* **2005**, *47*, 1391–1403. [CrossRef]
35. Nayak, A.K.; Anilkumar, C.; Behera, S.; Sah, R.P.; Lavanya, G.R.; Kumar, A.; Behera, L.; Tp, M.A. Genetic dissection of grain size traits through genome-wide association study based on genic markers in rice. *Rice Sci.* **2022**, *29*, 462–472. [CrossRef]
36. Yan, B.; Liu, R.; Li, Y.; Wang, Y.; Gao, G.; Zhang, Q.; Liu, X.; Jiang, G.; He, Y. QTL analysis on rice grain appearance quality, as exemplifying the typical events of transgenic or backcrossing breeding. *Breed. Sci.* **2014**, *64*, 231–239. [CrossRef]
37. Lin, S.; Liu, Z.; Zhang, K.; Yang, W.; Zhan, P.; Tan, Q.; Gou, Y.; Ma, S.; Luan, X.; Huang, C. *GL9* from *Oryza glumaepatula* controls grain size and chalkiness in rice. *Crop J.* **2023**, *11*, 198–207. [CrossRef]
38. Fujita, D.; Tagle, A.G.; Koide, Y.; Simon, E.V.; Fukuta, Y.; Ishimaru, T.; Kobayashi, N. Characterization of QTLs for grain weight from New Plant Type rice cultivars through the development of near-isogenic lines with an IR64 background. *Euphytica* **2022**, *218*, 50. [CrossRef]
39. Bian, J.M.; Jiang, L.; Liu, L.L.; Wei, X.J.; Xiao, Y.H.; Zhang, L.J.; Zhao, Z.G.; Zhai, H.Q.; Wan, J.M. Construction of a new set of rice chromosome segment substitution lines and identification of grain weight and related traits QTLs. *Breed. Sci.* **2010**, *60*, 305–313. [CrossRef]
40. Zhang, Y.D.; Zhang, Y.H.; Dong, S.L.; Tao, C.; Zhao, Q.Y.; Zhen, Z.; Zhou, L.H.; Shu, Y.; Ling, Z.; Xing, Y. QTL mapping for grain size traits based on extra-large grain rice line TD70. *Rice Sci.* **2013**, *20*, 400–406. [CrossRef]
41. Zhang, P.; Liu, X.; Tong, H.; Lu, Y.; Li, J. Association mapping for important agronomic traits in core collection of rice (*Oryza sativa* L.) with SSR markers. *PLoS ONE* **2014**, *9*, e111508. [CrossRef]
42. Zhao, D.; Li, P.; Wang, L.; Sun, L.; Xia, D.; Luo, L.; Gao, G.; Zhang, Q.; He, Y. Genetic dissection of large grain shape in rice cultivar 'Nanyangzhan'and validation of a grain thickness QTL (*qGT3. 1*) and a grain length QTL (*qGL3. 4*). *Mol. Breed.* **2017**, *37*, 42. [CrossRef]
43. Zhu, Y.; Zhang, Z.; Chen, J.; Fan, Y.; Mou, T.; Tang, S.; Zhuang, J. Fine mapping of *qTGW10-20.8*, a QTL having important contribution to grain weight variation in rice. *Crop J.* **2019**, *7*, 587–597. [CrossRef]
44. Yin, C.; Li, H.; Li, S.; Xu, L.; Zhao, Z.; Wang, J. Genetic dissection on rice grain shape by the two-dimensional image analysis in one japonica× indica population consisting of recombinant inbred lines. *Theor. Appl. Genet.* **2015**, *128*, 1969–1986. [CrossRef] [PubMed]
45. Shi, H.; Yun, P.; Zhu, Y.; Wang, L.; Li, P.; Lou, G.; Xia, D.; Zhang, Q.; Xiao, J.; Li, X. Fine mapping of *qTGW2b* and *qGL9*, two minor QTL conferring grain size and weight in rice. *Mol. Breed.* **2022**, *42*, 68. [CrossRef]
46. Zhang, H.; Zhu, Y.J.; Zhu, A.D.; Fan, Y.Y.; Huang, T.X.; Zhang, J.F.; Xie, H.A.; Zhuang, J.Y. Fine-mapping of *qTGW2*, a quantitative trait locus for grain weight in rice (*Oryza sativa* L.). *Peer J.* **2020**, *8*, e8679. [CrossRef]
47. Gao, F.Y.; Zeng, L.H.; Ling, Q.; Lu, X.J.; Ren, J.S.; Wu, X.T.; Su, X.W.; Gao, Y.M.; Ren, G.J. QTL mapping of grain appearance quality traits and grain weight using a recombinant inbred population in rice (*Oryza sativa* L.). *J. Integr. Agr.* **2016**, *15*, 1693–1702. [CrossRef]
48. Liu, W.; Li, X.; Zhou, K.; Pan, X.; Li, Y.; Lu, T.; Sheng, X. Mapping of QTLs controlling grain shape and populations construction derived from related residual heterozygous lines in rice. *J. Agric. Sci.* **2016**, *8*, 104–113. [CrossRef]
49. Qi, L.; Sun, Y.; Li, J.; Su, L.; Zheng, X.; Wang, X.; Li, K.; Yang, Q.; Qiao, W. Identify QTLs for grain size and weight in common wild rice using chromosome segment substitution lines across six environments. *Breed. Sci.* **2017**, *67*, 472–482. [CrossRef]
50. Ferdous, N.; Elias, S.M.; Howlader, Z.H.; Biswas, S.K.; Rahman, M.S.; Habiba, K.K.; Seraj, Z.I. Profiling Bangladeshi rice diversity based on grain size and amylose content using molecular markers. *Curr. Plant Biol.* **2018**, *14*, 56–65. [CrossRef]
51. Sanguinetti, C.J.; Simpson, A. Rapid silver staining and recovery of PCR products separated on polyacrylamide gels. *Biotechniques* **1994**, *17*, 914–921. [PubMed]

52. Wang, L.; Liu, W.; Xu, Y.; He, Y.; Luo, L.; Xing, Y.; Xu, C.; Zhang, Q. Genetic basis of 17 traits and viscosity parameters characterizing the eating and cooking quality of rice grain. *Theor. Appl. Genet.* **2007**, *115*, 463–476. [CrossRef]
53. Ma, M.; Meng, H.; Lei, E.; Wang, T.; Zhang, W.; Lu, B. De novo transcriptome assembly, gene annotation, and EST-SSR marker development of an important medicinal and edible crop, *Amomum tsaoko* (Zingiberaceae). *BMC Plant Biol.* **2022**, *22*, 467. [CrossRef] [PubMed]
54. Yeh, F.C.; Yang, R.; Boyle, T.B.; Ye, Z.; Mao, J.X. *Popgene Version 1.32, the User-Friendly Shareware for Population Genetic Analysis*; Molecular Biology and Biotechnology Centre, University of Alberta: Edmonton, AB, Canada, 1999.
55. Liu, K.; Muse, S.V. Powermarker: An integrated analysis environment for genetic marker analysis. *Bioinformatics* **2005**, *21*, 2128–2129. [CrossRef] [PubMed]
56. Pritchard, J.K.; Stephens, M.; Donnelly, P. Inference of population structure using multilocus genotype data. *Genetics* **2000**, *155*, 945–959. [CrossRef]
57. Earl, D.A.; VonHoldt, B.M. Structure harvester: A website and program for visualizing structure output and implementing the Evanno method. *Conserv. Genet. Resour.* **2012**, *4*, 359–361. [CrossRef]

plants

MDPI

Article

Pyramiding of Multiple Genes to Improve Rice Blast Resistance of Photo-Thermo Sensitive Male Sterile Line, without Yield Penalty in Hybrid Rice Production

Pei Peng [1], Haoyu Jiang [1], Lihua Luo [1], Changrong Ye [2] and Yinghui Xiao [1,*]

[1] College of Agronomy, Hunan Agricultural University, Changsha 410128, China
[2] Huazhi Biotech Co., Ltd., Changsha 410125, China
* Correspondence: xiaoyh@hunau.edu.cn; Tel.: +86-0731-84618076

Abstract: Rice blast caused by pathogenic fungus *Magnaporthe oryzae* is one of the most serious diseases in rice. The pyramiding of effective resistance genes into rice varieties is a potential approach to reduce the damage of blast disease. In this study, combinations of three resistance genes, *Pigm*, *Pi48* and *Pi49*, were introduced into a thermo-sensitive genic male sterile (PTGMS) line Chuang5S through marker-assisted selection. The results showed that the blast resistance of improved lines increased significantly compared with Chuang5S, and the three gene pyramiding lines (*Pigm* + *Pi48* + *Pi49*) had higher rice blast resistance level than monogenic line and digenic lines (*Pigm* +*Pi48*, *Pigm* + *Pi49*). The genetic backgrounds of the improved lines were highly similar (>90%) to the recurrent parent Chuang5S by using the RICE10K SNP chip. In addition, agronomic traits evaluation also showed pyramiding lines with two or three genes similar to Chuang5S. The yields of the hybrids developed from improved PTGMS lines and Chuang5S are not significantly different. The newly developed PTGMS lines can be practically used for the breeding of parental lines and hybrid varieties with broad spectrum blast resistance.

Keywords: hybrid rice; blast resistance; marker-assisted selection; thermo-sensitive genic male sterile line; gene pyramiding

Citation: Peng, P.; Jiang, H.; Luo, L.; Ye, C.; Xiao, Y. Pyramiding of Multiple Genes to Improve Rice Blast Resistance of Photo-Thermo Sensitive Male Sterile Line, without Yield Penalty in Hybrid Rice Production. *Plants* **2023**, *12*, 1389. https://doi.org/10.3390/plants12061389

Academic Editors: Xiangjin Wei, Yingxin Zhang, Weixun Wu and Guiai Jiao

Received: 23 February 2023
Revised: 17 March 2023
Accepted: 20 March 2023
Published: 21 March 2023

1. Introduction

Rice is one of the most important crops in the world, feeding more than half of the world's population. More than 60% of the population in China take rice as their staple food [1]. Rice blast, caused by pathogenic fungus *Magnaporthe oryzae*, is one of the most serious diseases affecting rice production [2]. The use of a resistance gene to create new cultivars with broad-spectrum disease resistance has been considered a cost-effective and environmentally friendly method to control the disease.

So far, more than 100 resistance loci have been identified, and among them, 38 R genes have been successfully cloned [3,4]. Most of the genes are distributed in the form of gene clusters on the other 11 chromosomes except the third chromosome. The *Piz* locus on chromosome 6, the *Pik* locus on chromosome 11 and the *Pita* locus on chromosome 12 are the hotspots of resistance gene clusters [5]. Most importantly, R genes in these hotspot regions often show broad-spectrum resistance to rice blast fungus, which provides important genetic resources for breeding rice varieties with broad-spectrum blast resistance [6].

The broad-spectrum rice blast resistance gene *Pigm* at the *Piz* locus has both leaf blast and panicle blast resistance effects and has been widely used in rice breeding [7,8]. Gene *Pi49* is a new broad-spectrum resistance gene at the *Pik* locus and discovered in the landrace Mowanggu, which has also been used in rice blast resistance breeding in recent years [9]. Genes *Pi47* and *Pi48* were identified in the *indica* cultivar Xiangzi 3150 and located at the *Pik* and *Pita* locus, respectively; they determined the stable broad-spectrum resistance of Xiangzi 3150, which conferred resistance to 95% of 303 blast isolates from China [10].

Chuang5S is an *indica* thermo-sensitive genic male sterile (TGMS) rice line bred by Hunan Agricultural University; it has excellent agronomic characteristics, such as lower critical sterility inducing temperature (refers to the critical temperature when a sterile line changes from a sterile state to a fertile state), stable sterility and a high combining ability and outcrossing rate [11]. More than ten hybrid combinations bred from Chuang5S have been approved and widely cultivated. However, it was found in the production practice that the rice blast resistance of Chuang5S was poor, and there is risk involved by planting the hybrids from this male sterile line in certain areas [12,13]. Therefore, in this study, three blast resistance genes (*Pigm, Pi48, Pi49*) were introgressed into Chuang5S by combining the traditional breeding method with MAS technology to develop new PTGMS lines with resistance to rice blast, aiming to improve the disease resistance of two-line hybrid rice bred from Chuang5S. Meanwhile, we also evaluated the yield of hybrid combinations derived from improved lines with one, two or three resistance genes in order to explore whether the yield was affected by the number of resistance genes.

2. Materials and Methods

2.1. Plant Materials

In this study, Chuang5S (C5S) and NIL-C5S carrying the R gene were used as the recurrent parents. Blast resistance gene *Pigm* was derived from a native variety Gumei 4 (GM4). Dominant blast resistance gene *Pi48* was derived from Xiangzi 3150 (XZ3150), a local rice variety in Hunan Province of China. Long-lasting blast resistance gene *Pi49* was derived from Mowanggu (MWG), a local variety in Yunnan Province of China. Rice variety CO39 was used as the susceptible control. A cross was made between Chuang5S and donor parents to generate the F_1 hybrids during the summer of 2011 at Changsha. After three backcross generations, the target BC_3F_3 lines were obtained. BC_3F_3 plants containing different resistance genes were crossed to generate new F_1 hybrids, then F_1 were self-pollinated to obtain the F_2 population. Then, the F_2 plants containing different resistance genes were crossed to generate the F_1 hybrids which harbored multiple resistance genes. After three self-pollination generations, the pedigree selection was followed to obtain the target MF_4 lines (Figure 1).

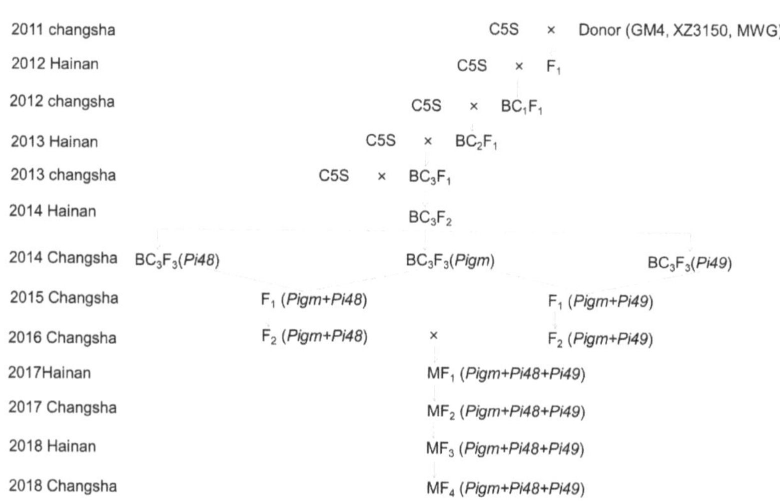

Figure 1. Breeding scheme for developing introgression lines of C5S carrying rice blast resistance genes *Pigm*, *Pi48* and *Pi49*.

2.2. Molecular Marker Selection and Genotyping

Molecular markers linked to the target genes were selected based on previous research; a set of molecular markers with polymorphism among parents were selected for foreground selection (Table 1). SSR markers RM7178 and RM7311 were used for *Pigm* detection, LY2 was used to detect *Pi48* and RM224 was used to detect *Pi49* [12,14].

Table 1. Molecular markers used to select individuals with the resistance genes.

Gene	Chr	Marker	Marker Position	Forward Primer	Reverse Primer
Pigm	6	RM7311	11046701-11046847	agtggtcgttgaactcggag	tcgtggcgcctttaatctc
		RM7178	10199892-10200042	taaccttcacagcgaacgtg	ccgtgagatgggctacctac
Pi48	12	LY2	11935927-11936129	attacgctcgatagtggc	ctagcgggaggttggaag
Pi49	11	RM224	27673251-27673353	atcgatcgatcttcacgagg	tgctataaaaggcattcggg

DNA samples were prepared by using the SDS-CTAB method and sucrose extraction method [15]. The PCR system (10 μL) for amplifcation was as follows: 1.0 μL 10 × buffer, 0.2 μL 5-mM dNTPs, 1.0 μL 2 pmol/μL primers, 0.1 μL 5 U/μL Taq polymerase, 1.0 μL DNA template (about 10 ng/μL) and 6.7 μL ddH_2O. The PCR was performed on the ABI PCR system 2700. The PCR procedure was 94 °C for 5 min, followed by 35 cycles of 94 °C for 30 s, 55 °C for 1 min, 72 °C for 30 s and finally 72 °C for 10 min. The PCR products were separated via electrophoresis on 8% non-denaturing polyacrylamide gel, and then visualized via silver staining and finally observed using a fluorescent light box.

2.3. Genetic Background Examination of Pyramided Lines

The whole-genome single-nucleotide polymorphism (SNP) array RICE10K was used to evaluate the similarity in genetic background between the improved lines and the recurrent parent C5S. This SNP array was developed based on a multiplex PCR and genotyping using targeted sequencing technology and comprises 9906 high-quality SNP and insertion-deletion (InDel) markers evenly distributed on the 12 chromosomes of rice with an average density of 27 SNPs per Mb. The pyramided lines of MF_6, F_8 and BC_3F_{11} were selected and genotyped for examination of the genetic background. For each improved line or C5S, the total DNA was extracted from the leaves of 30 plants. Genetic background similarity analysis was performed at China National Hybrid Rice Research and Development Center (Changsha, China) according to Genotyping by Targeted Sequencing Protocol.

2.4. Phenotyping of Blast Resistance in the Field

To evaluate the rice blast resistance of the newly improved TGMS lines, the blast resistance experiment was conducted under natural conditions in a rice blast disease hotspot location, Daweishan Village of Liuyang City, Hunan Province, China, where rice blast disease is epidemic every year. One hundred seeds of C5S and every newly improved PTGMS line were sown with the susceptible control CO39, and then CO39 was sown around them. About 30 days after seeding, the disease resistance was visually scored according to a 0–9 SES scale (Standard Evaluation System for Rice, IRRI, 2002) when the susceptible control CO39 showed more than 90% dead seedlings. Twenty random plants in the middle of each line were examined, and the average score was used to measure the disease resistance level of each line. Scores 0–3 were considered as resistant (R), 4 as moderately resistant (MR), 5 as moderately susceptible (MS) and 6–9 as susceptible (S).

2.5. Agronomic Performance Evaluation of Rice Blast Resistance Improved Lines and Chuang5S

The MF_3 plants of newly improved TGMS lines and plants of C5S (CK) were transplanted in the field with a space of 20 cm × 20 cm at the Jiangbei experimental field of Hunan Agricultural University, Hunan Province, in the summer of 2018. The main agronomic characters of five individual plants randomly selected from each line were examined, including plant height, panicle length, flag leaf length, number of primary branches on the panicle, number of grains per panicle and length of panicle exertion. One-way ANOVA was performed to detect the statistical differences. A *p*-value less than 0.05 was considered to be a significant difference.

2.6. Yield and Agronomic Performance Evaluation of Hybrid Rice Combinations Developed from the Improved Lines and Chuang5S

Thirty-five hybrid rice combinations crossed between 7 PTGMS lines and 5 restorer lines were evaluated during the summer of 2022. The evaluation of agronomic traits under the natural field condition was conducted in the field at the Jiangbei experimental field of Hunan Agricultural University, Hunan Province. Each combination was planted in a plot of 6 rows with 8 plants per row with a space of 20 cm × 20 cm. At maturity, five plants in the middle of each plot were taken randomly for measurements of the plant height, panicle length, number of panicles per plant, number of grains per panicle, spikelet fertility, 1000 grain weight and yield per plant. One-way ANOVA was performed to detect the statistical differences. A *p*-value less than 0.05 was considered to be a significant difference.

3. Results

3.1. Development of Gene Pyramided Lines of C5S through MAS

In this study, three backcrossing events were conducted to pyramid the rice blast resistance genes in PTGMS line C5S. At the first step, three rice blast resistance genes (*Pigm*, *Pi48* and *Pi49*) from three donor parents (GM4, XZ3150 and MWG) were introgressed into C5S via marker-assisted backcrossing. F_1 generation was obtained in the summer of 2011. Then, three successive backcrossing generations were conducted with C5S as the recurrent parent, and BC_3F_1 generation was developed in the summer of 2013. For each backcross generation, the presence of disease resistance genes in individual plants was detected using molecular markers; only the individuals harboring the target genes and showing similar morphological phenotypes to their recurrent parent C5S were selected for the following steps. Subsequently, the BC_3F_1 and its descendant pedigree individual plants selected via MAS and field phenotyping were handled with continuous irrigation of cold water with a temperature about of 20–22 °C in summer at Changsha, or were planted under low temperature conditions in winter at Sanya to restore the male fertility. Finally, after two times of self-pollination, the BC_3F_3 generation was obtained in the spring of 2014.

The second step was to pyramid two blast resistance genes. The hybridization between the BC_3F_3 population harboring different rice blast resistance genes of the first step was carried out at Changsha in the summer of 2014. The F_1 plants were genotyped using molecular markers, and the individuals with heterozygous genotype of target genes were selected for self-pollination; the selected plants were handled with cold water to recover the male fertility in Changsha. The F_2 populations carrying two rice blast resistance genes were obtained in the summer of 2016.

The third step was to pyramid three blast resistance genes. The F_2 plants with two rice blast resistance genes were selected for hybridization to obtain multiple F_1 populations at Changsha in the summer of 2016. In each subsequent generation, the rice blast resistant gene loci were detected using molecular markers, followed by agronomic traits selection in the field; finally, selected individuals were induced to restore fertility to harvest seeds. In the MF_1 generation, the plants with a heterozygous genotype of blast resistant genes were selected, and in the MF_2 or descendant generations, the plants with a homozygous genotype of blast resistant genes were selected.

3.2. Blast Resistance of the Pyramided Lines

The pyramided lines carrying homozygous blast resistance genes were selected for the evaluation of leaf blast resistance at the seedling stage and panicle blast resistance at the grain filling stage. The scores of the recurrent parent C5S and the susceptible control CO39 were 6.2 and 9.0, respectively. Three donor parents, GM4, XZ3150 and MWG, were resistant to leaf blast with scores of 1.6, 3.2 and 4.0, respectively. The introgression lines carrying single blast genes *Pigm*, *Pi48* and *Pi49* showed similar resistance to the donor, with resistance scores of 1.8, 3.6 and 4.8, respectively. In the two-gene pyramided lines, the resistance scores of *Pigm* + *Pi48* and *Pigm* + *Pi49* improved lines were 1.6 and 1.8, respectively, which were similar to *Pigm* monogenic lines and *Pigm* donor GM4, but significantly lower than *Pi48* and *Pi49* monogenic lines. The three-gene pyramided lines, including *Pigm*, *Pi49* and *Pi48*, showed comparable resistance to the donor GM4 with resistance grades of 1.6, which were significantly stronger than other monogenic lines and digenic lines (Table 2, Figure 2). At the grain filling stage, the disease reaction scores of all improved materials were 3–5, while the susceptible control CO39 and recurrent parent C5S showed more than 50% death (Table 2, Figure 3). The results revealed that the blast resistance of the newly developed lines was significantly improved in comparison to C5S.

Table 2. Blast resistance of C5S and improved lines.

Materials	Blast Resistance Gene	Leaf Blast Resistance at Seedling Stage	Panicle Blast Resistance at Grain Filling Stage
CO39	-	9.0 ± 0 **	9
C5S	-	6.2 ± 0.5	7
GM4	*Pigm*	1.6 ± 0.2 **	3
XZ3150	*Pi48*	3.2 ± 0.4 **	3
MWG	*Pi49*	4.0 ± 0.3 **	5
C5S-Pigm	*Pigm*	1.8 ± 0.4 **	3
C5S-Pi48	*Pi48*	3.6 ± 0.5 **	5
C5S-Pi49	*Pi49*	4.8 ± 0.4 **	5
C5S-Pigm + Pi48	*Pigm* + *Pi48*	1.6 ± 0.2 **	3
C5S-Pigm + Pi49	*Pigm* + *Pi49*	1.8 ± 0.4 **	5
C5S-Pigm + Pi48 + Pi49	*Pigm* + *Pi48* + *Pi49*	1.6 ± 0.2 **	3

** Means the disease severity score difference between improved lines and C5S was significant at 0.01 level. For blast resistance, scores 0–3 are resistant, score 4 is moderately resistant, score 5 is moderately susceptible and scores of 7–9 are susceptible.

Figure 2. Leaf blast resistance of C5S and the improved lines at seedling stage.

Figure 3. Panicle blast resistance of C5S and the improved lines at grain filling stage.

3.3. Genetic Background Examination of the Pyramided Lines

In this study, a 10K whole-genome SNP array was used to analyze the genetic background of the gene pyramided lines. The genetic background recovery rates of the improved lines 19RS00207 (*Pigm*) and 19RS00209 (*Pi49*) were 97.26% and 94.34%, respectively. The genetic background recovery rate of the digenic line 19RS00451 (*Pigm + Pi49*) was 97.53%. The genetic background recovery rate of the trigenic line 19RS00625 (*Pigm + Pi48 + Pi49*) was 92.68%. Fragments carrying target genes were substituted in the positions (red dots) of *Pigm* in chromosome 6, *Pi48* in chromosome 12 and *Pi49* in chromosome 11, indicating the successful introgression of blast resistance genes (Figure 4).

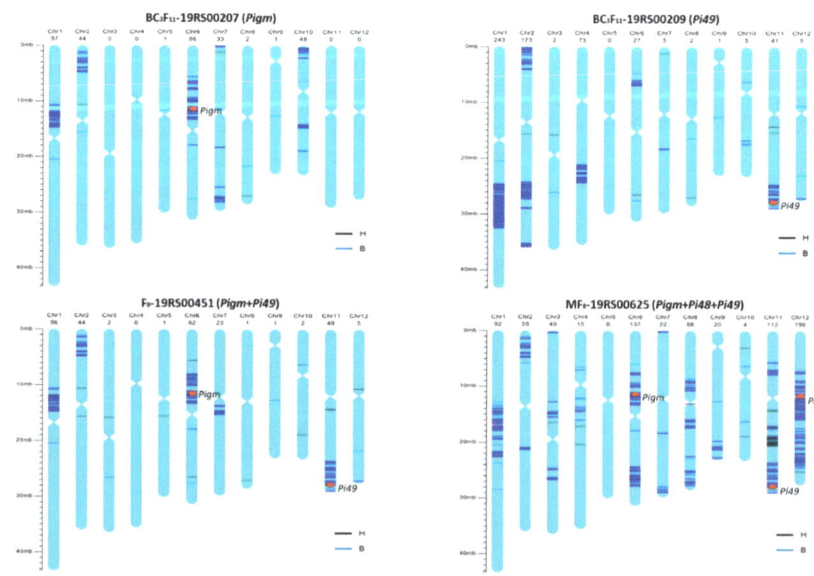

Figure 4. Genetic background of improved lines using the 10K array. The blue color represents homozygous genotype different from C5S, and the black color represents heterozygous genotype. The red dots indicate the positions of the target blast resistance genes.

3.4. Agronomic Traits of the Pyramided Lines

The main agronomic traits of the pyramided lines and their recurrent parent C5S were evaluated; the results showed that the agronomic traits such as plant height, number of grains per panicle and panicle length of the improved lines were lower than recurrent parent C5S. Among the improved lines, the agronomic traits of C5S-3R-4 were similar to the recurrent parent, and its plant height was not significantly different from C5S. The improved line C5S-3R-3 performed poorly in agronomic traits, and its number of grains per panicle was significantly lower than the recurrent parents (Table 3).

Table 3. Agronomic traits of C5S and the improved lines.

Lines	Genes	Plant Height (cm)	Panicle Length (cm)	Flag-Leaf Length (cm)	Number Primary Branches	Number of Grains per Spike	Length of the Upmost Internode (cm)	Length of Panicle Exertion (cm)
C5S		62.7 ± 0.7	22.1 ± 0.3	34.1 ± 1.4	11.0 ± 0.2	176.7 ± 7.7	20.7 ± 0.4	8.4
C5S-3R-1	*Pigm* + *Pi48* + *Pi49*	54.5 ± 0.8 **	19.9 ± 0.2 **	27.6 ± 1.0 **	9.1 ± 0.3 **	104.8 ± 3.9 **	16.2 ± 0.5 **	10.3
C5S-3R-2	*Pigm* + *Pi48* + *Pi49*	53.9 ± 0.6 **	18.7 ± 0.3 **	25.0 ± 0.6 **	9.8 ± 0.2 **	115.5 ± 4.6 **	16.7 ± 0.4 **	11.7
C5S-3R-3	*Pigm* + *Pi48* + *Pi49*	50.0 ± 0.5 **	17.5 ± 0.2 **	18.9 ± 1.0 **	6.9 ± 0.2 **	77.7 ± 2.8 **	16.8 ± 0.5 **	7.5
C5S-3R-4	*Pigm* + *Pi48* + *Pi49*	62.1 ± 0.8	19.7 ± 0.3 **	27.1 ± 1.0 **	10.0 ± 0.4 **	140.1 ± 6.9 **	21.9 ± 0.4 **	6.7
C5S-3R-5	*Pigm* + *Pi48* + *Pi49*	59.8 ± 0.9 **	18.9 ± 0.3 **	27.2 ± 0.9 **	9.3 ± 0.4 **	112.1 ± 6.1 **	20.6 ± 0.5	6.8

** means there is a significant difference at 0.01 level compared with C5S.

3.5. Agronomic Traits of the Hybrid Rice Combinations

In order to study the effect of gene introgression on the agronomic characters of hybrid rice combinations, seven male sterile lines were crossed with five restorer lines, and thirty-five hybrid combinations were obtained for investigation of agronomic traits. The results showed that most of the agronomic traits of the combinations derived from four restorer lines (R1128, HRZ, MFZ1 and YNSM) were similar to those combinations from C5S. In the SFSM group, although panicle length and number of grains per panicle of some combinations from improved lines was significantly higher than the combination from C5S, the 1000-grain weight of these lines were significantly lower than that of the control, and there was no significant difference in the plant height, number of panicles per plant and spikelet fertility, so there was no significant difference in yield (Table 4).

Table 4. Agronomic traits of the hybrids of C5S and the improved lines.

Materials	Restorer Lines	PTGMS Lines	Gene	Plant Height (cm)	Panicle Length (cm)	Panicles per Plant	Grain Number per Panicle	Spikelet Fertility (%)	1000-Grain Weight (g)	Yield per Plant (g)
22RC11	R1128	C5S	-	100.7 ± 1.0	20.8 ± 2.4	10.0 ± 1.0	178.7 ± 27.4	78.8 ± 2.9	25.1 ± 0.8	35.5 ± 8.4
22RC12		21YCS01	*Pi49*	115.2 ± 1.9 **	22.6 ± 0.4	9.0 ± 1.0	184.5 ± 10.4	81.6 ± 1.6	24.4 ± 0.2	33.1 ± 5.1
22RC13		21YCS22	*Pi48*	101.7 ± 2.4	23.3 ± 1.7	10.0 ± 3.0	216.8 ± 15.6	87.7 ± 4.6	23.9 ± 2.0	44.9 ± 12.1
22RC14		21YCS23	*Pigm*	106.1 ± 4.6	22.1 ± 1.3	9.7 ± 0.6	209.6 ± 11.8	82.6 ± 3.1	26.1 ± 2.7	43.7 ± 5.1
22RC15		21YCS24	*Pigm + Pi49*	101.9 ± 1.6	22.3 ± 1.9	9.3 ± 0.6	211.7 ± 26.3	79.2 ± 1.0	25.1 ± 1.8	39.2 ± 4.9
22RC16		21YCS61	*Pigm + Pi48*	111.1 ± 2.9 **	21.9 ± 1.0	8.7 ± 0.6	212.4 ± 33.0	83.1 ± 7.0	29.3 ± 1.6 **	45.1 ± 10.5
22RC17		21YCS25	*Pigm + Pi48 + Pi49*	102.3 ± 1.3	21.1 ± 0.8	7.3 ± 0.6	202.4 ± 33.7	88 ± 1.6	26.4 ± 0.5	34.5 ± 6.6
22RC21	HRZ	C5S	-	109.7 ± 2.7	23.1 ± 0.7	11.0 ± 0.0	229.0 ± 28.7	86.7 ± 1.7	24.0 ± 0.2	52.5 ± 7.9
22RC22		21YCS01	*Pi49*	106.0 ± 1.1	20.4 ± 0.8 *	15.3 ± 1.2 **	168.2 ± 18.9 *	89.4 ± 0.7	23.3 ± 0.5	53.4 ± 2.3
22RC23		21YCS22	*Pi48*	101.1 ± 5.3 *	21.7 ± 0.9	11.0 ± 1.0	206.7 ± 9.5	80.1 ± 6.2	22.9 ± 1.0	42.3 ± 10.4
22RC24		21YCS23	*Pigm*	106.7 ± 0.5	22.5 ± 0.9	10.7 ± 0.6	213.0 ± 20.0	85.6 ± 2.5	23.7 ± 0.9	46.3 ± 6.8
22RC25		21YCS24	*Pigm + Pi49*	113.1 ± 2.1	24.7 ± 0.2	9.0 ± 1.0	241.9 ± 14.7	92.0 ± 2.2	22.7 ± 0.5	45.4 ± 3.8
22RC26		21YCS61	*Pigm + Pi48*	108.7 ± 0.6	24.4 ± 0.3	11.3 ± 0.6	243.2 ± 10.8	88.3 ± 1.0	22.6 ± 0.2	55.1 ± 4.4
22RC27		21YCS25	*Pigm + Pi48 + Pi49*	112.4 ± 2.9	23.6 ± 0.8	9.0 ± 1.7	233.9 ± 25.6	87.9 ± 2.9	23.8 ± 0.1	44.2 ± 11.7
22RC31	MFZ1	C5S	-	101.8 ± 1.5	23.2 ± 0.5	13.0 ± 2.0	181.9 ± 11.8	82.8 ± 0.7	22.9 ± 0.8	44.5 ± 4.0
22RC32		21YCS01	*Pi49*	112.1 ± 4.8 *	24.7 ± 0.6	12.7 ± 1.2	209.7 ± 2.9	80.2 ± 1.5 *	23.1 ± 0.9	49.0 ± 2.5
22RC33		21YCS22	*Pi48*	103.5 ± 0.8	22.0 ± 0.9	14.0 ± 1.0	166.4 ± 16.4	86.8 ± 5.6	23.4 ± 0.3	47.2 ± 5.1
22RC34		21YCS23	*Pigm*	104.4 ± 6.2	22.9 ± 0.5	11.7 ± 0.6	208.9 ± 23.1	80.6 ± 5.5	21.5 ± 1.1	42.2 ± 6.3
22RC35		21YCS24	*Pigm + Pi49*	100.8 ± 2.3	23.5 ± 0.3	11.7 ± 0.6	207.8 ± 6.1 *	88.2 ± 1.9	20.8 ± 0.2 **	44.6 ± 3.1
22RC36		21YCS61	*Pigm + Pi48*	97.7 ± 3.9	22.4 ± 0.3	16.0 ± 1.7	191.3 ± 12.4	78.8 ± 1.2	22.8 ± 1.4	54.8 ± 3.3
22RC37		21YCS25	*Pigm + Pi48 + Pi49*	101.4 ± 2.9	23.0 ± 0.9	12.7 ± 0.6	199.0 ± 13.4	85.3 ± 4.0	22.9 ± 0.3	49.3 ± 5.3
22RC41	SFSM	C5S	-	111.9 ± 5.0	20.9 ± 0.7	15.0 ± 1.0	151.3 ± 19.9	87.9 ± 2.9	24.4 ± 0.9	55.6 ± 6.2
22RC42		21YCS01	*Pi49*	103.2 ± 2.6	21.5 ± 1.0	14.0 ± 1.0	165.0 ± 3.8	81.8 ± 1.2	22.8 ± 0.4 *	43.1 ± 4.3
22RC43		21YCS22	*Pi48*	110.3 ± 3.6	23.9 ± 0.4 **	12.3 ± 1.5	227.0 ± 6.1 **	79.8 ± 7.7	21.3 ± 0.4 **	47.4 ± 4.7
22RC44		21YCS23	*Pigm*	119.3 ± 0.8	23.2 ± 0.4 *	13.0 ± 1.0	220.4 ± 13.7 **	87.0 ± 0.8	22.5 ± 0.4 *	56.0 ± 2.6
22RC45		21YCS24	*Pigm + Pi49*	121.0 ± 1.0	24.0 ± 0.2 **	12.3 ± 0.6	208.0 ± 2.6 **	81.0 ± 0.2	22.7 ± 0.2 *	47.1 ± 2.5
22RC46		21YCS61	*Pigm + Pi48*	99.9 ± 7.5 *	23.4 ± 1.3 *	13.0 ± 1.7	214.5 ± 9.2 **	86.5 ± 3.3	21.7 ± 1.0 *	52.2 ± 4.6
22RC47		21YCS25	*Pigm + Pi48 + Pi49*	117.5 ± 1.6	24.2 ± 0.1 **	11.3 ± 1.2 *	238.5 ± 17.1 **	88.6 ± 3.7	22.1 ± 0.2 **	52.7 ± 5.3
22RC51	YNSM	C5S	-	104.2 ± 6.2	22.8 ± 0.7	15.7 ± 1.5	177.9 ± 22.4	79.8 ± 0.9	23.8 ± 0.2	52.6 ± 1.7
22RC52		21YCS01	*Pi49*	105.3 ± 3.3	24.3 ± 1.2	12.0 ± 1.0 *	263.3 ± 30.2 **	76.8 ± 4.6	23.4 ± 1.5	56.6 ± 5.8
22RC53		21YCS22	*Pi48*	106.0 ± 0.4	23.7 ± 0.8	11.7 ± 0.6 **	241.7 ± 25.6	78.2 ± 5.0	23.4 ± 0.2	51.7 ± 8.9
22RC54		21YCS23	*Pigm*	106.6 ± 1.4	24.6 ± 0.9	11.3 ± 0.6 **	230.7 ± 9.2	79.3 ± 5.0	24.5 ± 0.9	50.7 ± 1.7
22RC55		21YCS24	*Pigm + Pi49*	104.0 ± 1.6	23.6 ± 0.8	13.0 ± 1.0 *	226.3 ± 20.8	84.8 ± 1.8	22.3 ± 1.0	55.8 ± 9.6
22RC56		21YCS61	*Pigm + Pi48*	104.6 ± 3.5	23.0 ± 1.4	11.0 ± 0.0 **	241.9 ± 25.8	82.3 ± 7.3	22.7 ± 0.1	49.9 ± 9.0
22RC57		21YCS25	*Pigm + Pi48 + Pi49*	102.4 ± 2.2	22.7 ± 0.2	13.7 ± 0.6	212.8 ± 21.9	85.9 ± 3.4	22.8 ± 0.4	56.8 ± 5.9

* and ** means there is a significant difference at 0.05 and 0.01 level compared with hybrids from C5S.

4. Discussion

Marker-assisted selection is an effective way to improve the disease resistance of rice; the use of MAS can determine the resistance and fertility of individual plants as early as possible, and the selected individuals are treated with low temperature to restore fertility, thus greatly improving the selection efficiency and shortening the breeding procedure. The blast resistance gene *Pi2* was introgressed into TGMS line C815S through MAS and phenotypic selection approaches, and the agronomic and grain quality traits of four new TGMS lines met the requirement for two-line hybrid rice production [16]. Two PTGMS lines, Tai S and Wo S, were developed using the broad-spectrum resistance gene *Xa23* through MAS combined with phenotypic selection [17]. The blast and bacterial blight resistance of PTGMS line Guangzhan63-4S was improved by introducing the R genes *Pi2* and *Xa7* through MAS [18]. The rice blast and brown planthopper resistance of PTGMS line C815S was improved via the MAS of the *Pi9*, *Pi47*, *Pi48*, *Pi49*, *Bph14* and *Bph15* genes [19]. Similar to previous studies, this study combined a traditional backcross method with MAS to pyramid three blast resistance genes into sterile lines of C5S through several generations of hybridization and backcrossing. The genotyping results indicate that target genes were successfully introgressed into the improved lines. The new developed lines conferred high resistance to rice blast in the field. The genetic background recovery rates were 97.26% and 94.34% for the improved monogenic lines 19RS00207 (*Pigm*) and 19RS00209 (*Pi49*), respectively, and 97.53% for the digenic line 19RS00451 (*Pigm* + *Pi49*) and 92.68% for the trigenic line 19RS00625 (*Pigm* + *Pi48* + *Pi49*). There were more fragments introduced into the recurrent parent when more genes were pyramided, and these fragments may carry genes affecting agronomic traits. Thus, some of the agronomic traits such as the plant height, number of grains per panicle and panicle length of the trigenic lines were lower than recurrent parent C5S (Table 3). However, the effects of the introduced fragments on the performance of the hybrids were dependent on the restore lines. Among the five restore lines used, the 1000-grain weight of hybrids from SFSM were lower than those from the control C5S, while no difference was observed in hybrids from other restore lines (Table 4). Importantly, the hybrid rice combinations of improved lines showed no significant difference in yield compared to the control. The developed PTGMS lines with blast resistance and a similar yield as C5S could be used for the hybrid rice seeds production.

In the practice of crop disease resistance breeding, repeated use of a single gene for a long time can easily lead to the loss of resistance, thus pyramiding broad-spectrum resistance genes is generally considered as an effective way to solve this problem. Multi-gene pyramiding, which is conducive for broadening the resistance spectrum and improving crop resistance, is an important approach for rice variety improvement [20]. Genes *Pi46* and *Pita* were pyramided into an elite restorer line HH179 to improve its blast resistance using the marker-assisted backcrossing procedure, and the resistance spectrum of the three improved lines was markedly broader than that of HH179 [21]. The blast resistances of a japonica rice variety 07GY31 was improved by pyramiding R genes *Pi9*, *Pizt* and *Pi54* [22]. The blast resistances of four sterile lines were improved by the introgression of broad-spectrum blast resistance genes *Pi37*, *Pit*, *Pid3*, *Pigm*, *Pi36*, *Pi5*, *Pi54*, *Pikm* and *Pb1* [23]. The disease resistances of two cultivars (ASD 16 and ADT 43) were enhanced through the introgression of bacterial blight (*xa5*, *xa13*, and *Xa21*), blast (*Pi54*) and sheath blight (*qSBR7-1*, *qSBR11-1* and *qSBR11-2*) resistance genes [24]. In our previous study, we pyramided two rice blast resistance genes, *Pi9* and *Pi49*, into C5S, and the results showed that the resistances of C5S-Pi9/Pi49 lines were improved when different blast strains were inoculated [12]. In this study, we developed lines containing two or three blast resistance genes that were resistant to leaf and panicle blast under natural conditions. The three-gene pyramiding lines of C5S (*Pigm* + *Pi48* + *Pi49*) have higher rice blast resistance levels than the monogenic line and digenic lines. In addition, the present study results also reveal that the gene combination with *Pigm* had better resistance than the remaining improved lines. These results were consistent with the results of previous studies. The improved lines containing one, two or three R genes can be used to breed two-line hybrid rice with broad-spectrum resistance and can also be used to polymerize more resistance genes into intermediate materials via MAS to obtain new materials with a broader resistance spectrum, wider adaptation range and durable resistance. In addition, they can be pyramided with more genes of important agronomic traits, such as the low accumulation of heavy metals and superior grain quality.

In this study, molecular markers linked to the target gene were used for foreground selection, but genome-wide markers were not used for background selection in each generation due to the cost. The genetic background examination results showed that some non-target genomic regions of the higher generation materials still retained genotypes of the donor, especially the large chromosomal segment of *Pi48*. Similar results have been reported in previous studies. A large flanking fragment remained in the recurrent parent when introgressing gene *Pita* to improve rice blast resistance [25]. The *Pi48* gene was the allele of *Pita*, and the *Pita* locus is close to the centromere region, which has a low recombination rate. The background selection is considered to be an effective way to improve the efficiency of recombination selection and reduce the transmission of linkage drag. With the advances in high-throughput genotyping technology and the reduction in genotyping costs, it has become easier to obtain high-quality whole genome genotype. New genotyping technology such as KASP, whole genome sequencing, genotyping via target sequencing and SNP breeding chips have been widely applied in recent years [26–30]. The use of an SNP chip for genetic background selection in each generation can select individuals with higher genomic background recovery as early as possible and shorten the breeding process, and, therefore, it has been widely employed in breeding programs. The RICE6K SNP array was used for background detection in backcross breeding, and obtained improved lines, with a genome recovery of 99.67% [31]. In the case of genomics-assisted breeding, the background analysis using the 56K SNP chip revealed that the selected BC_2F_3 line has a background recovery rate of 96.8% comparing to the recurrent parent [32]. The use of the SNP chip for background selection in the process of backcross breeding can reduce the near isogenic line development by 2–3 years, which will further shorten the breeding cycle [31]. Using a liquid chip based on targeted sequencing technology [33] to further improve breeding efficiency is our future research direction on the basis of the improved lines in this study. Using a genomics-assisted breeding strategy to polymerize more important genes and create a series of better C5S is our next research focus.

Author Contributions: Conceptualization, Y.X. and P.P.; methodology, H.J. and L.L.; validation, H.J. and P.P.; formal analysis, H.J., C.Y. and P.P.; resources, L.L.; writing—original draft preparation, P.P.; writing—review and editing, P.P., C.Y. and Y.X.; supervision, Y.X.; funding acquisition, Y.X. All authors have read and agreed to the published version of the manuscript.

Funding: This research was funded by the National Key Research and Development Program of China, grant number 2016YFD0101100.

Institutional Review Board Statement: Not applicable.

Informed Consent Statement: Not applicable.

Data Availability Statement: Not applicable.

Acknowledgments: The authors thank Na Xu and Gai zeng of Hunan Agricultural University for all their help during the experiment.

Conflicts of Interest: The authors declare no conflict of interest.

References

1. Saud, S.; Wang, D.; Fahad, S.; Alharby, H.F.; Bamagoos, A.A.; Mjrashi, A.; Alabdallah, N.M.; AlZahrani, S.S.; AbdElgawad, H.; Adnan, M.; et al. Comprehensive impacts of climate change on rice production and adaptive strategies in china. *Front. Microbiol.* **2022**, *13*, 926059. [CrossRef] [PubMed]
2. Skamnioti, P.; Gurr, S.J. Against the grain: Safeguarding rice from rice blast disease. *Trends Biotechnol.* **2009**, *27*, 141–150. [CrossRef] [PubMed]

3. Ning, X.; Yunyu, W.; Aihong, L. Strategy for use of rice blast resistance genes in rice molecular breeding. *Rice Sci.* **2020**, *27*, 263–277. [CrossRef]
4. Devanna, B.N.; Jain, P.; Solanke, A.U.; Das, A.; Thakur, S.; Singh, P.K.; Kumari, M.; Dubey, H.; Jaswal, R.; Pawar, D.; et al. Understanding the dynamics of blast resistance in rice-magnaporthe oryzae interactions. *J. Fungi.* **2022**, *8*, 584. [CrossRef]
5. Wang, X.; Lee, S.; Wang, J.; Ma, J.; Bianco, T.; Ji, Y. Current advances on genetic resistance to rice blast disease. In *Rice-Germplasm, Genetics and Improvement*; InTech: Rang-Du-Fliers, France, 2014; pp. 196–216.
6. Ying, Z.; Tao, W.; Bin, Y.; Fang, L.; Meijuan, C.; Qiong, W.; Ping, H.; Shuyan, K.; Wenxiu, Q.; Li, L. Improving rice blast resistance by mining broad-spectrum resistance genes at pik locus. *Rice Sci.* **2022**, *29*, 133–142. [CrossRef]
7. Deng, Y.; Zhu, X.; Shen, Y.; He, Z. Genetic characterization and fine mapping of the blast resistance locus pigm(t) tightly linked to pi2 and pi9 in a broad-spectrum resistant chinese variety. *Appl. Genet.* **2006**, *113*, 705–713. [CrossRef]
8. Deng, Y.; Zhai, K.; Xie, Z.; Yang, D.; Zhu, X.; Liu, J.; Wang, X.; Qin, P.; Yang, Y.; Zhang, G.; et al. Epigenetic regulation of antagonistic receptors confers rice blast resistance with yield balance. *Science* **2017**, *355*, 962–965. [CrossRef]
9. Sun, P.; Liu, J.; Wang, Y.; Jiang, N.; Wang, S.; Dai, Y.; Gao, J.; Li, Z.; Pan, S.; Wang, D.; et al. Molecular mapping of the blast resistance gene pi49 in the durably resistant rice cultivar mowanggu. *Euphytica* **2012**, *192*, 45–54. [CrossRef]
10. Huang, H.; Huang, L.; Feng, G.; Wang, S.; Wang, Y.; Liu, J.; Jiang, N.; Yan, W.; Xu, L.; Sun, P.; et al. Molecular mapping of the new blast resistance genes pi47 and pi48 in the durably resistant local rice cultivar xiangzi 3150. *Phytopathology* **2011**, *101*, 620–626. [CrossRef] [PubMed]
11. Wang, Y.; Hao, M.; Lei, D.; Tang, W.; Cheng, L. Character of early growth and quick development of genic male sterility lines. *Crop. Res.* **2014**, *28*, 341–344.
12. Zhang, J.; Hao, M.; Zeng, G.; Cao, Z.; Jiang, H.; Huang, X.; Xiao, Y. Polymerization of pi9 and pi49 loci by marker assisted selection to improve blast resistance of dual-purpose genic sterile rice chuang 5s. *Mol. Plant Breed.* **2018**, *16*, 7372–7379.
13. Chen, Q.; Tang, W.; Zeng, G.; Sheng, H.; Shi, W.; Xiao, Y. Reduction of cadmium accumulation in the grains of male sterile rice chuang-5s carrying pi48 or pi49 through marker-assisted selection. *3 Biotech* **2020**, *10*, 539. [CrossRef] [PubMed]
14. Cao, Z.; Zheng, G.; Hao, M.; Sheng, H.; Ye, N.; Xiao, Y. Improving blast resistance of dual-purpose genic sterile line c815s by using molecular marker-assisted selection. *Mol. Plant Breed.* **2015**, *13*, 1193–1200.
15. Cao, Z.; Zeng, G.; Sheng, H.; Xiao, Y. A simple approach for rapid preparation of rice genomic DNA for pcr analysis. *J. Hunan. Agric. Univ.* **2013**, *39*, 13–16.
16. Jiang, J.; Mou, T.; Yu, H.; Zhou, F. Molecular breeding of thermo-sensitive genic male sterile (tgms) lines of rice for blast resistance using pi2 gene. *Rice* **2015**, *8*, 11. [CrossRef] [PubMed]
17. Wang, S.; Liu, W.; Lu, D.; Lu, Z.; Wang, X.; Xue, J.; He, X. Distribution of bacterial blight resistance genes in the main cultivars and application of xa23 in rice breeding. *Front. Plant Sci.* **2020**, *11*, 555228. [CrossRef] [PubMed]
18. Mi, J.; Yang, D.; Chen, Y.; Jiang, J.; Mou, H.; Huang, J.; Ouyang, Y.; Mou, T. Accelerated molecular breeding of a novel p/tgms line with broad-spectrum resistance to rice blast and bacterial blight in two-line hybrid rice. *Rice* **2018**, *11*, 11. [CrossRef]
19. Chen, Q.; Zeng, G.; Hao, M.; Jiang, H.; Xiao, Y. Improvement of rice blast and brown planthopper resistance of ptgms line c815s in two-line hybrid rice through marker-assisted selection. *Mol. Breed.* **2020**, *40*, 21. [CrossRef]
20. Yin, J.; Zou, L.; Zhu, X.; Cao, Y.; He, M.; Chen, X. Fighting the enemy: How rice survives the blast pathogen's attack. *Crop. J.* **2021**, *9*, 543–552. [CrossRef]
21. Xiao, W.; Luo, L.; Wang, H.; Guo, T.; Liu, Y.; Zhou, J.; Zhu, X.; Yang, Q.; Chen, Z. Pyramiding of pi46 and pita to improve blast resistance and to evaluate the resistance effect of the two r genes. *J. Integr. Agric.* **2016**, *15*, 2290–2298. [CrossRef]
22. Xiao, N.; Wu, Y.; Pan, C.; Yu, L.; Chen, Y.; Liu, G.; Li, Y.; Zhang, X.; Wang, Z.; Dai, Z.; et al. Improving of rice blast resistances in japonica by pyramiding major r genes. *Front. Plant Sci.* **2016**, *7*, 1918. [CrossRef]
23. Jiang, H.; Li, Z.; Liu, J.; Shen, Z.; Gao, G.; Zhang, Q.; He, Y. Development and evaluation of improved lines with broad-spectrum resistance to rice blast using nine resistance genes. *Rice* **2019**, *12*, 29. [CrossRef] [PubMed]
24. Ramalingam, J.; Raveendra, C.; Savitha, P.; Vidya, V.; Chaithra, T.L.; Velprabakaran, S.; Saraswathi, R.; Ramanathan, A.; Pillai, M.P.A.; Arumugachamy, S.; et al. Gene pyramiding for achieving enhanced resistance to bacterial blight, blast, and sheath blight diseases in rice. *Front. Plant Sci.* **2020**, *11*, 591457. [CrossRef] [PubMed]
25. Xiao, W.; Yang, Q.; Huang, M.; Guo, T.; Liu, Y.; Wang, J.; Yang, G.; Zhou, J.; Yang, J.; Zhu, X.; et al. Improvement of rice blast resistance by developing monogenic lines, two-gene pyramids and three-gene pyramid through mas. *Rice* **2019**, *12*, 78. [CrossRef]
26. Semagn, K.; Babu, R.S.; Hearne, S.J.; Olsen, M. Single nucleotide polymorphism genotyping using kompetitive allele specific pcr (kasp): Overview of the technology and its application in crop improvement. *Mol. Breed.* **2013**, *33*, 1–14. [CrossRef]
27. Majeed, U.; Darwish, E.; Rehman, S.U.; Zhang, X. Kompetitive allele specific pcr (kasp): A singleplex genotyping platform and its application. *J. Agric. Sci.* **2018**, *11*, 1. [CrossRef]
28. Abe, A.; Takagi, H.; Yaegashi, H.; Natsume, S.; Utsushi, H.; Tamiru, M.; Terauchi, R. Next-generation breeding of rice by whole-genome approaches. In *Rice Genomics, Genetics and Breeding*; Sasaki, T., Ashikari, M., Eds.; Springer: Singapore, 2018; pp. 511–522.
29. Xu, Y.; Yang, Q.; Zheng, H.; Sang, Z.; Zhang, J. Genotyping by Target Sequencing (gbts) and Its Applications. *Sci. Agric. Sin.* **2020**, *53*, 2983–3004.
30. Rasheed, A.; Hao, Y.; Xia, X.; Khan, A.; Xu, Y.; Varshney, R.K.; He, Z. Crop breeding chips and genotyping platforms: Progress, challenges, and perspectives. *Mol. Plant* **2017**, *10*, 1047–1064. [CrossRef]

31. Yang, D.; Tang, J.; Yang, D.; Chen, Y.; Ali, J.; Mou, T. Improving rice blast resistance of feng39s through molecular marker-assisted backcrossing. *Rice* **2019**, *12*, 70. [CrossRef]
32. He, Z.; Xin, Y.; Wang, C.; Yang, H.; Xu, Z.; Cheng, J.; Li, Z.; Ye, C.; Yin, H.; Xie, Z.; et al. Genomics-assisted improvement of super high-yield hybrid rice variety "super 1000" for resistance to bacterial blight and blast diseases. *Front. Plant Sci.* **2022**, *13*, 881244. [CrossRef]
33. Guo, Z.; Yang, Q.; Huang, F.; Zheng, H.; Sang, Z.; Xu, Y.; Zhang, C.; Wu, K.; Tao, J.; Prasanna, B.M.; et al. Development of high-resolution multiple-snp arrays for genetic analyses and molecular breeding through genotyping by target sequencing and liquid chip. *Plant Commun* **2021**, *2*, 100230. [CrossRef] [PubMed]

plants

MDPI

Article

Isolation and Characterization of *SPOTTED LEAF42* Encoding a Porphobilinogen Deaminase in Rice

Lin Liu [1], Yunpeng Wang [1], Yunlu Tian [1], Shuang Song [1], Zewan Wu [1], Xin Ding [1], Hai Zheng [1], Yunshuai Huang [1], Shijia Liu [1], Xiaoou Dong [1], Jianmin Wan [1,2] and Linglong Liu [1,*]

1 State Key Laboratory for Crop Genetics and Germplasm Enhancement, Jiangsu Plant Gene Engineering Research Center, Nanjing Agricultural University, Nanjing 210095, China
2 National Key Facility for Crop Gene Resources and Genetic Improvement, Institute of Crop Science, Chinese Academy of Agricultural Sciences, Beijing 100081, China
* Correspondence: liulinglong@njau.edu.cn

Abstract: The formation and development of chloroplasts play a vital role in the breeding of high-yield rice (*Oryza sativa* L.). Porphobilinogen deaminases (PBGDs) act in the early stage of chlorophyll and heme biosynthesis. However, the role of PBGDs in chloroplast development and chlorophyll production remains elusive in rice. Here, we identified the *spotted leaf 42* (*spl42*) mutant, which exhibited a reddish-brown spotted leaf phenotype. The mutant showed a significantly lower chlorophyll content, abnormal thylakoid morphology, and elevated activities of reactive oxygen species (ROS)-scavenging enzymes. Consistently, multiple genes related to chloroplast development and chlorophyll biosynthesis were significantly down-regulated, whereas many genes involved in leaf senescence, ROS production, and defense responses were upregulated in the *spl42* mutant. Map-based cloning revealed that *SPL42* encodes a PBGD. A C-to-T base substitution occurred in *spl42*, resulting in an amino acid change and significantly reduced PBGD enzyme activity. SPL42 targets to the chloroplast and interacts with the multiple organelle RNA editing factors (MORFs) OsMORF8-1 and OsMORF8-2 to affect RNA editing. The identification and characterization of *spl42* helps in elucidating the molecular mechanisms associated with chlorophyll synthesis and RNA editing in rice.

Keywords: chlorophyll biosynthesis; chloroplast; rice (*Oryza sativa* L.); RNA editing; spotted leaf; porphobilinogen deaminase (PBGD)

Citation: Liu, L.; Wang, Y.; Tian, Y.; Song, S.; Wu, Z.; Ding, X.; Zheng, H.; Huang, Y.; Liu, S.; Dong, X.; et al. Isolation and Characterization of *SPOTTED LEAF42* Encoding a Porphobilinogen Deaminase in Rice. *Plants* **2023**, *12*, 403. https://doi.org/10.3390/plants12020403

Academic Editors: Xiangjin Wei, Yingxin Zhang, Weixun Wu and Guiai Jiao

Received: 16 December 2022
Revised: 9 January 2023
Accepted: 12 January 2023
Published: 15 January 2023

1. Introduction

The chloroplast is an important organelle for photosynthesis in higher plants. Its main function is to convert light energy into chemical energy through chlorophyll. Chlorophyll is involved in the absorption, transmission, and conversion of light energy. It is a long-term breeding objective to improve the chloroplast and chlorophyll photosynthesis capability.

So far, many genes related to chlorophyll synthesis in rice (*Oryza sativa*) have been cloned, including glutamyl-tRNA synthetase (OsGluRS), magnesium ion chelatase (Chl), protochlorophyll ester oxidoreductase (OsPORB), porphobilinogen deaminase (PBGD), and so on. OsGluRS is the first enzyme in the chlorophyll synthesis pathway in plants. When OsGluRS is mutated, plant leaves turn yellowish green while plant growth and development are inhibited. [1]. Chl contains three subunits, ChlH, ChlD, and ChlI [2,3]. Mutations in the former two gene results in a yellow-green phenotype, whereas mutating the last one (ChlH) affects the chloroplast structure and chlorophyll biosynthesis, resulting in plant death four weeks after germination [4,5]. Likewise, the abnormal expression of OsPORB produces yellow, white, and necrotic spots on the leaves [6]. PBGD is an enzyme that functions during the early stage of chlorophyll and heme biosynthesis. PBGD catalyzes the deamination of four porphobilinogen molecules and their gradual polymerization into linear hydroxymethylbilane. A loss-of-function mutant of PBGD in maize (cf1) exhibits yellowing and even necrotic leaves. The necrotic spotted leaf phenotype of the cf1 mutant is caused by the decreased activity of tetrapyrrole synthetase, accompanied by reactive oxygen species (ROS) burst and the disturbance of the ROS scavenge system [7]. Another loss-of-function mutant of PBGD in Arabidopsis, rugosa1, showed spontaneous lesions reminiscent of those seen in lesion-mimic mutants (LMM), such as irregularly shaped leaves and a reduced growth. [8,9].

An LMM often exhibits hypersensitive responses in the absence of pathogens, resulting in the production of disease-resistant substances such as ROS, plant hormones, and phenolic complexes [10,11]. By screening rice brown necrotic spot mutants, Wang et al. identified an LMM-related gene, *OsCNGC9*. This gene relies on calcium ion channels to stimulate the plant's own immune system, thereby regulating innate immunity in rice [10]. Another rice LMM, *apoptosis leaf and sheath 1* (*als1*), showed the accumulation of jasmonic acid (JA) and salicylic acid (SA) and an increased resistance to rice fungus, indicating that *ALS1* is involved in the defense responses mediated by SA and JA [11]. Moreover, most of the rice LMMs, such as *spl2*, *spl7*, *spl11*, *dj-lm*, and *pls3*, accumulate ROS (such as H_2O_2 and O^{2-}) in the lesion sites, thereby negatively affecting the chloroplast development [12–17]. Despite the clarification of the various PBGDs in the chloroplast development of maize and Arabidopsis, the role of PBGD in rice defense responses and chloroplast development remains elusive.

As a common processing event of organellar transcripts in higher plants, RNA editing is essential for gene repair and plant development [18–21]. The mutation of the RNA recognition motif (RRM) protein ORRM1, which was involved in the RNA editing of plant organelles, leads to the complete loss of the functions of 12 editing sites in Arabidopsis [22]. ORRM1 interacts with multiple organelle RNA editing factors (MORFs), MORF2 and MORF8/RIP1, to regulate the development of chloroplasts and mitochondria [22]. The RanBP2 zinc finger protein family member OZ1 affects the expression of the MORF protein by interacting with ORRM1 [23]. Although MORF and RRM play crucial roles in RNA editing in mitochondria and plastids [24,25], other factors such as PBGD have not been studied in detail.

In this study, we identified the *spotted leaf 42* mutant (*spl42*) in rice. The *spl42* mutant displayed an irregular reddish-brown blotchy phenotype at the third to fourth leaf stage, which was accompanied throughout the whole growth period. Map-based cloning revealed that *SPL42* locates on the short arm of chromosome 2 and encodes a PBGD family protein. Transgenic experiments confirmed that the functional defect of PBGD was the cause of the *spl42* mutant phenotype. The aim of this study is to reveal the role of PBGD in chloroplast development and RNA editing in rice. At the same time, the characterization of the lesion-mimic mutant *spl42* helps to decipher the molecular mechanism of the defense response. This study helps in stabilizing the production of rice, a major staple food crop for more than half of the world's population.

2. Results

2.1. Phenotypic Characteristics of the spl42 Mutant

The *spl42* mutant was isolated from the ethyl methane sulfonate (EMS)-mutagenized japonica variety Ningjing6 and designated according to the leaf phenotype. Under field conditions, *spl42* exhibited an irregular reddish-brown spotted leaf phenotype at the third or fourth leaf stage, which then spread to the whole leaf (Figure 1A,B). Two weeks after transplanting in the field, the old leaves of the *spl42* mutant were full of lesion spots, but the new leaves showed fewer spots (Figure 1C,D). As the plant developed, the spots on the blade of the *spl42* mutant continued to spread, and the mature leaves of *spl42* demonstrated an early senescence symptom during anthesis (Figure 1E–H). To further characterize the *spl42* mutant, the chlorophyll and carotenoid contents were measured at the seedling and heading stages. The contents of chlorophyll a and b and carotenoids in the *spl42* mutant were significantly lower than those of the wild type at both seedling and heading stages (Figure 1I,J). The mature *spl42* mutant plants displayed a significant reduction in the plant height, panicle length, flag leaf length, and seed setting rate compared with the wild type (Table A1).

Figure 1. Phenotypic characterization of the *spl42* mutant. (**A**,**B**) Individual images of wild type (WT, left) and *spl42* (right) at the four-leaf stage. Scale Bar = 5 cm. (**C**,**D**) Field phenotypes of wild type (left) and *spl42* (right) two weeks after transplanting into the field. (**E**,**F**) Images of individual plants (left) and flag leaves (right) of wild type and *spl42* at heading stage. Scale Bar = 10 cm. (**G**,**H**) Images of individual plants (left) and flag leaves (right) of wild type and *spl42* at the mature stage. Scale Bar = 10 cm. (**I**,**J**) The chlorophyll contents of wild type and *spl42* at the seedling (left) and heading stages (right). Error bars represent ± SD ($n = 3$). (**K**) Seedling phenotypes of wild type and *spl42* under different temperature treatments. Scale Bar = 5 cm. (**L**) The chlorophyll content of wild type and *spl42* under different temperature treatments. Chl a: Chlorophyll a; Chl b: Chlorophyll b; Car: total carotenoids. Error bars represent ± SD ($n = 3$). Asterisks indicate a significant difference between the wild type and *spl42* plants by Student's *t*-test, ** $p < 0.01$.

Since temperature contributes to the formation of lesion-like symptoms [13], we investigated the phenotype of the *spl42* mutant under three temperature gradients (20 °C, 25 °C, and 30 °C; see the Materials and Methods section for the treatment conditions) and measured the chlorophyll contents. The *spl42* mutant under high-temperature conditions (30 °C) exhibited more severe red-brown spots than the wild type, accompanied by growth inhibition. In agreement with the lesion phenotype, chlorophyll a, chlorophyll b, and the carotenoids in the mutant were decreased significantly (Figure 1K,L). On the contrary, the *spl42* exhibited a leaf color similar to that of the wild type at low temperatures (20 °C and 25 °C) with no difference in the chlorophyll content (Figure 1K,L). These results suggested that *spl42* was sensitive to high-temperature stresses.

2.2. Chloroplast Structures and Related Gene Expression of the spl42 Mutant

Considering that the *spl42* spotted leaf phenotype might be related to the abnormal development of chloroplasts, we observed the chloroplast ultrastructure of four-leaf seedlings by transmission electron microscopy (TEM). In wild-type seedlings, the chloroplasts developed normally and the thylakoid structures were intact and closely arranged (Figure 2A,B). However, two types of chloroplasts were observed in mutant leaves: the first type was from the green area of the spotted leaf where the chloroplasts were larger than those of the wild type, and the thylakoids were relatively loosely arranged (Figure 2C,D). The second type was the spotted leaf part of the *spl42* mutant, where the chloroplasts were broken and degraded (Figure 2E,F).

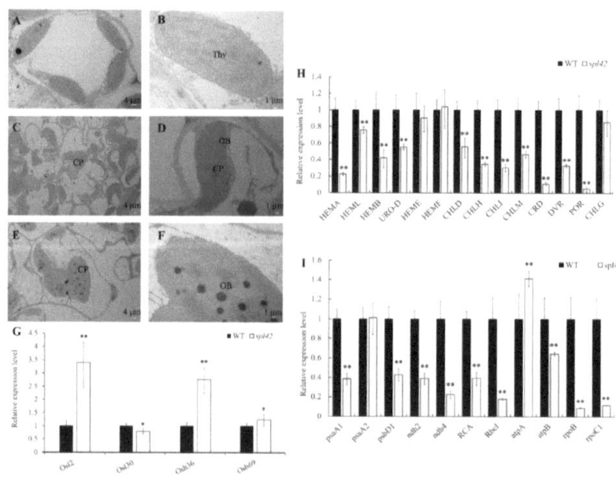

Figure 2. Transmission electron microscopic observation of the wild type (WT) and the *spl42* mutant and related gene expression analysis. (**A**,**B**) Chloroplast ultrastructure of wild-type leaves at four-leaf stage. Scale Bar = 4 μm in (**A**) and 1 μm in (**B**). (**C**,**D**) Ultrastructure of the chloroplast in green part of leaves of *spl42* at four-leaf stage. Scale Bar = 4 μm in (**C**) and 1 μm in (**D**). (**E**,**F**) Ultrastructure of the chloroplast in the spotted part of *spl42* at the four-leaf stage. Thy: thylakoid lamellar; CP: chloroplast; OB: osmophilic. Scale Bar = 4 μm in (**E**) and 1 μm in (**F**). (**G**) Relative expression analysis of the genes related to leaf senescence in the wild type and the *spl42* mutant. Error bars represent ± SD (n = 3). (**H**) Relative expression analysis of the genes related to chloroplast development in the wild type and the *spl42* mutant. Error bars represent ± SD (n = 3). (**I**) Relative expression analysis of the genes related to chlorophyll biosynthesis in the wild type and the *spl42* mutant. Error bars represent ± SD (n = 3). Asterisks indicate a significant difference between WT and the *spl42* plants by Student's *t*-test, ** $p < 0.01$; * $p < 0.05$.

Based on the spotted leaf phenotype of the *spl42* mutant, we used qRT-PCR to analyze the expression of genes related to leaf senescence in wild-type and *spl42* mutant plants at the four-leaf seedling stage. Except for *Osl30*, the expression levels of other leaf senescence genes were significantly up-regulated (Figure 2G). We also analyzed the expression of genes related to the chloroplast development. Most of the genes involved in the chloroplast development were significantly down-regulated, such as *psaA1* (encoding photosystem I subunit A1), *psbD1* (encoding photosystem I subunit D1), *ndh2* (encoding NADH dehydrogenase subunit 2), *ndh4* (encoding NADH dehydrogenase subunit 4), *RCA* (encoding rubisco activase), *RbcL* (encoding rubisco large subunit), *atpB* (encoding ATP synthase subunit B), *rpoB* (encoding RNA polymerase subunit B), and *rpoC1* (encoding RNA polymerase subunit C1) (Figure 2H). These results were consistent with the inhibition of chloroplast development in the *spl42* mutant. In addition, the reduced expression levels of the genes involved in the chlorophyll synthesis indicated that the chlorophyll synthesis was severely blocked in the *spl42* mutant (Figure 2I).

2.3. ROS Accumulation Is Elevated in the spl42 Mutant

Previous studies demonstrated that leaf spot symptoms were associated with a hypersensitive response (HR), accompanied by the accumulation of ROS [13]. Diaminobenzidine (DAB) could react with hydrogen peroxide (H_2O_2) to form a dark brown polymerization product, while nitro-blue tetrazolium (NBT) could be restored to a dark blue insoluble substance by O^{2-} [10,11]. Therefore, we used DAB and NBT staining to detect the ROS accumulation level in the leaves of wild type and *spl42* mutant seedlings. The DAB staining showed that the H_2O_2 in the mutant was obviously accumulated (Figure 3A,B). Similarly, the NBT staining indicated that the mutant had a higher level of superoxide anions (O^{2-}) accumulation compared with the wild type (Figure 3C,D). In addition, qRT-PCR was performed to analyze the expression of some genes involved in the ROS scavenging system, including *APX1* and *APX2* (encoding two isoforms of ascorbate peroxidase genes), *SOD-A* (encoding superoxide dismutase gene), *CATA* and *CATB* (encoding two isoforms of catalase genes), *AOX1a*, *AOX1b*, and *AOX1c* (encoding three isoforms of alternative oxidase gene). Our results showed that the expression levels of *CATB*, *AOX1a*, and *AOX1b* were significantly increased compared to the wild type (Figure 3E), indicating that the ROS levels in the *spl42* mutant was much higher than those in the wild type. The up-regulated expression of genes related to the antioxidant system might be a protective mechanism for the normal plant growth against physiological stress conditions.

At the same time, plant leaves were stained with trypan blue for the detected cell activity. The results showed that the *spl42* mutant displayed a lot of lesion spots, while the wild type did not show any lesions (Figure A1A,B). The gene expression associated with programmed cell death indicated that two metacaspase gene family members were significantly upregulated (Figure A1C). Correspondingly, the content of relevant reactive oxygen removal enzymes and reactive oxygen related substances were mostly elevated in the *spl42* mutant (Table A2).

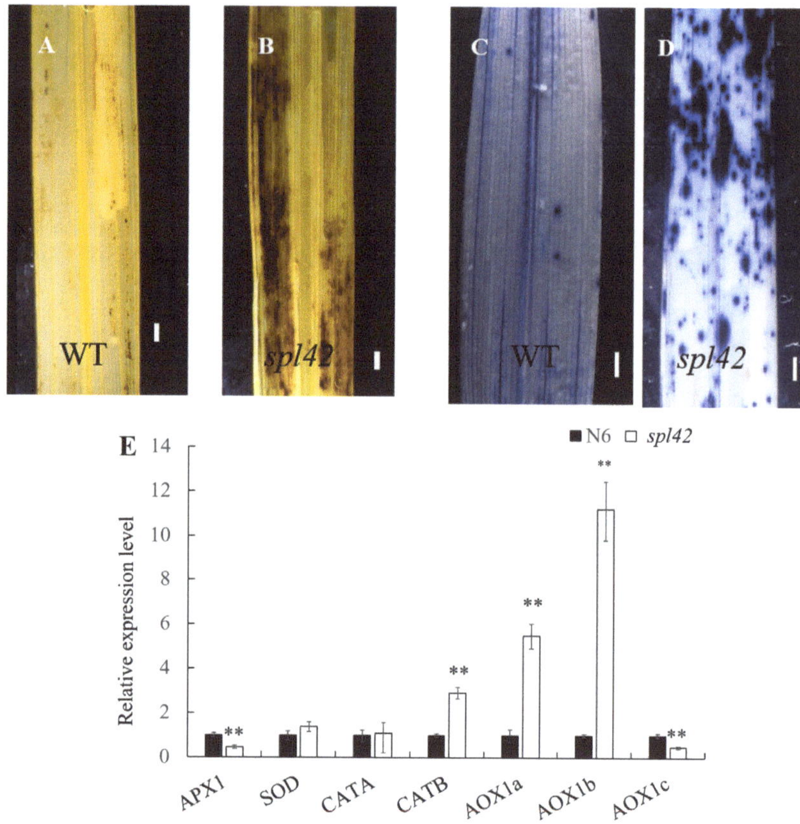

Figure 3. Comparison of reactive oxygen species (ROS) activity between wild-type and *spl42* mutants and expression analysis of genes related to ROS scavenging. (**A**,**B**) Hydrogen peroxide (H_2O_2) was visualized by diaminobenzidine (DAB) staining. Scale Bar = 0.1 cm. (**C**,**D**) Superoxide anion (O^{2-}) was visualized by nitro-blue tetrazolium (NBT) staining. Scale Bar = 0.1 cm. (**E**) qRT-PCR analysis of the genes related to ROS scavenging systems. Error bars represent ± SD (n = 3). Asterisks indicate a significant difference between NingJing6(N6) and the *spl42* plants by Student's *t*-test, ** $p < 0.01$.

2.4. Elevated Expression of Defense Response Genes

The activation of defense responses and an enhanced disease resistance have been reported in a number of lesion-mimic mutants [26,27]. We first analyzed the expression levels of endoplasmic reticulum (ER) chaperone genes consisting of *OsCNX1* (encoding cofactor for nitrate reductase and xanthine dehydrogenase1), *OsBip1* (encoding an ER molecular chaperone protein), and *OsPDIL1-1* (a marker indicator of unfolded protein accumulation in the ER caused by ROS accumulation) [28]. The expression levels of these genes in the *spl42* mutant increased by 1.7-fold, 0.8-fold, and 2.9-fold, respectively, compared with the wild type (Figure 4A).

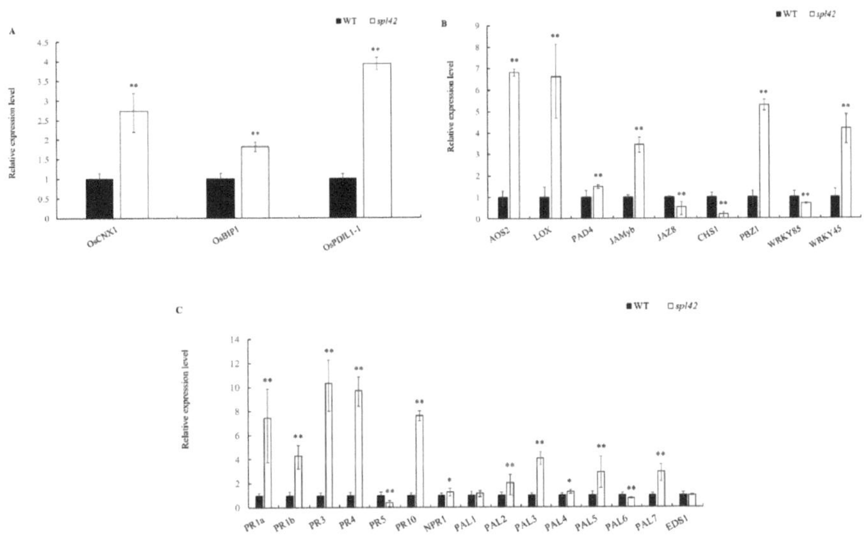

Figure 4. Expression analysis of defense-related gene families in wild type (WT) and *spl42*. (**A**) Endoplasmic reticulum chaperone gene qRT-PCR analysis. (**B**) qRT-PCR analysis of JA signaling genes. (**C**) qRT-PCR analysis of SA signaling genes. Error bars represent ± SD (n = 3). Asterisks indicate a significant difference between the WT and *spl42* plants by Student's *t*-test, ** $p < 0.01$; * $p < 0.05$.

We further analyzed the changes in the expression of the defense-related genes involved in jasmonic acid (JA) and salicylic acid (SA) signaling and/or biosynthesis. Compared with the wild type, the expression levels of JA signaling genes including *AOS2* (encoding propylene oxide synthase), *LOX* (encoding lipoxygenase), *PAD4* (encoding plant anti-toxin protein), *JAMyb* (encoding R2R3-type MYB transcription factor), *PBZ1* (encoding disease-course related protein), and *WRKY45* (encoding WRKY transcription factor) were significantly elevated in *spl42*, whereas the expression of *JAZ8* (encoding a jasmonate ZIM structural domain protein), *CHS1* (encoding Chalcone synthetase1), and *WRKY85* (encoding a WRKY transcription factor) was significantly lower (Figure 4B). In contrast, most SA signaling genes *PR1a*, *PR1b*, *PR3*, *PR4*, and *PR10* (encoding pathogenesis-related proteins), *NPR1* (non-expressor of pathogenesis-related genes 1), and *PAL2*, *PAL3*, *PAL4*, *PAL5*, and *PAL7* (encoding phenylalanine deaminase genes) were highly up-expressed in *spl42* compared with the wild type (Figure 4C).

2.5. Map-Based Cloning of the SPL42 Gene

An F2 population derived from the cross between the *spl42* mutant and N22 was used to map the *SPL42* locus. The segregation in the F2 population was 754 plants with normal (wild-type) leaves, and 217 with reddish-brown spotted leaves, corresponding to a 3:1 ratio (754:217, $\chi^2 = 1.89 < \chi^2_{0.05,1} = 3.84$; $p = 0.1692 > 0.05$), indicating that the spotted leaf phenotype in *spl42* was controlled by a single recessive-nuclear gene. The *SPL42* locus was initially mapped by 10 recessive reddish-brown spotted leaf individuals between the markers RM6357 and RM5622 on the short arm of chromosome 2 (Figure 5A). Based on an analysis of 217 plants with a recessive (mutated) phenotype, *SPL42* was delimited to a 123 kb genomic region between markers 2-3.6-3 and 2-3.7-3 containing fourteen open reading frames (ORFs) (Figure 5A). The specific data and information of the fourteen open reading frames were listed in the table through a website prediction (http://plants.ensembl.org/info/about/collaborations/gramene.html, accessed on 15 December 2022) (Table A3). Sequencing analysis showed that only the second exon of ORF9 (*Os02g0168800*) had a single-base C to T substitution, resulting in the conversion of alanine to valine

(Figure 5B–D). The sequencing of other thirteen candidate genes revealed no base changes between the *spl42* mutant and its wild type. Taken together, these results indicated that *Os02g0168800* was the *SPL42* candidate gene.

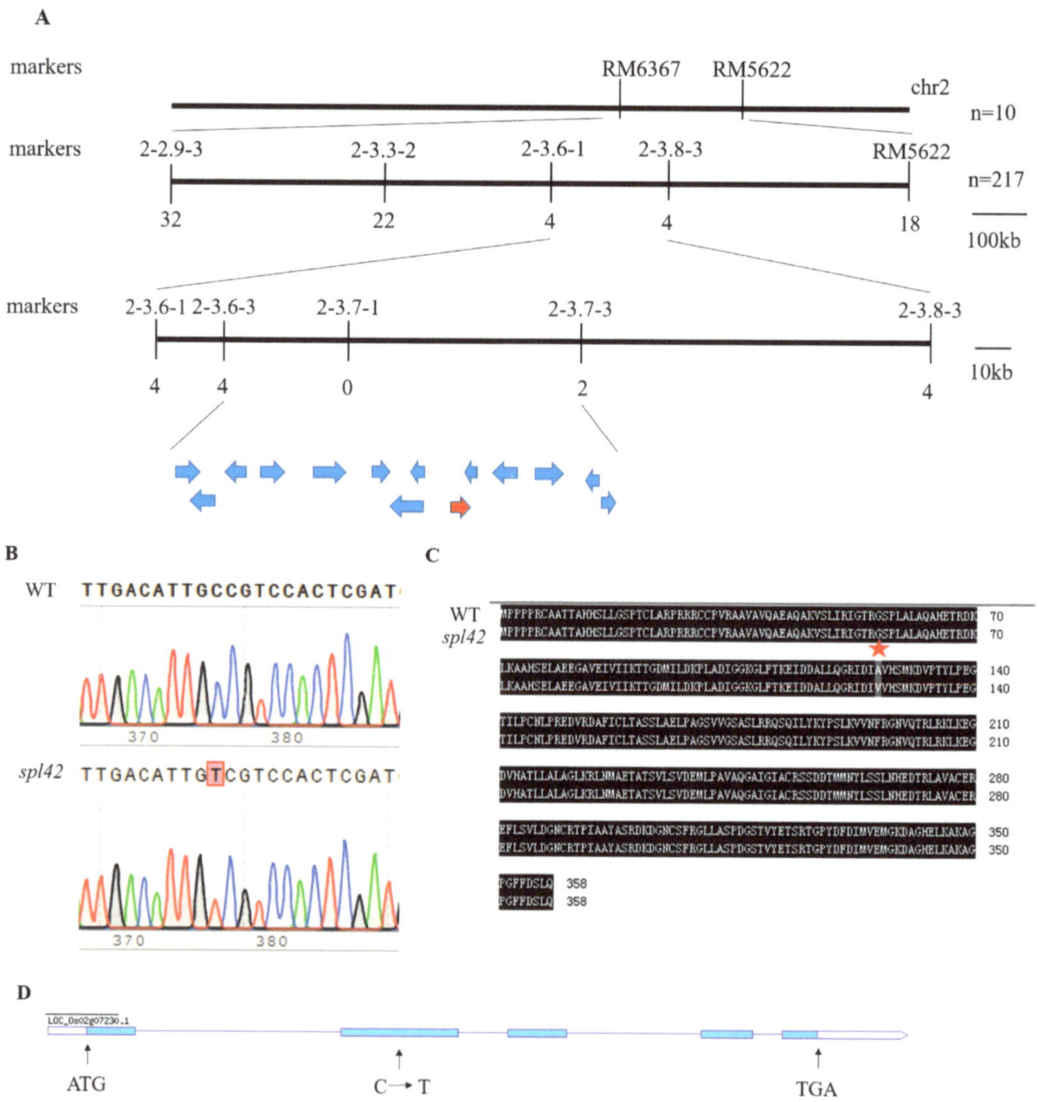

Figure 5. Map-based cloning of the *SPL42* gene. (**A**) The *SPL42* locus was delimited to a 123 kb genomic region between markers 2-3.6-3 and 2-3.7-3 on the short arm of chromosome 2. Fourteen open reading frames (indicating by arrow) were predicted in the region. Numbers below the chromosome indicate the numbers of recombinant individuals in gene mapping. (**B**) Sequencing results of the wild type (WT) and the *spl42* mutant. The mutated nucleotide was boxed. (**C**) Amino acid sequence alignment of the wild type and *spl42* mutant. A star indicates the putative amino acid change between the wild type and *spl42* mutant. (**D**) *SPL42* gene structure. ATG represent the start codon and TAG is stop codon, and the blue boxes represent exons, while the lines between the blue boxes represent introns. The single base substitution of C-T (arrow) causes amino acid changes.

The genomic sequence of the *Os02g0168800* gene is 2993 bp and contains five exons and four introns. To further confirm that the single-base substitution in *Os02g0168800* was responsible for the spotted leaf phenotype of the *spl42*, we generated *SPL42* transgenic complementary plants in the *spl42* mutant background. *SPL42* transgenic complementary plants recovered the green leaves phenotype similar to that of the wild type at the seedling stage (Figure 6A). In order to further clarify the function of *SPL42*, a knockout mutant of *SPL42* was generated by CRISPR/Cas9 technology in the NingJing6 background, which showing a spotted leaf phenotype similar to that of the mutant (Figure 6B). These results demonstrated that *SPL42* was a positive regulator of the leaf development in rice.

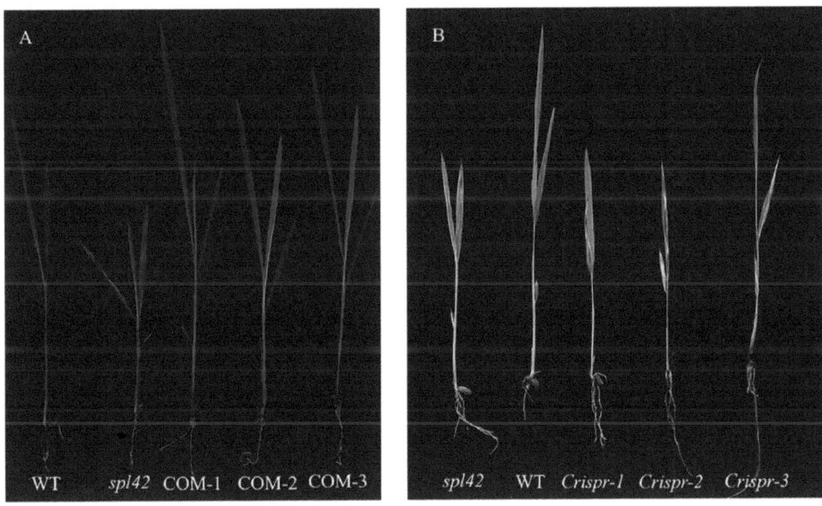

Figure 6. *SPL42* complementation and knockout transgene verification. (**A**) Phenotypes of wild type, *spl42* mutant, and transgenic complementary plants at the four-leaf stage. (**B**) Phenotypes of *spl42* mutant, wild type, and three knockout mutants Crispr-1/2/3.

2.6. Bioinformatics Analysis of SPL42

According to the functional prediction of the *SPL42* protein sequence on the NCBI website, *SPL42* encodes a PBGD, which is a key enzyme in the chlorophyll synthesis pathway. The homologues of *SPL42* were previously reported in Arabidopsis and maize species, both of which showed a lesion mutant phenotype once their normal functions were disrupted [7,8]. We compared the *SPL42* protein sequences with other proteins from different species via the NCBI website and constructed phylogenetic trees to study the evolutionary relationships between *SPL42* and its homologues (Figure 7A). The results showed that *SPL42* was 93%, 91%, 90%, and 88% homologous to its counterparts in *Eragrostis curvula* (TVU32733.1), maize (GRMZM2G041159), sorghum (Sb04g004630), and proso millet (RLM78486.1), respectively. The phylogenetic tree was divided into three branches, including monocots, dicots, bacteria, and yeast, and *SPL42* was highly conserved among monocots. Furthermore, we found only one purine choline deaminase in the rice genome, suggesting that *spl42* was a weak mutation in a single copy gene.

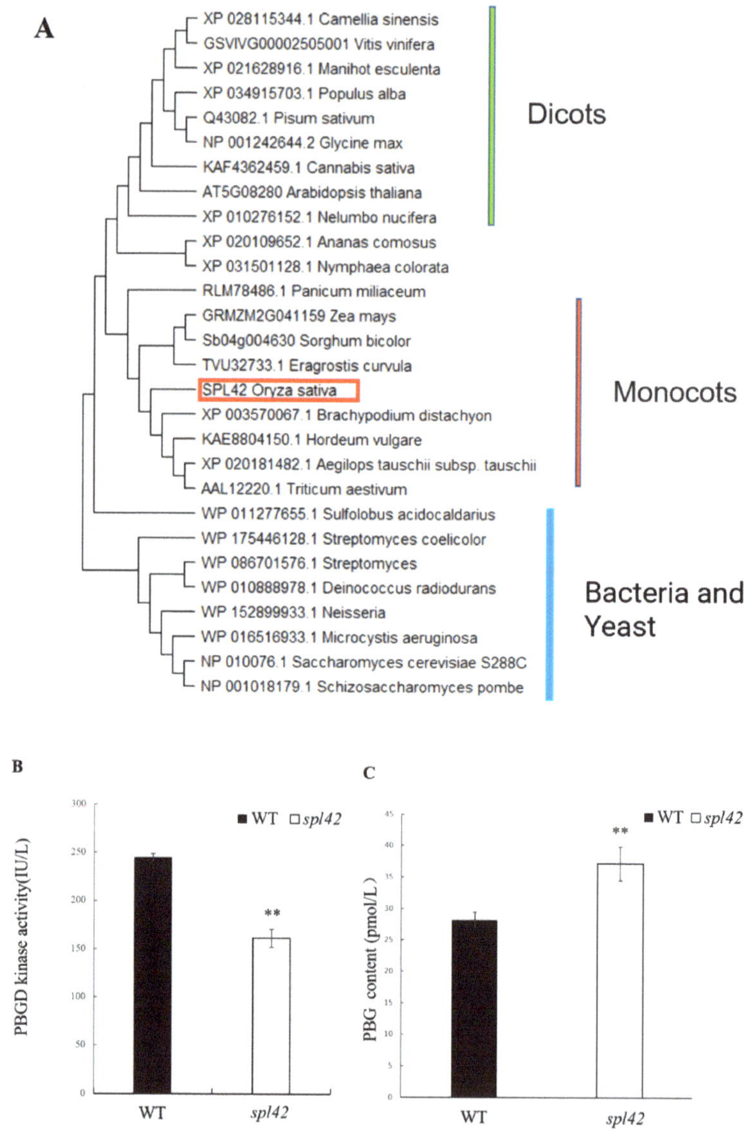

Figure 7. Phylogenetic tree analysis of *SPL42* and enzyme properties of porphobilinogen deaminase (PBGD). (**A**) Phylogenetic tree analysis of *SPL42* protein (boxed) with homologous proteins. The rooted tree using percentage identities is based on a multiple sequence alignment generated with the program MEGA-X. Scale represents percentage substitution per site. (**B**) Enzyme activity analysis of porphobilinogen deaminase (PBGD) in wild type (WT) and *spl42* mutants. Error bars represent ± SD (n = 3). (**C**) porphobilinogen (PBG) content of the precursor of enzyme reactions in wild type and *spl42* mutants. Error bars represent + SD (n = 3). Asterisks indicate a significant difference between WT and the *spl42* plants by Student's *t*-test, ** $p < 0.01$.

Considering that *SPL42* encoded a PBGD, the enzyme activity of *SPL42* and the content of the precursors for enzymatic reactions were measured. Our results showed that the enzyme activity of the *spl42* mutant was decreased significantly, whereas the content of the precursors for enzymatic reactions was increased significantly (Figure 7B,C). These results indicated that the single base change in *spl42* inhibited PBGD enzyme activity and subsequently resulted in the decrease in the chlorophyll content.

2.7. Expression Analysis of SPL42

The *SPL42* transcripts were detected in the roots, stems, young leaves, mature leaves, leaf sheaths, and developing panicles, indicating that *SPL42* was a constitutively expressed gene (Figure 8A). The expression level of *SPL42* was the highest in mature leaves and relatively low in root tissues, consistent with its role in photosynthesis.

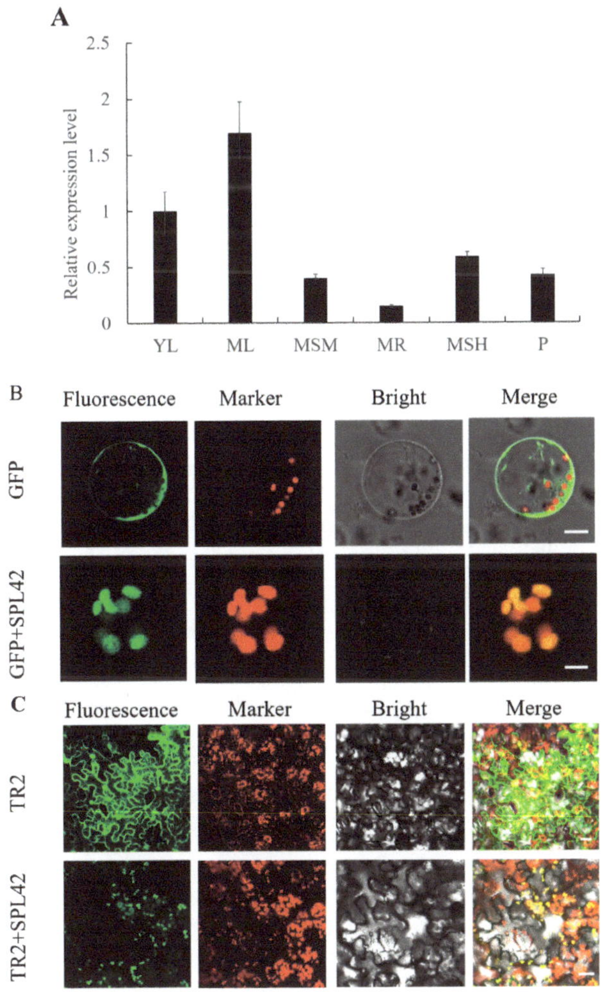

Figure 8. Expression pattern and subcellular localization of *SPL42*. (**A**) Relative expression levels of *SPL42* in the different tissues. YL: young leaf at the four-leaf stage; ML: mature leaf; MSM: mature stem; MR: mature root; MSH: mature sheath; P: panicle. Error bars represent + SD (n = 3). (**B**) Subcellular localization of SPL42 in rice protoplasts. (**C**) Subcellular localization of SPL42 in tobacco leaves. TR2 indicates SPL42-GFP fusion vector was driven by an octopine type Ti-plasmid right T-DNA gene 2′ promoter (TR 2). Signals from GFP fluorescence, chlorophyll autofluorescence, bright field, and merged images were shown. Scale Bar = 2 μm in (**B**) and 20 μm in (**C**).

We use the online tool ChloroP (http://www.cbs.dtu.dk/services/ChloroP/, accessed on 15 December 2022) to predict the *SPL42* protein and found that its N-terminal harbored a chloroplast transit peptide, suggesting that *SPL42* may locate in the chloroplast. To confirm the subcellular localization, the full-length *SPL42* cDNA was amplified from wild-type cDNA, and was inserted in the *PAN580* plasmid and transformed into rice protoplasts. Confocal microscopy showed that the green fluorescence of the *GFP-SPL42* fusion protein colocalized with the autofluorescence signal of chlorophyll, indicating that *SPL42* was indeed located in chloroplasts (Figure 8B). Further, the *SPL42* localization to chloroplasts was verified by the transient expression of the *TR2- SPL42* fusion protein in tobacco leaves (Figure 8C).

2.8. SPL42 Interacts with OsMORF8-1 and OsMORF8-2 and Is Involved in RNA Editing

RNA editing is a correction mechanism for pyrimidine-cytidine mutations in plant organelle genomes. It was previously reported that some chlorophyll synthesis-related enzymes, including protoporphyrinogen IX oxidase 1 (PPO1) and HEMC (an Arabidopsis homologous protein of SPL42), participate in RNA editing in Arabidopsis [29,30]. Since SPL42 is localized to chloroplasts, we examined 27 reported RNA editing sites in rice chloroplasts, as reported previously [21,31]. The results showed that six loci were significantly different, including atpA-C1148, ndhD-C878, ndhF-C62, rpoB-C560, rps14-C80, and ropC2 (Figure A2). However, only one of them, i.e., ndhF-C62, caused an amino acid change in the encoding protein. These results suggested that SPL42, similar to HEMC, not only affected the chlorophyll synthesis but also participated in RNA editing in rice.

The multiple organelle RNA editing factor (MORF)/RNA-editing factor interacting protein (RIP) extensively influenced the RNA editing sites in mitochondria and plastids [24,25]. In rice, there are a total of seven MORFs, among which OsMORF1(Os11g11020) and OsMORF3(Os03g38490) proteins are localized in mitochondria, and OsMORF2-1(Os04g51280), OsMORF2-2(Os06g02600), and OsMORF9(Os08g04450) proteins are localized in chloroplasts. However, OsMORF8-1(Os09g04670) and OsMORF8-2(Os09g33480) are a mitochondrial and chloroplast dual localization [21]. We thus investigated whether SPL42 interacted with MORFs. A yeast two-hybrid (Y2H) experiment demonstrated that SPL42 physically interacted with OsMORF8-1 and OsMORF8-2 (Figure 9A). A bimolecular fluorescence complementation (BiFC) assay revealed that SPL42, OsMORF8-1, and OsMORF8-2 all localized to overlapping cellular compartments and interacted with each other to generate a GFP fluorescence in the nuclei of tobacco leaf cells (Figure 9B).

Previous studies showed that HEMC interacted with the PPR-protein OTP85 to influence RNA editing in Arabidopsis [24,30]. The OTP85 homolog in rice, i.e., OsOTP85 (Os03g0795200), also showed the interaction with SPL42 by Y2H and BiFC experiments (Figure 9A,B). These results demonstrated that the SPL42 was involved in RNA editing through an association with MORFs and PPR(s).

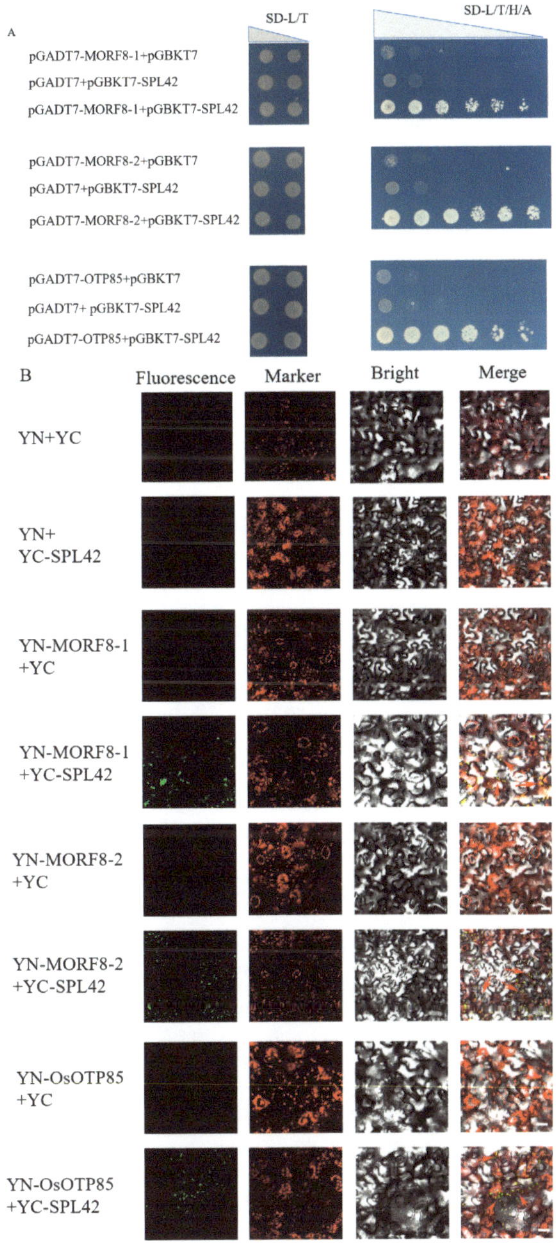

Figure 9. SPL42 physically interacts with OsMORF8-1 and OsMORF8-2. (**A**) SPL42 physically interacts with OsMORF8-1 and OsMORF8-2 in yeast. The indicated construct pairs were co-transformed into yeast strain AH109. Interactions between bait and prey were examined on control medium 2 (SD/-Leu/-Trp) and selective medium 4 (SD/-Leu/-Trp/-His/-Ade). AD: activation domain; BD: binding domain. (**B**) Bimolecular fluorescence complementation assay showing the interactions between SPL42 and OsMORF8-1 as well as OsMORF8-2 in tobacco leaf epidermal cells. Signals from GFP fluorescence, chlorophyll autofluorescence, bright field, and merged images are shown. The arrows indicate signal areas with a strong interaction. Scale Bar = 20 μm.

2.9. RNA-Sequencing Analysis

To comprehensively understand the effect of the *spl42* mutant on the gene expression, we selected the leaves of wild-type and *spl42* mutant plants at the four-leaf stage for RNA-seq analysis. A summary of the transcriptome sequencing data can be found in the attached Figure A3. It was noteworthy that the RNA-seq results were consistent with the aforementioned RT-PCR analyses. A total of 2772 differentially expressed genes (DEG) were detected, of which 1777 genes were up-regulated and 995 genes were down-regulated in the *spl42* mutant (Figure 10A). In order to further verify the reliability of RNA-seq data, we randomly selected 10 up-regulated and 10 down-regulated genes for qRT-PCR analysis. As predicted, the results of qRT-PCR were generally consistent with those of RNA-seq analysis (Figure 10B,C). The clustering analysis of differentially expressed genes showed that most significant gene expression differences were concentrated in the light harvesting and pigment synthesis pathways in photosystem I (Figure 10D). It was reasonable that the expression changes in these genes in *spl42* led to the failure of the photosynthesis pathway. Furthermore, we found that the expression levels of ten genes related to porphyrin, the chlorophyll metabolism, and the carotenoid biosynthesis pathway, twenty-five genes related to the photosynthesis pathway were down-regulated, whereas seventeen genes related to peroxisome (a primary organelle associated with ROS production) were significantly up-regulated (Figure A3). Moreover, our results of the RNA-seq indicated that most of the genes associated with a disease resistance were up-regulated in *spl42*. Collectively, the RNA-seq analysis revealed that the *spl42* might promote a ROS production by affecting the efficiency of chlorophyll synthesis and photosynthesis.

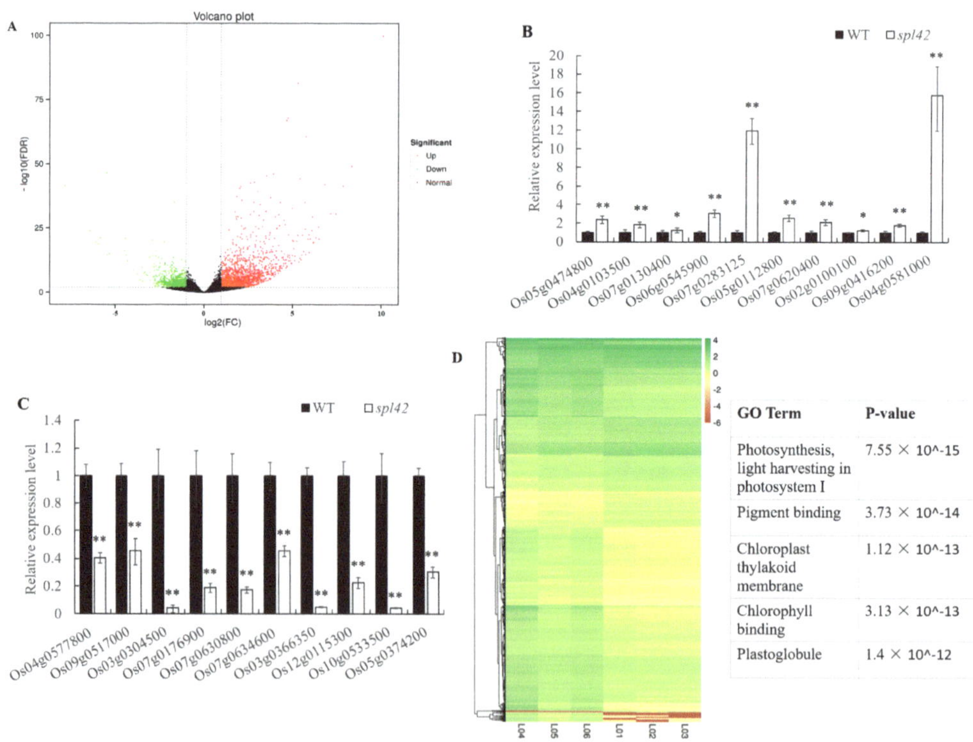

Figure 10. RNA-sequencing related data analysis of the wild type (WT) and the *spl42* mutant. (**A**) Distribution of volcano plot of wild type and *spl42* mutant. Volcano plot shows the differential expression of genes in accordance with fold difference and significant levels, and the criteria for significant differentially

expressed gene screening were: $|\log^2\text{Ratio}| \geq 1$ and q-value ≤ 0.05. (**B**,**C**) qRT-PCR analysis of 10 up-regulated and 10 down-regulated genes which are selected randomly from RNA-sequencing results. Asterisks indicate a significant difference between WT and the *spl42* plants by Student's *t*-test, ** $p < 0.01$; * $p < 0.05$. (**D**) Cluster analysis of differentially expressed genes in wild type and *spl42* mutants.

3. Discussion

3.1. *spl42* Is a Temperature-Sensitive Mutant

The normal formation and development of chloroplasts play a decisive role in breeding high-yield crops. The discovery of leaf color mutants provides an ideal material for studying the chloroplast metabolism, development, and photosynthesis, as well as molecular mechanisms of chlorophyll biosynthesis [32]. Among the reported mutants of rice leaf spots, the lesion-like phenotype is largely affected by environmental conditions such as the temperature, light, and humidity [33]. For example, the formation of diseased spots in the *spl7* mutant was induced by a high temperature and ultraviolet radiation [14]. Conversely, leaf necrosis was inhibited by a high temperature and long day in *OsLSD1* antisense plants [33]. In this study, we identified the high-temperature inducible mutant *spl42*, which is similar to *spl7*. The phenotype of *spl42* was the most obvious under high temperature conditions (30 °C) (Figure 1K) and, correspondingly, the chlorophyll content was decreased significantly (Figure 1L). In addition, when *spl42* was growth in the field of Hainan, China (a natural low temperature condition), the leaf spot formation was later than its wild type (data not shown). The lesion spots in *spl42* also resulted in a significant decrease in the photosynthetic-related parameters such as the light quantum yield and photosynthetic rate. Moreover, *spl42* exhibited other abnormal phenotypes, including the reductions in the plant height, panicle length, flag leaf length, and seed setting rate compared with the wild type (Table A1). These results suggest that *SPL42* plays a pleotropic role in rice.

3.2. Disruption of Chlorophyll Synthesis Pathway Leads to the Production of Abnormal Chloroplast

Imbalances in metabolic pathways can lead to the accumulation of intermediate metabolites, thereby inducing autoimmunity and cell death. For example, *SPL18* encodes an acyltransferase [34], and *OsSSI2* encodes a fatty acid desaturase. Mutations in both genes produced a necrotic plaque phenotype on the leaves [35]. Likewise, the mutation of the protoporphyrogen III oxidase gene *RLIN1* in the tetrapyrrole biosynthesis pathway led to a leaf spot formation and programmed cell death [36,37]. In our study, the defect of *SPL42* caused a significant reduction in not only enzyme activity but also the precursor deposition of the PBGD enzyme reactions (i.e., porphobilinogen) in rice plants (Figure 7B,C). Since porphobilinogen is one of the tetrapyrrole, we hypothesized that the tetrapyrrole intermediates in *spl42* were photoactivated and oxidized by light, resulting in the accumulation of reactive oxygen intermediates and photodynamic destruction. Moreover, the TEM analysis showed that there were two types of chloroplasts in *spl42*: one kind of chloroplasts was bigger in size than the wild type, and its thylakoid arrangement was relatively loose, while the chloroplast of the other one was broken and degraded and could not produce the complete thylakoid lamella structure (Figure 2A–F). These results suggested that the phenotype of leaf color mutants was closely related to the development of chloroplast and chlorophyll biosynthesis. RNA-sequencing results further confirmed the down-regulation of porphyrin, the chlorophyll metabolism, and carotenoid biosynthesis genes in *spl42*, consistent with our results that chloroplast numbers are crucial for ensuring the chloroplast's structure.

3.3. The Disorder of ROS Scavenging System May Result in the Enhancement of Defense Response in *spl42*

Maintaining the homeostasis of ROS is pivotal for normal plant growth and defense response. In many spotted leaf mutants, the balance of ROS was broken, leading to the disturbance of a plant development [15–17]. Under light conditions, chloroplasts in plant cells are a major source of ROS and have been shown to control programmed cell death in response to ROS [38]. Chloroplast damage is caused by oxidative bursts that lead to the excessive deposition of material, such as plastoglobules and callose, in the chloroplast [38]. In our study, the levels of ROS were abnormal in the *spl42* mutant. DAB and NBT staining experiments confirmed the excessive accumulation of H_2O_2 and O^{2-} in the mutant (Figure 3A,B). In addition, the expression of SOD, POD, and MDA of the ROS scavenging system were mostly up-regulated in *spl42* (Table A2). These results suggested that the disruption of ROS homeostasis might be one of the reasons for the formation of the spotted leaf. More than 80% of the reported spotted leaf mutants showed an enhanced resistance to rice blast or bacterial blight [11,39–42]. For example, the *spl40* showed a higher resistance to bacterial blight strains PXO79, PXO145, C5, and OS225 than its wild type [43]. The resistance of the *spl35* to grisea and bacterial strains CH43, CH680, and PXO61 was also significantly enhanced [27]. The RNA-Seq analysis revealed that the genes related to the disease resistance were upregulated in *spl42*, which was consistent with the changes in the expression levels of the defense genes related to the SA and JA pathways (Figure 4A–C).

3.4. *spl42 Is a Temperature-Sensitive Mutant*

Previous reports indicated that Arabidopsis HEMC was involved in RNA editing by interacting with MORF family proteins, and PPO1 is involved in RNA editing as an enzyme of the chlorophyll synthesis pathway [29,30]. We examined 27 reported RNA editing genes in rice chloroplasts. Of these, six differed in nucleotides, including *atpA-C1148*, *ndhD-C878*, *ndhF-C62*, *rpoB-C560*, *rps14-C80*, and *ropC2*, but only one of these *(ndhF-C62)* underwent a non-synonymous base change (Figure A2). In addition, yeast two-hybrid experiments and bimolecular fluorescence complementation experiments showed that SPL42 interacted with OsMORF8-1 and OsMORF8-2, as well as OsOTP85 (Figure 9A,B). Thus, SPL42 might be involved in RNA editing with OsMORFs and PPR protein(s). Although a model was proposed to decipher the association of MORFs, PPRs, and HEMC [30], the detailed molecular mechanism between SPL42 and RNA editing needs to be uncover in future.

4. Materials and Methods

4.1. *Plant Materials and Growth Conditions*

The rice *spl42* mutant was isolated from rice cultivar (cv.) NingJing6. An F2 population derived from a cross between *spl42* and N22 was used for the genetic analysis. Wild-type and *spl42* mutant plants were grown in Nanjing (Jiangsu province) and Lingshui (Hainan province) under natural conditions. For temperature-sensitive treatments, plants in the growth chamber (GXM-258B, Ningbo) were treated with a 16 h light/8 h darkness photoperiod at a constant temperature of 20 °C, 25 °C, or 30 °C, respectively [44]. The thousand-grain weight, seed length, and width were examined by a seed phenotyping system (SC-G automatic test seed analysis software, Hangzhou). At the maturity of the wild type and *spl42*, the plant height, number of tillers, spike length, and leaf length and width were investigated. At seed harvest, agronomic traits such as the number of branches, seed set, number of grains per spike, grain length, grain width, and thousand-grain weight were investigated. The photosynthetic rate and other photosynthetic indicators of the wild type and *spl42* mutant were measured at the maximum tillering stage using a portable photosynthesizer (Li6400XT, LI-COR, USA) and an ultra-portable modulated chlorophyll fluorometer (MINI-PAM, WALZ, Germany). Twenty plants were surveyed for each trait and the mean was taken.

4.2. *Measurement of Chlorophyll Contents*

About 0.03 g of fresh leaves of the wild type and *spl42* mutant plants at the four-leaf stage were collected to determine the chlorophyll contents. The cut leaves were soaked in 95% ethanol and left out of the light for 48 h. After centrifugation, the supernatant was collected and the optical density at 665, 649, and 470 nm was measured by using a spectrophotometer. Three replicates of different samples were set up. The contents of chlorophyll a, chlorophyll b, and carotenoids were calculated according to the method described previously [45].

4.3. TEM Analysis

For TEM analysis, normal green and spotted leaves from the fourth leaf of 2-week-old wild-type and mutant plants grown in the field were cut into small pieces of 0.5 cm. The detailed method was described in the previous studies [44] with minor modifications. The leaves were fixed with 2.5% glutaraldehyde solution and vacuumed until fully sinking to the bottom, then rinsed and incubated in 1% OsO4 (PH = 7.2) overnight at 4 °C. After staining with uranyl acetate, the tissues were further dehydrated in an ethanol series and then embedded in Spurr's medium prior to ultrathin sectioning. Finally, the samples were stained again and observed with a Hitachi H-7650 transmission electron microscope (HITACHI, Tokyo, Japan).

4.4. DAB and NBT Staining

DAB can react with H_2O_2 to form a dark brown polymerization product, while NBT can be restored to a dark blue insoluble substance by O^{2-}. Therefore, the detection of O^{2-} and H_2O_2 was conducted by staining with NBT and DAB. We selected the third leaves of the *spl42* mutant and wild type at the four-leaf stage to determine the accumulation level of reactive oxygen species. The third leaves were soaked in the prepared dye, then vacuumed and stained overnight at 37 °C in darkness. The next day, the tissues were decolorized with 95% ethanol until the color no longer changed.

4.5. Trypan Blue Staining Test

Trypan blue, a staining reagent for testing the cell membrane integrity, was used to stain the third leaf of the four-leaf stage of wild type and mutant *spl42* leaves grown normally under field conditions. Dead cells were stained dark blue and normal cells were barely stained, as previously reported [28].

4.6. Determination of Reactive Oxygen Scavenger Enzyme Activity and Related Indexes

In order to study the accumulation of ROS in the mutant *spl42*, the third leaves of the *spl42* and wild type were selected under field conditions, and the enzymatic activities of CAT, SOD, and POD, as well as the content of H_2O_2 and MDA, were measured using kits from Jiancheng Biological Co., Ltd., Nanjing, Jiangsu, China (http://www.njjcbio.com/, accessed on 15 December 2022).

4.7. Map-Based Cloning of SPL42

For gene mapping, F1 was obtained by hybridization with the *spl42* mutant as the female parent and indica cultivar N22 as the male parent. Then, F1 plants were self-crossed to obtain F2 generation. The F2 populations were used for mapping the *SPL42* gene. Ten *spl42*-like individual DNA of the F2 population were selected for linkage analysis of the whole genome. Then, additional individuals were used to narrow down the *spl42* locus. Primers for a fine localization were listed in Table A4. An SSR primer amplification system (10 μL): template DNA (20 ng/μL) 1 μL, primer (2 mmol/L) 1 μL, mix 4 μL, ddH_2O 4 μL. The amplification procedure was as follows: pre-denaturation at 94 °C for 10 s, denaturation at 98 °C for 30 s, annealing at 58 °C for 30 s, and extension at 72 °C for 40 s. The denaturation–annealing–extension steps were carried out for 34 cycles, and the last step was 72 °C for 10 min and storing at 4 °C.

4.8. Transgene Constructs and Plant Transformation

For the complementation experiments, a fragment including the promoter region (2 kb) and full-length cDNA of *SPL42* in the wild type was amplified with the HB-F and HB-R prime pairs. Then, the correct PCR product was inserted to the pCUBi1390-Flag vector. The resultant recombinant construct *pCUbi1390-SPL42* was introduced into the *spl42* mutant by *Agrobacterium*-mediated transformation using the *Agrobacterium tumefaciens* strain EHA105 [46]. For the construction of transgenic knockout, a 20 bp fragment (CCTTGCCACCTATATCTGCC) targeted to *SPL42* was designed using the CRISPR-Cas9 website tool (http://cbi.hzau.edu.cn/cgi-bin/CRISPR, accessed on 15 December 2022). The targeted fragment was inserted into the CRISPR/Cas9 binary vector pCAMBIA1305.1 carrying the CaMV35S promoter. The sequenced construct was introduced into the Japonica rice variety NingJing6. The primers used for vector construction were listed in Table A4.

4.9. Enzyme Activity Assay In Vivo

About 0.5 g of fresh leaves from the wild-type and the *spl42* mutant seedlings were weighed and mixed with 10 mL of phosphate buffer (PBS, pH = 7.4) after grinding in liquid nitrogen. Then, after ultrasonic crushing for 5–10 min, the supernatant was aspirated by centrifugation and prepared for the determination of the enzyme activity. The test method was according to the manual of plant CAT, SOD, PAD, and PBGD ELISA (enzyme-linked immunosorbent assay) kits (Shanghai Zhen Ke Biological Tecchnology Co., Ltd., http://www.shzkbio.com/, accessed on 15 December 2022).

4.10. RNA Preparation and Reverse Transcription Quantitative PCR (RT-qPCR)

Total RNA was extracted from the wild type and the *spl42* mutant at different leaf stages (the three-leaf to four-leaf stage) or different tissues. The detailed extraction protocol followed the instruction of an RNA Prep Pure Plant kit (TIANGEN Biotech, Beijing, China, http://www.tiangen.com/en/, accessed on 15 December 2022). The total RNA (2 μg) was used for a single-strand cDNA synthesis using Superiorscript Reverse Transcriptase (Enzynomics, http://www.enzynomics.com, accessed on 15 December 2022) and oligo(dT) primers. Three technical replicates on three biological replicates were carried out for each sample. qRT-PCR was conducted using PikoReal real-time PCR (ThermoFisher Scientific, https://www.thermofisher.com, accessed on 15 December 2022) with SYBR Green Premix Ex TaqII (Takara, https://www.takarabio.com, accessed on 15 December 2022) according to the manufacturer's instructions. The primers for the qRT-PCR were listed in Table A4, and the *UBQ5* (ubiquitin5) was used as an internal control to normalize the RT-qPCR results. Data were analyzed and processed using the $2^{-\Delta\Delta CT}$ method [44,47].

4.11. Validation of Intrachloroplast RNA Editing

The total RNA was isolated using the RNAprep pure Plant Kit (Tengen Biochemistry Ltd.) and first-strand cDNA was synthesized using the PrimeScript II (TaKaRa Inc., Dalian, China) reverse transcription kit and random hexamer primers (TaKaRa). The editing efficiency of RNA was primarily measured by sequencing the RT-PCR product according to previously reported procedures [48,49]. The level of RNA editing at each site was measured by the relative height of the nucleotide peaks in the sequencing results. A statistical analysis was performed using four replicates.

4.12. Subcellular Localization Experiment

For a subcellular localization, the cDNA sequence of *SPL42* was amplified from the wild type and introduced into the vector pCAMBIA1305.1-GFP in frame with the N terminus of GFP. The recombinant vector was transformed into Agrobacterium EHA105 and subsequently transformed into the 5- to 6-week-old tobacco leaves by an A. tumefaciens-mediated method as described previously [50]. Similarly, the full-length cDNA as cloned into the transient expression vector pAN580 with the CaMV35S promoter. The recombinant vector was used to transform rice protoplasts as described previously [51]. GFP fluorescence signals were observed with a confocal laser scanning microscope (LSM780, Carl Zeiss, Germany) [50]. The primers for a subcellular localization were listed in Table A4.

4.13. Bimolecular Fluorescence Complementation Assay

The full-length cDNA fragments of *SPL42* and *OsMORF8-1* and *OsMORF8-2* and *OsTP85* were cloned into the linearized pSPYNE173 and Pspyce (M) vectors, respectively. The primers used are listed in Table A4. Different combinations of Agrobacterium containing the above plasmids were co-transformed into *Nicotiana benthamiana* leaves [50]. After 48 h, YFP fluorescence was observed in the leaves using an LSM780 (Carl Zeiss, Oberkoche, Germany) confocal laser scanning microscope.

4.14. Yeast Two-Hybrid Assay

The full-length coding region of *SPL42* was cloned into the pGBKT7 vector as prey. Then, the full-length coding region of *OsMORF8-1*, *OsMORF8-2*, and *OsTP85* were cloned into the pGADT7 vector as the bait plasmid vectors. For the interaction test, each bait construct was co-transformed with each prey construct into the yeast strain AH109, plated on SD/-Trp-Leu medium, and grown at 30 °C for 2 days. Then, co-transformed yeast clones were serially diluted (1:10, 1:100, 1:1000) and spotted and grown on SD/-Leu/-Trp/-His/-Ade medium at 30 °C for 4 days. Empty vectors were co-transformed as the negative controls. The primers used are listed in Table A4.

4.15. RNA-Sequencing Analysis

mRNA was purified from the total RNA that was extracted from the third leaves of the wild-type and the *spl42* seedlings and sent to Beijing Biomarker Biotechnology Co. Ltd. (http://www.biomarker.com.cn/, accessed on 15 December 2022) for sequencing and analysis. The criteria for significant differentially expressed gene screening were: $|\log^2\text{Ratio}| \geq 1$ and q-value ≤ 0.05. The gene ontology (GO) functional analysis was performed on the Blast2GO program [52]. A pathway enrichment analysis was conducted using the Kyoto Encyclopedia of Genes and Genomes database [53]. All requirements for the RNA-sequencing reads will be provided in a timely manner by the corresponding author.

Author Contributions: Conceptualization, methodology, software, formal analysis: L.L. (Linglong Liu) and L.L. (Lin Liu); investigation, L.L. (Lin Liu), Y.W., Y.T., S.S., Z.W., X.D. (Xin Ding), H.Z., Y.H., S.L. and L.L. (Linglong Liu); validation, L.L. (Lin Liu) and L.L. (Linglong Liu); resources, L.L. (Lin Liu) and L.L. (Linglong Liu); data curation, L.L. (Lin Liu) and L.L. (Linglong Liu); writing—original draft preparation, L.L. (Lin Liu) and Y.W.; writing—review and editing, L.L. (Linglong Liu), X.D. (Xiaoou Dong) and J.W. supervision, L.L. (Linglong Liu) and J.W.; project administration, L.L. (Linglong Liu) and J.W.; funding acquisition, L.L. (Linglong Liu) and J.W. All authors have read and agreed to the published version of the manuscript.

Funding: This research was funded by Jiangsu key R& D project (BE2021360), National Agricultural key Sci & Tech project (NK2022050104), Key Laboratory of Biology, Genetics, and Breeding of Japonica Rice in Midlower Yangtze River, Ministry of Agriculture, P. R. China, Jiangsu Collaborative Innovation Center for Modern Crop Production.

Data Availability Statement: Sequence of the *spl42* mutant gene has been deposited in NCBI (https://www.ncbi.nlm.nih.gov, accessed on 15 December 2022) with GenBank accession number OQ186605.

Conflicts of Interest: The authors declare no conflict of interest.

Appendix A

Table A1. Agronomic traits of wild type (WT) and *spl42*.

Agronomic Traits	WT	*spl42*
Plant height (cm)	89.75 ± 1.58	85.70 ± 2.10 **
Panicle length (cm)	18.64 ± 1.54	16.92 ± 1.21 **
Effective tiller number	5.8 ± 1.33	5.75 ± 1.92
Flag leaf length (cm)	28.99 ± 2.87	24.13 ± 0.83 **
Flag leaf width (cm)	1.8 ± 0.13	1.71 ± 0.07 *
Setting rate (%)	0.92 ± 0.05	0.83 ± 0.07 **
Grain length (mm)	7.58 ± 0.37	7.57 ± 0.37
Grain width (mm)	3.1 ± 0.2	3.1 ± 0.23
Thousand grain weight (g)	26.12 ± 0.06	26.35 ± 0.24
Maximum light quantum efficiency	0.81 ± 0.01	0.80 ± 0.003
Quantum yield	626 ± 64.78	496 ± 28.28 **
Photosynthetic rate (μmol CO^2 m^{-2} s^{-1})	9.24 ± 0.57	5.45 ± 0.77 **
Conductance to H_2O (mol H_2O m^{-2} s^{-1})	0.27 ± 0.04	0.18 ± 0.01 **
Transpiration rate (mmol H_2O m^{-2} s^{-1})	5.27 ± 0.46	3.84 ± 0.07 **

For 20 replicates, analysis of difference significance is based on Student's *t*-test, * $p < 0.05$; ** $p < 0.01$.

Table A2. Peroxide related index content of wild type and *spl42*.

Peroxide Related Index	Wild Type	*spl42*
H_2O_2 content (μmol/g)	611.67 ± 7.70	1044.60 ± 13.38 **
CAT activity (U/g)	1436.05 ± 139.42	326.61 ± 18.33 **
SOD activity (U/g)	1755.19 ± 23.39	2321.24 ± 28.36 **
POD activity (U/g)	12,306.67 ± 231.71	29,966.67 ± 124.72 **
MDA content (nmol/g)	5.04 ± 0.62	8.65 ± 1.23 **

For 3 replicates, the analysis of difference significance is based on Student's *t*-test, ** $p < 0.01$.

Table A3. The putative function of fourteen different genes in 123 kb mapping region.

ORFs	Locus	Functional Description
ORF1	*Os02g0168000*	Conserved hypothetical protein
ORF2	*Os02g0168100*	Similar to 4-hydroxyphenylpyruvate dioxygenase (EC 1.13.11.27) (4HPPD) (HPD) (HPPDase)
ORF3	*Os02g0168200*	Similar to transfactor-like protein
ORF4	*Os02g0168300*	Hypothetical protein
ORF5	*Os02g0168400*	Hypothetical protein
ORF6	*Os02g0168500*	Conserved hypothetical protein
ORF7	*Os02g0168600*	Ovarian tumor, otubain domain containing protein
ORF8	*Os02g0168700*	Peptidylprolyl isomerase, FKBP-type domain containing protein
ORF9	*Os02g0168800*	Similar to Porphobilinogen deaminase (Fragment)
ORF10	*Os02g0168900*	Conserved hypothetical protein
ORF11	*Os02g0169000*	Conserved hypothetical protein
ORF12	*Os02g0169300*	Similar to Phosphoglycerate kinase, cytosolic (EC 2.7.2.3)
ORF13	*Os02g0169400*	Similar to Argonaute 4 protein
ORF14	*Os02g0169700*	Glycine cleavage H-protein family protein

Table A4. List of primer sequences (5′→3′) used in this study.

Primers for map-based cloning	
RM6367-F	CAGACAGAACAGCGGTCAAG
RM6367-R	GATGGATGGATGGATTGGAG
RM5622-F	CGAAACAAGCAGCTCTAGGG
RM5622-R	CATGTCGAATTGTGGTGAGG
2-2.9-3-F	AGGGCATGAAAGCAGCTCAA
2-2.9-3-R	TGATCGTTCCCAAGCATCCC
2-3.3-2-F	AGAACCCGGTCCTTCTGAAC
2-3.3-2-R	AAGGGAAGTCTCTGCTTGCC
2-3.6-1-F	CTGCTAGCCTCACACCTACG
2-3.6-1-R	CATCTGATGCATGCACGTCG
2-3.6-3-F	AGCCAAAGTTACAGTAGCATGGA
2-3.6-3-R	GATTGGCGCCACCAGAAAAT
2-3.7-1-F	TTGCAGGTAGCCACAGGATT
2-3.7-1-R	GTTCGTGCAGCTTGAGACCA
2-3.7-3-F	AAATCTGAGTTTATTGGGGGAGTTT
2-3.7-3-R	CAGCAACCTGGAGGCTAAAAG
2-3.8-3-F	TGTGTGTTGACGATTGCTGAA
2-3.8-3-R	AAGGCCTAAGGGGCGAAAAA
Primers for transgenic constructs	
HB-F	CCGGCGCGCCAAGCTTATCTCACTGCCAAGTACTCCT
HB-R	GAATTCCCGGGGATCCGACAGCTCAACATTGTGGTACA
Crispr-F	GGCACCTTGCCACCTATATCTGCC
Crispr-R	AAACGGCAGATATAGGTGGCAAGG
Primers for yeast two hybrid	
SPL42-BD-F	CATGGAGGCCGAATTCATGCCGCCGCCGCCGAGATG
SPL42-BD-R	GCAGGTCGACGGATCCTCATTGCAAGCTATCAAAGA
MORF8-1-AD-F	GGAGGCCAGTGAATTCATGGTGTCGGCGTCGC
MORF8-1-AD-R	CGAGCTCGATGGATCCCTACTGGTAATTCCTCCCTGGCT
MORF8-2-AD-F	GGAGGCCAGTGAATTCATGGCGTCGGCGTCG
MORF8-2-AD-R	CGAGCTCGATGGATCCCTACTGGTAATTCCTCCCTTGTCCG
OsOTP85-AD-F	GGAGGCCAGTGAATTCATGTCGTCCTCGCCTCTAACC
OsOTP85-AD-R	CGAGCTCGATGGATCCCTACCAGTAGTCACCACAAGAGCA
Primers for bimolecular fluorescence complementation (BiFC)	
YC-SPL42-F	CATTTACGAACGATAGTTAATTAAATGCCGCCGCCGCCGAGATG
YC-SPL42-R	CACTGCCACCTCCTCCACTAGTTTGCAAGCTATCAAAGAAGC
YN-MORF8-1-F	CATTTACGAACGATAGTTAATTAAATGGTGTCGGCGTCGC
YN-MORF8-1-R	CACTGCCACCTCCTCCACTAGTCTGGTAATTCCTCCCTGGCT
YN-MORF8-2-F	CATTTACGAACGATAGTTAATTAAATGGCGTCGGCGTCG
YN-MORF8-2-R	CACTGCCACCTCCTCCACTAGTCTGGTAATTCCTCCCTTGTCCG
YN-OsOTP85-F	CATTTACGAACGATAGTTAATTAAATGTCGTCCTCGCCTCTAACC
YN-OsOTP85-R	CACTGCCACCTCCTCCACTAGTCCAGTAGTCACCACAAGAGCA
Primers for subcellular localization	
PAN580-SPL42-F	CGGAGCTAGCTCTAGAATGCCGCCGCCGCCGAG
PAN580-SPL42-R	TGCTCACCATGGATCCTTGCAAGCTATCAAAGAAGCCAGGG
TR2M-SPL42-F	CACCAAATCGTCTAGAATGCCGCCGCCGCCGAGATG
TR2M-SPL42-R	TCGAGACGTCTCTAGATTGCAAGCTATCAAAGAAGC
Primers for quantitative real-time RT-PCR	
SPL42-F	TGCTTCTCGTGACAAGGATGGG
SPL42-R	ACTGTAGATCCATCAGGTGAAGCC
UBQ5-F	ACCACTTCGACCGCCACTACT
UBQ5-R	ACGCCTAAGCCTGCTGGTT
HEMA-F	GATGCAATCACTGCTGGAAAGCGT
HEMA-R	CCATCTTGCCAGCACCAATCAACA
HEML-F	AGAACAAAGGGCAGATTGCTGCTG
HEML-R	TGTTTCGTCAAGTCACGGAGAGCA
HEMB-F	TGGCATTGTCAGGGAAGATGGAGT
HEMB-R	CCAAAGCAGCACGTATTGCTCCAA

Table A4. *Cont.*

HEMC-F	TCATTCCGAGGGCTATTGGCTTCA
HEMC-R	ACACTCTAGTTGGCCAATGGTGGA
URO-D-F	AGGCTTCCACTGACAGGTGTTGAT
URO-D-R	AAAGAACGCCAGGGTCAACATTCC
HEME-F	AATGGAGGCTTGCTTGAGCGAATG
HEME-R	TTGTTACCAAGGCGTCTCCTTCCA
HEMF-F	ACTGACTGCACGATGGCAGTATGA
HEMF-R	AGAGATCGAGCCATTCCTTTGGGT
CHLD-F	TAGCACAGCTGTCAGAGTGGGTTT
CHLD-R	TTGCCAGCCACCTCAAGTATCTCA
CHLH-F	GCACGGGAACTTGGCGTTTCATTA
CHLH-R	ACATGTCCTGGAGCTGCTTCTCAT
CHLI-F	AGGGATGCTGAACTCAGGGTGAAA
CHLI-R	AAGTAGGACTCACGGAACGCCTTT
CHLM-F	GCTTCATCTCCACGCAGTTCTACT
CHLM-R	GCAATGACGAATCGAAGACGCACA
CRD-F	TGGATCTAACATGACACGCACCCA
CRD-R	ACTGTAACGGCATTCTTCTCCGGT
DVR-F	TTCTTCGAGAGGGTGATCAGGGAA
DVR-R	GAAACTGGCAATGGCAGCCAAGAA
POR-F	TCGTCGGCCTCGTCTGAGTTTATT
POR-R	AGGCCTCTCTCACTGAAAGCTGAA
CHLG-F	CCAGCCACTGATGAAAGCAGCAAT
CHLG-R	AGAGCGCTAATACACTCGCGAACA
psaA1-F	GGGAGGTGGCGAGTTAGTAG
psaA1-R	AATGCGTGAATGTGATGGAC
PsaA2-F	TTATCTTCAACGAGCGGT
PsaA2-R	TATCTCCAGGTCCTATTGTT
psbD1-F	CTGCTACTGCTGTTTTCT
psbD1-R	GATGTTATGCTCTGCCTG
ndh2-F	ATCACTGTAGGACTTGGGTT
ndh2-R	TTTCCAGAAGAAGATGCC
ndh4-F	TCCTTATTGCTTATGCTGTC
ndh4-R	CCGTATGCTCCCATCTTTA
RCA-F	AAGATGACGATCGAAAAGC
RCA-R	AGTACTGCTCAGCCAGCTG
RbcL-F	AATGCGACTGCAGGTACA
RbcL-R	GTCATGCATTACAATAGGAA
atpA-F	TCAAAAAGGGCAAGATGT
atpA-R	TTGTAATGTAGCAGGGGAAT
atpB-F	ATGAATGTTATTGGTGAGCCA
atpB-R	CTGTTCTGTGGCTTGCTCAA
rpoB-F	GTGGGGAACTTGCTTTAGG
rpoB-R	GCTTGTTGTATCCGTCTGA
rpoC1-F	CATAGATTAGGCATACAGGC
rpoC1-R	AATAGCGGGAGATAGGAG
APX1-F	AGGTGCCACAAGGAAAGATCTGGT
APX1-R	TCAGCAGGGCTTTGTCACTAGGAA
APX2-F	TCAGCAGGGCTTTGTCACTAGGAA
APX2-R	TCCGCAGCATATTTCTCCACCAGT
SOD-A-F	ATCTGGATGGGTGTGGCTAGCTTT
SOD-A-R	AGTACGCATGCTCCCAGACATCAA
CATA-F	CAACCGCAACGTCGACAACTTCTT
CATA-R	TTCACCGGCAGCATCAGGTAGTTT
CATB-F	GCTTGCTTTCTGCCCAGCGATAAT
CATB-R	AAATAGTTTGGGCCAAGACGGTGC
AOX1a-F	CTTCGCATCGGACATCCATTA

Table A4. *Cont.*

AOX1a-R	TCCTCGGCAGTAGACAAACATC
AOX1b-F	CCTGCTCAGTTCATCACCATCA
AOX1b-R	GCATAAAACGGAGTGACAATAGC
AOX1c-F	GCGTCGGATGTTCATTTTCA
AOX1c-R	CCGCCGTAAGATTTCTTCAGT
Osl2-F	GCAGACAACAAATCGCCAAAT
Osl2-R	TCTCCAGCAACTCTAACCAGCAT
Osl30-F	GAGAAATCCCTTGAAGCCAA
Osl30-R	CACAAAGCAGTGAAAGCACA
Osl43-F	ATTCAGCATTCCACTGCAAG
Osl43-R	ATTCAGCATTCCACTGCAAG
Osh36-F	GTGCACCATGCACTTAATCC
Osh36-R	CACCGACCCTTCCTGTAGTT
Osh69-F	ACGAGCTACACGCCTACCTT
Osh69-R	ACTTCCTTGCCAGAAGCACT
OsMC1-F	GCTTCATCAAGGCGGTGGAGTG
OsMC1-R	AAGTTGGCGACCTTGCGGATG
OsMC2-F	CGACCCGTACAGGGTGCCGA
OsMC2-R	GCACAGCGCCTCGTCGTAGC
OsMC3-F	GGCTCCTTCGTCCGCAAGAT
OsMC3-R	CACAGGAGAAACGGTTTCCTGT
OsMC4-F	TCGACGTTCGTGGAGATGCTC
OsMC4-R	ATTCACGAGCCGCCTGATCTT
OsMC5-F	GTGCCAGACCGACCAGACAT
OsMC5-R	CCGCTCTTCTCCGACAGGAT
OsMC6-F	CCACACCGCAGGGTTCTTCAT
OsMC6-R	GTCCAGGCTGCTGAGTGTATCC
OsMC7-F	ATACAGACCGTGCTGGCGTC
OsMC7-R	AGGAATGGCGTCTCGGCGTT
OsMC8-F	TCCGGCAAGTGCCTCGTAAC
OsMC8-R	CAATGCGGTCGGTCACAGGAT
Realtime PCR primers to verification RNA-seq	
Os05g0474800-F	GACGTCCAACTCCTTCATGC
Os05g0474800-R	CATGGTCGGTTCTTGGTAGC
Os04g0103500-F	CTCCATGCTCTGCTGCAAAT
Os04g0103500-R	ACTTGCCGTTGTTCGAGATG
Os07g0130400-F	GGATACCTAGCGCCAGAACT
Os07g0130400-R	TCTTCGCCCACAAGTAACCT
Os06g0545900-F	AAGGCTTCTGCAAATGGCAA
Os06g0545900-R	CCATTCGCGCTGATCAGAAT
Os07g0283125-F	TCCTCCTGCAGCTCCTATAC
Os07g0283125-R	TGGTCAGGTTAGCACCAGAG
Os05g0112800-F	CGTCTGGTACGACTACTGCT
Os05g0112800-R	TTCTGCGTGTTGAGGAGGAT
Os07g0620400-F	CTGGCTAGGCATCTTCACCT
Os07g0620400-R	CGGTGCGTTTACTCCTTTCC
Os02g0100100-F	GCGTTCCCAAGGACTACAAC
Os02g0100100-R	CGAAGCAGCAGCAATCTCAT
Os09g0416200-F	GGTGTTCCGGAAGAAGAACG
Os09g0416200-R	GAGGTAGAGGGAGGAGGTGA
Os04g0581000-F	AGCTGTATTCCGATGAGCCA
Os04g0581000-R	TCGGCAGTACGTGCTCATAA
Os04g0577800-F	GGCTCCATCTTTGGAGGACT
Os04g0577800-R	TCCAAATCCAACAGCGTCAC
Os09g0517000-F	GCCAGAGGAGAGGAAGGAC
Os09g0517000-R	CTCATCTTGGCCACGAACTC
Os03g0304500-F	TCCAAGAATCCCAACACGGA
Os03g0304500-R	GAAGAGAAGCTGTCCTCCGA

Table A4. *Cont.*

Os07g0176900-F	GTTGTGCAGTTCTGCTGGAA
Os07g0176900-R	GTCGGTAACGTAGGGTTTGC
Os07g0630800-F	ATCTTGGACACTGCAACACG
Os07g0630800-R	GTTGATGCCGAACAGGTCAT
Os07g0634600-F	CCAATGCCAACTCTGGTAGC
Os07g0634600-R	CCTCGGTGTAGGTGATACCC
Os03g0366350-F	AAGGCCAACAACAACAGCAA
Os03g0366350-R	TGACGACGTTCTTCCTCTCC
Os12g0115300-F	GATGGTGATGGCGATGGTG
Os12g0115300-R	CCGTCACGTAAGAGATGCAC
Os10g0533500-R	ATCGAGAACGTGCCCTACTT
Os10g0533500-F	CATGGCCTCCTTCCTCTTGA
Os05g0374200-F	GCTCCTGTTACTTGCCCTTG
Os05g0374200-R	ATCGTGGAAGCTGAAATGGC
Primers for RNA editing site in chloroplast	
atpA-F	CCCAGGGGATGTTTTTATT
atpA-R	TGAAAAAAGCGTCCATTGTC
ndhD-F	ATTTTGGCTTCCTTATTGC
ndhD-R	GCCTCTACCCTGTCAACG
ndhF-F	ATATGCATGGGTAATCCCTC
ndhF-R	AGTGGCTCCTAAGAAAAGTG
ropB-F	ACTAAGCGTGCTATTCTCAA
ropB-R	TTTATGGTCTAATTCCGAGC
rps14-F	ATGGCAAAAAAAAAGTTTGATTC
rps14-R	TTACCAACTGGATCTTGTTGCA
rpoC2-F	GGTCCTTGGGGATTCTTGAT
rpoC2-R	TCTTGTTTTGTGGGTAACGG

Appendix B

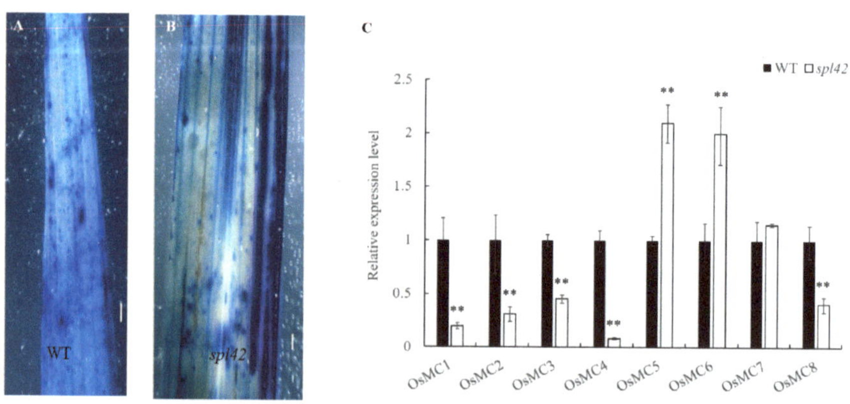

Figure A1. Trypan blue staining experiment and meta caspase gene family expression analysis. (**A**,**B**) Trypan blue staining in wild type (WT) and mutant *spl42* leaves. Scale bars = 0.1 cm. (**C**) qRT-PCR analysis for meta caspase family genes. Error bars represent ± SD (n = 3). Asterisks indicate a significant difference between WT and the *spl42* plants by Student's t-test, ** $p < 0.01$.

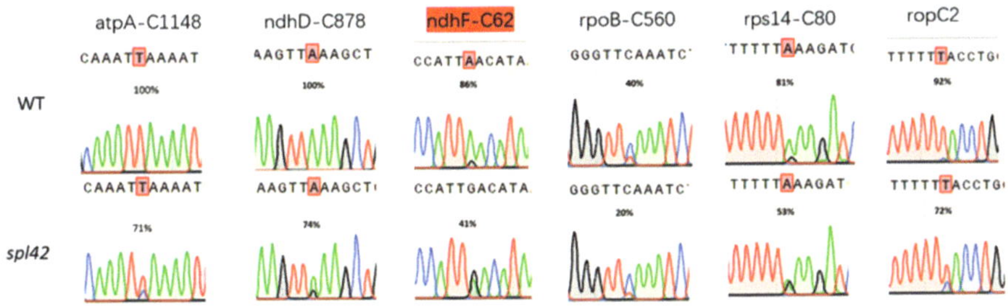

Figure A2. Changes in RNA editing efficiency in chloroplasts of the wild type and the *spl42* mutant. The red box represents the only RNA editing site that changes amino acid of encoding protein.

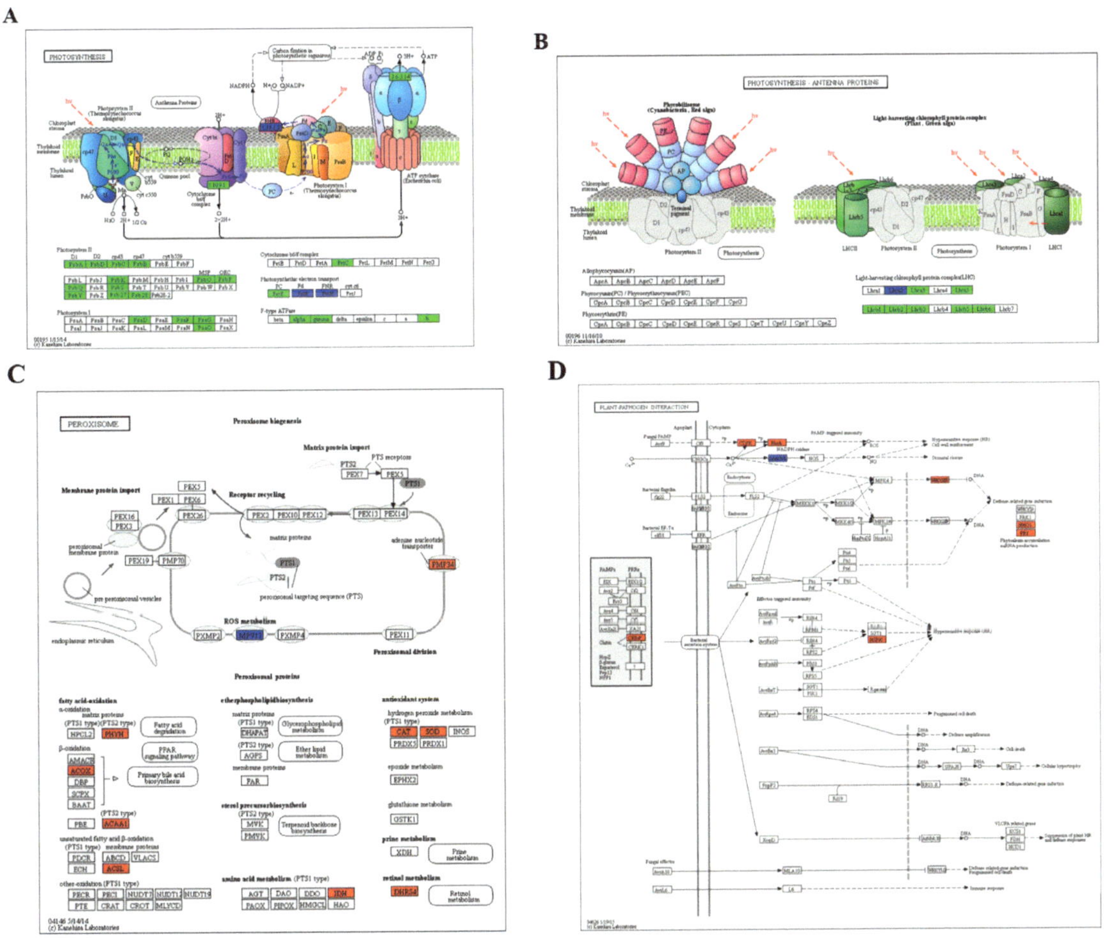

Figure A3. Differential map of genes in different pathways between wild type and *spl42* mutant. (**A**) Differential genes in photosynthesis pathway. (**B**) Differential gene of photosynthesis-antenna protein. (**C**) Differential genes of the peroxisome pathway. (**D**) Differential genes for the interaction between plants and pathogens.

References

1. Liu, W.; Fu, Y.; Hu, G.; Si, H.; Zhu, L.; Wu, C.; Sun, Z. Identification and fine mapping of a thermo-sensitive chlorophyll deficient mutant in rice (*Oryza sativa* L.). *Planta* **2007**, *226*, 785–795. [CrossRef]
2. Goh, C.H.; Satoh, K.; Kikuchi, S.; Kim, S.C.; Ko, S.M.; Kang, H.G.; Jeon, J.S.; Kim, C.; Park, Y.I. Mitochondrial activity in illuminated leaves of chlorophyll-deficient mutant rice (OsCHLH) seedlings. *Plant Biotechnol. Rep.* **2010**, *4*, 281–291. [CrossRef]
3. Ruan, B.; Hua, Z.; Zhao, J.; Zhang, B.; Ren, D.; Liu, C.; Yang, S.; Zhang, A.; Jiang, H.; Yu, H.; et al. OsACL-A2 negatively regulates cell death and disease resistance in rice. *Plant Biotechnol. J.* **2019**, *17*, 1344–1356. [CrossRef] [PubMed]
4. Zhang, H.; Li, J.; Yoo, J.H.; Yoo, S.C.; Cho, S.H.; Koh, H.J.; Seo, H.S.; Paek, N.C. Rice Chlorina-1 and Chlorina-9 encode ChlD and ChlI subunits of Mg-chelatase, a key enzyme for chlorophyll synthesis and chloroplast development. *Plant Mol. Biol.* **2006**, *62*, 325–337. [CrossRef] [PubMed]
5. Jung, K.H.; Hur, J.; Ryu, C.H.; Choi, Y.; Chung, Y.Y.; Miyao, A.; Hirochika, H.; An, G. Characterization of a rice chlorophyll-deficient mutant using the T-DNA gene-trap system. *Plant Cell Physiol.* **2003**, *44*, 463–472. [CrossRef]
6. Sakuraba, Y.; Rahman, M.L.; Cho, S.H.; Kim, Y.S.; Koh, H.J.; Yoo, S.C.; Paek, N.C. The rice faded green leaf locus encodes protochlorophyllide oxidoreductase B and is essential for chlorophyll synthesis under high light conditions. *Plant J.* **2013**, *74*, 122–133. [CrossRef]
7. Huang, M.; Slewinski, T.L.; Baker, R.F.; Janick-Buckner, D.; Buckner, B.; Johal, G.S.; Braun, D.M. Camouflage patterning in maize leaves results from a defect in porphobilinogen deaminase. *Mol. Plant* **2009**, *2*, 773–789. [CrossRef]
8. Quesada, V.; Sarmiento-Mañús, R.; González-Bayón, R.; Hricová, A.; Ponce, M.R.; Micol, J.L. Porphobilinogen deaminase deficiency alters vegetative and reproductive development and causes lesions in *Arabidopsis*. *PLoS ONE* **2013**, *8*, e53378. [CrossRef] [PubMed]
9. Ishikawa, A.; Okamoto, H.; Iwasaki, Y.; Asahi, T. A deficiency of coproporphyrinogen III oxidase causes lesion formation in *Arabidopsis*. *Plant J.* **2001**, *27*, 89–99. [CrossRef]
10. Wang, J.; Liu, X.; Zhang, A.; Ren, Y.; Wu, F.; Wang, G.; Xu, Y.; Lei, C.; Zhu, S.; Pan, T.; et al. A cyclic nucleotide-gated channel mediates cytoplasmic calcium elevation and disease resistance in rice. *Cell Res.* **2019**, *29*, 820–831. [CrossRef]
11. Du, D.; Liu, M.; Xing, Y.; Chen, X.; Zhang, Y.; Zhu, M.; Lu, X.; Zhang, Q.; Ling, Y.; Sang, X.; et al. Semi-dominant mutation in the cysteine-rich receptor-like kinase gene, *ALS1*, conducts constitutive defence response in rice. *Plant Biol.* **2019**, *21*, 25–34. [CrossRef] [PubMed]
12. Yoshimura, A.; Ideta, O.; Iwata, N. Linkage map of phenotype and RFLP markers in rice. *Plant Mol. Biol.* **1997**, *35*, 49–60. [CrossRef] [PubMed]
13. Yin, Z.; Chen, J.; Zeng, L.; Goh, M.; Leung, H.; Khush, G.S.; Wang, G.L. Characterizing rice lesion mimic mutants and identifying a mutant with broad-spectrum resistance to rice blast and bacterial blight. *Mol. Plant-Microbe Interact.* **2000**, *13*, 869–876. [CrossRef] [PubMed]
14. Yamanouchi, U.; Yano, M.; Lin, H.; Ashikari, M.; Yamada, K. A rice spotted leaf gene, *Spl7*, encodes a heat stress transcription factor protein. *Proc. Natl. Acad. Sci. USA* **2002**, *99*, 7530–7535. [CrossRef] [PubMed]
15. Shirsekar, G.S.; Vega-Sanchez, M.E.; Bordeos, A.; Baraoidan, M.; Swisshelm, A.; Fan, J.; Park, C.H.; Leung, H.; Wang, G.L. Identification and characterization of suppressor mutants of *spl11*-mediated cell death in rice. *Mol. Plant-Microbe Interact.* **2014**, *27*, 528–536. [CrossRef] [PubMed]
16. Li, Z.; Ding, B.; Zhou, X.; Wang, G.L. The rice dynamin-related protein OsDRP1E negatively regulates programmed cell death by controlling the release of cytochrome c from Mitochondria. *PLoS Pathogens.* **2017**, *13*, e1006157. [CrossRef]
17. Hong, Y.; Zhang, Y.; Sinumporn, S.; Yu, N.; Zhan, X.; Shen, X.; Chen, D.; Yu, P.; Wu, W.; Liu, Q.; et al. Premature leaf senescence 3, encoding a methyltransferase, is required for melatonin biosynthesis in rice. *Plant J.* **2018**, *95*, 877–891. [CrossRef] [PubMed]
18. Takenaka, M.; Zehrmann, A.; Verbitskiy, D.; Härtel, B.; Brennicke, A. RNA editing in plants and its evolution. *Annu. Rev. Genet.* **2013**, *47*, 335–352. [CrossRef] [PubMed]
19. Oldenkott, B.; Yang, Y.; Lesch, E.; Knoop, V.; Schallenberg-Rüdinger, M. Plant-type pentatricopeptide repeat proteins with a DYW domain drive C-to-U RNA editing in *Escherichia coli*. *Commun. Biol.* **2019**, *10*, 85. [CrossRef]
20. Chateigner-Boutin, A.; Small, I. Plant RNA editing. *RNA Biology*. **2010**, *7*, 213–219. [CrossRef]
21. Wang, Y.; Wang, Y.; Ren, Y.; Duan, E.; Zhu, X.; Hao, Y.; Zhu, J.; Chen, R.; Lei, J.; Teng, X.; et al. *white panicle2* encoding thioredoxin z, regulates plastid RNA editing by interacting with multiple organellar RNA editing factors in rice. *New Phytol.* **2021**, *229*, 2693–2706. [CrossRef]
22. Sun, T.; Germain, A.; Giloteaux, L.; Hammani, K.; Barkan, A.; Hanson, M.R.; Bentolila, S. An RNA recognition motif-containing protein is required for plastid RNA editing in *Arabidopsis* and maize. *Proc. Natl. Acad. Sci. USA* **2013**, *110*, E1169–E1178. [CrossRef]
23. Sun, T.; Shi, X.; Friso, G.; Van, W.; Bentolila, S.; Hanson, M.R. A zinc finger motif-containing protein is essential for chloroplast RNA editing. *PLoS Genet.* **2015**, *11*, e1005028. [CrossRef]
24. Takenaka, M.; Zehrmann, A.; Verbitskiy, D.; Kugelmann, M.; Härtel, B.; Brennicke, A. Multiple organellar RNA editing factor (MORF) family proteins are required for RNA editing in mitochondria and plastids of plants. *Proc. Natl. Acad. Sci. USA* **2012**, *109*, 5104–5109. [CrossRef] [PubMed]
25. Bentolila, S.; Heller, W.P.; Sun, T.; Babina, A.M.; Friso, G.; Van Wijk, K.J.; Hanson, M.R. RIP1, a member of an *Arabidopsis* protein family, interacts with the protein RARE1 and broadly affects RNA editing. *Proc. Natl. Acad. Sci. USA* **2012**, *109*, E1453–E1461. [CrossRef] [PubMed]

26. Tang, J.; Zhu, X.; Wang, Y.; Liu, L.; Xu, B.; Li, F.; Fang, J.; Chu, C. Semi-dominant mutations in the CC-NB-LRR-type R gene, *NLS1*, lead to constitutive activation of defense responses in rice. *Plant J.* **2011**, *66*, 996–1007. [CrossRef]
27. Ma, J.; Wang, Y.; Ma, X.; Meng, L.; Jing, R.; Wang, F.; Wang, S.; Cheng, Z.; Zhang, X.; Jiang, L.; et al. Disruption of gene *SPL35*, encoding a novel CUE domain-containing protein, leads to cell death and enhanced disease response in rice. *Plant Biotechnol. J.* **2019**, *17*, 1679–1693. [CrossRef] [PubMed]
28. Cui, Y.; Peng, Y.; Zhang, Q.; Xia, S.; Ruan, B.; Xu, Q.; Yu, X.; Zhou, T.; Liu, H.; Zeng, D.; et al. Disruption of *EARLY LESION LEAF 1*, encoding a cytochrome P450 monooxygenase, induces ROS accumulation and cell death in rice. *Plant J.* **2021**, *105*, 942–956. [CrossRef]
29. Zhang, F.; Tang, W.; Hedtke, B.; Zhong, L.; Liu, L.; Peng, L.; Lu, C.; Grimm, B.; Lin, R. Tetrapyrrole biosynthetic enzyme protoporphyrinogen IX oxidase 1 is required for plastid RNA editing. *Proc. Natl. Acad. Sci. USA* **2014**, *111*, 2023–2028. [CrossRef]
30. Huang, C.; Yu, Q.B.; Li, Z.R.; Ye, L.S.; Xu, L.; Yang, Z.N. Porphobilinogen deaminase HEMC interacts with the PPR-protein AtECB2 for chloroplast RNA editing. *Plant J.* **2017**, *92*, 546–556. [CrossRef]
31. Zhang, Q.; Xu, Y.; Huang, J.; Zhang, K.; Xiao, H.; Qin, X.; Zhu, L.; Zhu, Y.; Hu, J. The Rice Pentatricopeptide Repeat Protein PPR756 Is Involved in Pollen Development by Affecting Multiple RNA Editing in Mitochondria. *Front. Plant Sci.* **2020**, *11*, 749. [CrossRef]
32. Shin, N.H.; Trang, D.T.; Hong, W.J.; Kang, K.; Chuluuntsetseg, J.; Moon, J.K.; Yoo, Y.H.; Jung, K.H.; Yoo, S.C. Rice senescence-induced receptor-like kinase (OsSRLK) is involved in phytohormone-mediated chlorophyll degradation. *Int. J. Mol. Sci.* **2019**, *21*, 260. [CrossRef]
33. Wang, L.; Pei, Z.; Tian, Y.; He, C. OsLSD1, a rice zinc finger protein, regulates programmed cell death and callus differentiation. *Mol. Plant-Microbe Interact.* **2005**, *18*, 375–384. [CrossRef] [PubMed]
34. Mori, M.; Tomita, C.; Sugimoto, K.; Hasegawa, M.; Hayashi, N.; Dubouzet, J.G.; Ochiai, H.; Sekimoto, H.; Hirochika, H.; Kikuchi, S. Isolation and molecular characterization of a *Spotted-leaf 18* mutant by modified activation-tagging in rice. *Plant Mol. Biol.* **2007**, *63*, 847–860. [CrossRef]
35. Jiang, C.J.; Shimono, M.; Maeda, S.; Inoue, H.; Mori, M.; Hasegawa, M.; Sugano, S.; Takatsuji, H. Suppression of the rice fatty-acid desaturase gene *OsSSI2* enhances resistance to blast and leaf blight diseases in rice. *Mol. Plant-Microbe Interact.* **2009**, *22*, 820–829. [CrossRef] [PubMed]
36. Sun, C.; Liu, L.; Tang, J.; Lin, A.; Zhang, F.; Fang, J.; Zhang, G.; Chu, C. *RLIN1*, encoding a putative coproporphyrinogen III oxidase, is involved in lesion initiation in rice. *J. Genet. Genom.* **2011**, *38*, 29–37. [CrossRef]
37. Wang, J.; Ye, B.; Yin, J.; Yuan, C.; Zhou, X.; Li, W.; He, M.; Wang, J.; Chen, W.; Qin, P.; et al. Characterization and fine mapping of a light-dependent *leaf lesion mimic mutant 1* in rice. *Plant Physiol. Biochem.* **2015**, *97*, 44–51. [CrossRef]
38. Kang, S.G.; Lee, K.E.; Singh, M.; Kumar, P.; Matin, M.N. Rice Lesion Mimic Mutants (LMM): The current understanding of genetic mutations in the failure of ROS scavenging during lesion formation. *Plants* **2021**, *10*, 1598. [CrossRef]
39. Xiao, G.; Zhou, J.; Lu, X.; Huang, R.; Zhang, H. Excessive UDPG resulting from the mutation of *UAP1* causes programmed cell death by triggering reactive oxygen species accumulation and caspase-like activity in rice. *New Phytol.* **2018**, *217*, 332–343. [CrossRef]
40. Liu, J.; Li, W.; Ning, Y.; Shirsekar, G.; Cai, Y.; Wang, X.; Dai, L.; Wang, Z.; Liu, W.; Wang, G.L. The U-Box E3 ligase SPL11/PUB13 is a convergence point of defense and flowering signaling in plants. *Plant Physiol.* **2012**, *160*, 28–37. [CrossRef] [PubMed]
41. Yuan, Y.; Zhong, S.; Li, Q.; Zhu, Z.; Lou, Y.; Wang, L.; Wang, J.; Wang, M.; Li, Q.; Yang, D.; et al. Functional analysis of rice *NPR1*-like genes reveals that *OsNPR1/NH1* is the rice orthologue conferring disease resistance with enhanced herbivore susceptibility. *Plant Biotechnol. J.* **2007**, *5*, 313–324. [CrossRef]
42. Zeng, L.R.; Qu, S.; Bordeos, A.; Yang, C.; Baraoldan, M.; Yan, H.; Xie, Q.; Nahm, B.H.; Leung, H.; Wang, G.L. *Spotted leaf11*, a negative regulator of plant cell death and defense, encodes a U-box/armadillo repeat protein endowed with E3 ubiquitin ligase activity. *Plant Cell* **2004**, *16*, 2795–2808. [CrossRef]
43. Sathe, A.P.; Su, X.; Chen, Z.; Chen, T.; Wei, X.; Tang, S.; Zhang, X.B.; Wu, J.L. Identification and characterization of a spotted-leaf mutant *spl40* with enhanced bacterial blight resistance in rice. *Rice* **2019**, *12*, 68. [CrossRef]
44. Liu, L.L.; You, J.; Zhu, Z.; Chen, K.Y.; Hu, M.M.; Gu, H.; Liu, Z.W.; Wang, Z.Y.; Wang, Y.H.; Liu, S.J.; et al. *WHITE STRIPE LEAF8*, encoding a deoxyribonucleoside kinase, is involved in chloroplast development in rice. *Plant Cell Rep.* **2020**, *39*, 19–33. [CrossRef]
45. Lichtenthaler, H.K. Chlorophylls and carotenoids: Pigments of photosynthetic biomembranes. *Methods Enzymol.* **1987**, *148*, 350–382.
46. Hiei, Y.; Ohta, S.; Komari, T.; Kumashiro, T. Efficient transformation of rice (*Oryza sativa* L.) mediated by Agrobacterium and sequence analysis of the boundaries of the T-DNA. *Plant J.* **1994**, *6*, 271–282. [CrossRef]
47. Livak, K.J.; Schmittgen, T.D. Analysis of relative gene expression data using real-time quantitative PCR and the $2^{-\Delta\Delta C}{}_T$ method. *Methods* **2001**, *25*, 402–408. [CrossRef]
48. Robbins, J.C.; Heller, W.P.; Hanson, M.R. A comparative genomics approach identifies a PPR-DYW protein that is essential for C-to-U editing of the *Arabidopsis* chloroplast accD transcript. *RNA* **2009**, *15*, 1142–1153. [CrossRef] [PubMed]
49. Du, L.; Zhang, J.; Qu, S.; Zhao, Y.; Su, B.; Lv, X.; Li, R.; Wan, Y.; Xiao, J. The pentratricopeptide repeat protein pigment-defective mutant 2 is involved in the regulation of chloroplast development and chloroplast gene expression in *Arabidopsis*. *Plant Cell Physiol.* **2017**, *58*, 747–759. [CrossRef] [PubMed]

50. Waadt, R.; Kudla, J. In planta visualization of protein interactions using bimolecular fluorescence complementation (BiFC). *Cold Spring Harb. Protoc.* **2008**, *2008*, pdb.prot4995. [CrossRef]
51. Yoo, S.D.; Cho, Y.H.; Sheen, J. *Arabidopsis* mesophyll protoplasts: A versatile cell system for transient gene expression analysis. *Nat. Protoc.* **2007**, *2*, 1565–1572. [CrossRef] [PubMed]
52. Conesa, A.; Götz, S.; García-Gómez, J.M.; Terol, J.; Talón, M.; Robles, M. Blast2GO: A universal tool for annotation, visualization and analysis in functional genomics research. *Bioinformatics* **2005**, *21*, 3674–3676. [CrossRef] [PubMed]
53. Kanehisa, M.; Araki, M.; Goto, S.; Hattori, M.; Hirakawa, M.; Itoh, M.; Katayama, T.; Kawashima, S.; Okuda, S.; Tokimatsu, T.; et al. KEGG for linking genomes to life and the environment. *Nucleic Acids Res.* **2008**, *36*, D480–D484. [CrossRef] [PubMed]

plants

Article

Mapping and Validation of *qHD7b*: Major Heading-Date QTL Functions Mainly under Long-Day Conditions

Amir Sohail [1], Liaqat Shah [1,2], Ling Liu [1], Anowerul Islam [1,3], Zhengfu Yang [1,4], Qinqin Yang [1], Galal Bakr Anis [1,5], Peng Xu [1], Riaz Muhammad Khan [1,6], Jiaxin Li [1], Xihong Shen [1], Shihua Cheng [1], Liyong Cao [1,7], Yingxin Zhang [1,*] and Weixun Wu [1,*]

1 State Key Laboratory of Rice Biology, China National Rice Research Institute, Hangzhou 310006, China
2 Department of Agriculture, Mir Chakar Khan Rind University, Sibi 82000, Pakistan
3 Department of Agricultural Extension, Ministry of Agriculture, Dhaka 1215, Bangladesh
4 State Key Laboratory of Subtropical Silviculture, Zhejiang A&F University, Hangzhou 311300, China
5 Rice Research and Training Center, Field Crops Research Institute, Agriculture Research Center, Kafrelsheikh 33717, Egypt
6 Cereal Crops Research Institute (CCRI) Pirsabak Nowshera, Agriculture Research System, Nowshera 24100, Pakistan
7 Northern Center of China National Rice Research Institute, Shuangyashan 155600, China
* Correspondence: zhangyingxin@caas.cn (Y.Z.); wuweixun@caas.cn (W.W.)

Abstract: Heading date (HD) is one of the agronomic traits that influence maturity, regional adaptability, and grain yield. The present study was a follow-up of a previous quantitative trait loci (QTL) mapping study conducted on three populations, which uncovered a total of 62 QTLs associated with 10 agronomic traits. Two of the QTLs for HD on chromosome 7 (*qHD7a* and *qHD7b*) had a common flanking marker (RM3670) that may be due to tight linkage, and/or weakness of the statistical method. The objectives of the present study were to map QTLs associated with HD in a set of 76 chromosome segment substitution lines (CSSLs), fine map and validate one of the QTLs (*qHD7b*) using 2997 $BC_5F_{2:3}$ plants, and identify candidate genes using sequencing and expression analysis. Using the CSSLs genotyped with 120 markers and evaluated under two short-day and two long-day growing conditions, we uncovered a total of fourteen QTLs (*qHD2a*, *qHD4a*, *qHD4b*, *qHD5a*, *qHD6a*, *qHD6b*, *qHD7b*, *qHD7c*, *qHD8a*, *qHD10a*, *qHD10b*, *qHD11a*, *qHD12a*, and *qHD12b*). However, only *qHD6a* and *qHD7b* were consistently detected in all four environments. The phenotypic variance explained by *qHD6a* and *qHD7b* varied from 10.1% to 36.1% (mean 23.1%) and from 8.1% to 32.8% (mean 20.5%), respectively. One of the CSSL lines (CSSL52), which harbored a segment from the early heading XieqingzaoB (XQZB) parent at the *qHD7b* locus, was then used to develop a $BC_5F_{2:3}$ population for fine mapping and validation. Using a backcross population evaluated for four seasons under different day lengths and temperatures, the *qHD7b* interval was delimited to a 912.7-kb region, which is located between RM5436 and RM5499. Sequencing and expression analysis revealed a total of 29 candidate genes, of which *Ghd7* (*Os07g0261200*) is a well-known gene that affects heading date, plant height, and grain yield in rice. The *ghd7* mutants generated through CRISPR/Cas9 gene editing exhibited early heading. Taken together, the results from both the previous and present study revealed a consistent QTL for heading date on chromosome 7, which coincided not only with the physical position of a known gene, but also with two major effect QTLs that controlled the stigma exertion rate and the number of spikelets in rice. The results provide contributions to the broader adaptability of marker-assisted breeding to develop high-yield rice varieties.

Keywords: rice (*Oryza sativa* L.); quantitative trait locus; chromosome segment substitution lines; *qHD7b*; fine-mapping

Citation: Sohail, A.; Shah, L.; Liu, L.; Islam, A.; Yang, Z.; Yang, Q.; Anis, G.B.; Xu, P.; Khan, R.M.; Li, J.; et al. Mapping and Validation of *qHD7b*: Major Heading-Date QTL Functions Mainly under Long-Day Conditions. *Plants* **2022**, *11*, 2288. https://doi.org/10.3390/plants11172288

Academic Editors: Kassa Semagn and Tika Adhikari

Received: 28 June 2022
Accepted: 30 August 2022
Published: 1 September 2022

1. Introduction

Rice is a staple food for more than 50% of the world's population, with its production expected to increase by about 25% in 2030 to keep pace with population growth. Rice is a facultative, short-day crop that flowers earlier under short-day (SD) conditions and later under long-day (LD) conditions [1]. Heading date (HD) is a crucial trait affecting rice adaption to diverse cultivation areas, cropping seasons, maturity, and grain yield [2]. The development of early- or late-maturing cultivars depends on ecological conditions. In the regions where growing seasons are short, the aim is to develop early maturing varieties to escape frost damage, but there may be a yield penalty. However, in the regions where growing seasons are long, the aim is to develop late-maturing varieties with all of the assimilates efficiently transmitted to the grains, thereby enhancing grain weight and yield. Generally, there was a trade-off between flowering time and yield, which aimed to maximize production [3].

Numerous studies have been conducted to map and characterize 712 HD genes and quantitative trait loci (QTLs) that have been documented in the Gramene database (http://archive.gramene.org/qtl/ (accessed on 29 August 2022)). The *E1/Ghd7* was the first HD QTL reported in rice, which possesses a functional dominant *E1* allele and a non-functional *e1* allele [4]. The allelic variation of *Ghd7* contributes to the geographic distribution of cultivated rice [5], which has been investigated for photoperiod sensitivity and regional adaptability [6]. The functional *Ghd7* alleles (e.g., *Ghd7-1*, *Ghd7-2*, and *Ghd7-3*) delay heading, while the non-functional alleles (e.g., *Ghd7-0* and *Ghd7-0a*) shorten heading date in the different genetic backgrounds of rice. Both the *Ghd7-1* and *Ghd7-3* alleles were found in rice varieties grown in the tropics, subtropics, and areas with hot summers and long growing seasons in China and Southeast Asia. The *Ghd7-2* allele was found in temperate *japonica* varieties from Japan and northern China and had a smaller phenotypic effect than *Ghd7-1* [7]. *Se1/Hd1* was the first cloned heading-date QTL, an ortholog of *Arabidopsis CONSTANS* that promotes and suppresses flowering under short- and long-day growing conditions, respectively [8]. *Heading date 6* (*Hd6*) [9], *Heading date 3a* (*Hd3a*) [10], *Early heading date 1* (*Ehd1*) [11], *Days to heading 8* (*DTH8*)/*Ghd8* [12], *Heading date 17* (*Hd17*) [13], *RICE FLOWERING LOCUS T 1* (*RFT1*) [14], and *Days to heading 2* (*DTH2*) [3] are other HD QTLs in rice that have been cloned using a map-based approach. The analysis of these genes exhibited two main photoperiodic flowering pathways in rice: *Hd1-Hd3a* and *Ghd7-Ehd1-Hd3a/RFT1*. Major QTLs associated with the late heading, such as *Ghd7*, *Hd1*, *DTH8/Ghd8*, and *DTH7/Ghd7.1* [5,15], showed a strong correlation with an increase in grain yield, which suggests that the use of such types of HD QTLs can significantly influence rice's productivity and adaptability to specific growing conditions.

Recent progress in molecular technology and statistical methodology has provided researchers an opportunity to map and characterize HD QTLs in diverse types of populations, including F_2, recombinant inbred lines (RILs), and doubled haploid lines (DHLs) in rice [16,17]. However, these populations may not be ideal for the precise mapping of QTLs due to the simultaneous segregation of multiple loci originating from the two parents. Moreover, it would be more challenging to determine the actual genetic actions of the QTLs and differentiate the QTL effects from background noise [18,19]. Chromosome segment substitution lines (CSSLs) are genetic stocks that consist of overlapping segments of the complete genome of any genotype. CSSLs have been widely used to map QTLs accurately, evaluate gene interactions, discover new alleles, and compare the phenotypic effect of genes or QTLs [19,20]. Further fine mapping of QTLs of interest can be done by constructing segregating populations obtained from crossing one of the CSSLs and their recurrent parent [21].

In a previous study, our group identified 9 HD QTLs in a RIL population (Figure S1) and 2 BCF_1 populations derived from a cross between XieqingzaoB (XQZB) and Zhonghui9308 (ZH9308), which individually accounted for 2.6–18.6% of the phenotypic variance [22]. Three of the nine HD QTLs were mapped on chromosome 7 between RM3670 and RM2 markers (*qHD7a*), between RM5436 and RM3670 (*qHD7b*), and between RM118 and RM3555 (*qHD7c*), explaining 18.6%, 12.1%, and 5.6%, respectively. The *qHD7a* and the *qHD7b* QTL were physically located between 13,439,924–16,022,676 bp and 9,075,636–13,439,924 bp, respectively. RM3670, located at 13,439,924 bp, was a common flanking marker in both *qHD7a* and *qHD7b*, suggesting that the two QTLs are either tightly linked or the statistical method wrongly identified them as two independent QTLs. Both issues may be resolved using CSSLs, which form the basis of the present study. Therefore, the objectives of the present study were to understand the phenotypic variation of the CSSLs for heading date, fine map the HD QTL on chromosome 7, and identify candidate genes associated with HD under short- and long-day rice-growing conditions. Furthermore, we were also interested in determining the proportion of phenotypic variance explained by one of the QTLs on chromosome 7, validating and fine mapping its position using the $BC_5F_{2:3}$ population derived from a cross between one of the CSSLs and the recurrent parent, identifying candidate genes near the target QTL, and determining its actual effect in mutants generated through CRISPR/Cas9 gene editing.

2. Results

2.1. Phenotypic and Genotypic Analysis of CSSLs

Seventy-six CSSLs and the two parents were evaluated for HD under natural SD (NSD) at Hainan in 2015–2016 and under natural LD (NLD) at Hangzhou in 2014–2015 for two seasons. The XQZB matured about 32 and 20 days earlier than the ZH9308 in the Hangzhou and Hainan growing conditions, respectively (Figure 1A,B; Table 1). The days to heading of 76 CSSLs exhibited 59 to 122 days in NLDs and from 93 to 124 days in NSDs (Figure 1C–F; Table 1). Overall, HD showed continuous variation in both growing conditions but skewed distribution (Table 1, Figure 1). The broad-sense heritability was computed from all four environments, Hainan, and Hangzhou, and were 0.83, 0.82, and 0.79, respectively (Table 1). PC1 and PC2 from the principal component analysis (PCA) accounted for 75.7% and 15.5%, respectively (Figure S2), with most CSSLs showing an average heading date clustered together at the origin. Highly-significant positive correlations were observed among the tested environments for heading date (Figure S2).

Table 1. Summary of heading dates of parents and 76 chromosome segment substitution lines (CSSLs) evaluated under two natural short-day conditions at Hainan and long-day conditions at Hangzhou.

	Parents [b]		76 CSSLs [a]						
Year/Location	ZH9308	XQZB	Min	Max	Mean	SD	Kurtosis	Skewness	H
2014 Hangzhou	91.83 ± 1.11 **	60.17 ± 1.10	60.40	119.22	87.12	8.89	5.21	−0.23	0.79
2015 Hangzhou	88.33 ± 0.50 **	56.67 ± 0.54	57.67	124.28	88.21	9.42	4.45	−0.33	
2015 Hainan	110.99 ± 1.52 **	91.00 ± 1.31	96.57	124.94	110.87	3.94	4.33	0.63	0.82
2016 Hainan	109.67 ± 1.24 **	91.44 ± 1.67	90.00	122.80	103.59	6.10	1.54	0.36	
			Combined						
	Trait	Min	Max	Mean	σ^2G	σ^2GxE	σ^2E	CV	H^2
	HD	80.77	119.78	97.45	34.86	24.91	8.98	3.08	0.83

[a] Values for CSSLs are minimum (Min), maximum (Max), mean, standard deviation (SD), and repeatability (H^2). [b] Data of parents are presented as mean ± SD with ** indicating significant differences between ZH9308 and XQZB at the $p < 0.01$.

2.2. QTL Analysis of the CSSLs

The genetic linkage map of the CSSL population was constructed using 87 simple sequence repeats (SSRs) and 33 insertion/deletion (InDel) markers that followed the 1:1 Mendelian segregation pattern (Figure S3). A total of 120 markers were distributed across the whole genome with total coverage of 1311.02 cM using the Kosambi function of IciMapping software (Figure S3; Table S1). Inclusive composite interval mapping conducted on the BLUP HD data of the four environments using a LOD threshold value ≥ 2.5 identified 14 QTLs associated with heading date on Chr2 (*qHD2a*), Chr4 (*qHD4a* and *qHD4b*), Chr5 (*qHD5a*), Chr6 (*qHD6a* and *qHD6b*), Chr7 (*qHD7b* and *qHD7c*), Chr8 (*qHD8a*), Chr10 (*qHD10a* and *qHD10b*), Chr11 (*qHD11*), and Chr12 (*qHD12a* and *qHD12b*) (Table 2). The proportion of phenotypic variance (PVE) explained by each QTL ranged from 0.4% to 36.1%, and the additive effect from −15.4 to 18.2. Of the fourteen QTLs, only *qHD6a* and *qHD7b* were consistently detected in all four environments, from 10.1% to 36.1% (mean 23.14%) and from 8.1% to 32.8% (mean 20.5%), respectively. The remaining 12 QTLs were detected only in 1 or 2 environments (Table 2). We then focused on *qHD7b* for further research due to its consistent detection not only in all four environments in the present study, but also in two types of mapping populations in our previous study [23]. The favorable alleles of all QTL with a negative additive effect originated from the XQZB, while those with a positive allele originated from ZH9308. The XQZB allele at the *qHD7b* leads to early heading under both NSD and NLD conditions (Figure 1A,B; Table 2).

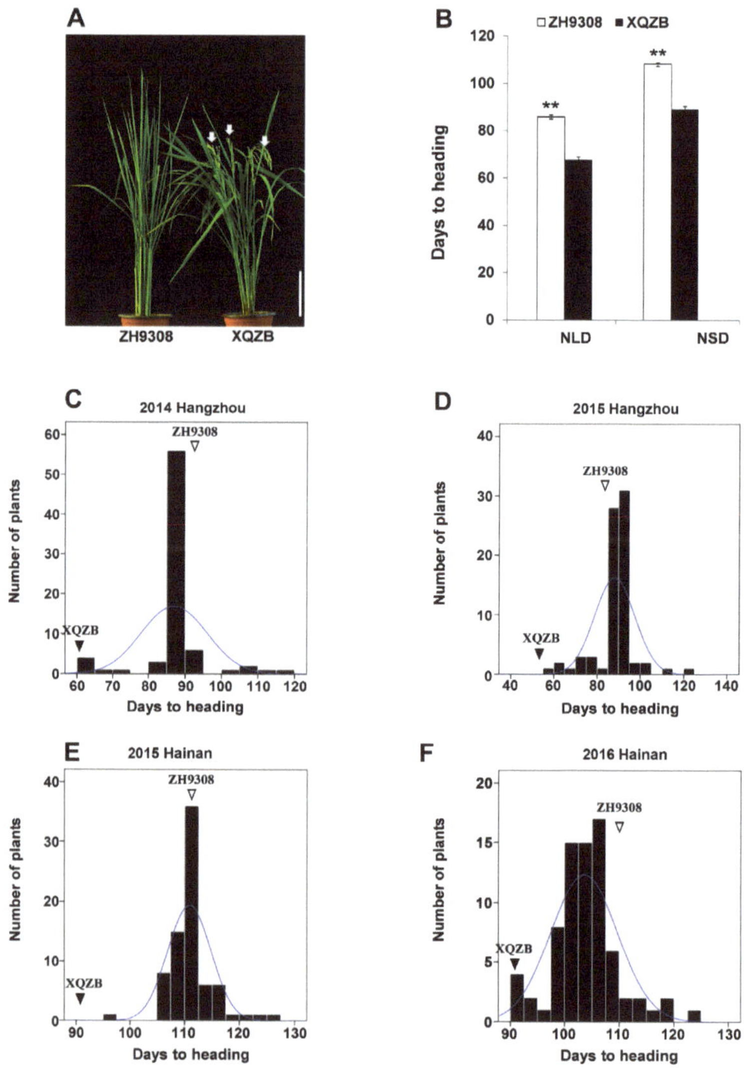

Figure 1. Comparison of ZH9308 and XQZB parents with ** indicating significant differences at $p < 0.01$ (**A**,**B**) and heading date distribution of the 76 chromosome segment substitution lines (CSSLs) based on the best linear unbiased prediction (BLUP) evaluated at the Hangzhou and Hainan growing conditions (**C**–**F**).

Table 2. Chromosomal locations of putative HD QTLs under four different environments using 76 CSSL populations.

QTLs	Year/Location [a]	Chr.	Region (cM)	Flanking Markers	LOD [b]	PVE (%) [c]	Add [d]
qHD2a	2015HZ	2	19.77	RM6424-InD31	2.80	2.54	2.37
qHD4a	2014HZ	4	6.58	InD62-RM1205	38.89	9.87	13.13
	2015HZ	4	10.54	RM1205-RM5979	6.46	4.50	1.75
qHD4b	2015HZ	4	21.08	RM3839-RM241	11.33	9.14	8.79
qHD5a	2014HZ	5	25.20	RM3638-RM6841	2.59	0.45	−0.47
	2016HN	5	14.52	InD79-RM3638	2.91	7.78	−3.91
qHD6a	2014HZ	6	6.58	RM5754-RM5963	49.67	19.03	18.21
	2015HN	6	5.27	RM510-RM5754	11.12	36.14	6.78
	2015HZ	6	6.58	RM5754-RM5963	13.14	10.13	9.90
	2016HN	6	5.27	RM510-RM5754	8.83	25.01	8.67
qHD6b	2014HZ	6	11.85	RM20069-InD94	3.77	0.33	0.36
qHD7b	2014HZ	7	27.69	RM3859-RM5875	54.42	26.54	−15.44
	2015HN	7	27.69	RM3859-RM5875	3.24	8.10	−2.61
	2015HZ	7	27.69	RM3859-RM5875	26.56	32.84	−12.90
	2016HN	7	27.69	RM3859-RM5875	7.55	20.33	−6.76
qHD7c	2015HZ	7	21.13	RM1132-RM455	2.56	2.79	0.88
qHD8a	2014HZ	8	7.91	RM5556-RM22529	46.40	16.12	−12.03
	2015HZ	8	7.91	RM5556-RM22529	15.91	13.30	−8.22
qHD10a	2015HZ	10	2.63	InD133-InD135	12.27	9.14	−13.28
	2015HN	10	2.63	InD133-InD135	6.67	18.74	−7.37
qHD10b	2016HN	10	6.58	RM6142-RM5620	3.60	8.89	−5.96
qHD11a	2014HZ	11	10.54	RM7463-RM26652	18.14	1.94	−7.76
	2015HZ	11	17.16	RM26652-InD151	2.51	3.10	−2.72
qHD12a	2014HZ	12	2.63	InD156-RM7003	15.42	2.21	−6.10
qHD12b	2014HZ	12	23.82	InD165-RM1300	2.85	0.43	−0.44

[a] Year/Location refers to the year of the experiment (2014, 2015, and 2016) followed by the location (Hangzhou—HZ/Hainan—HN). [b] Logarithm of odd, [c] the proportion of the phenotypic variance explained by the QTL effect, [d] the sign of the additive effects shows the parental origin of the favorable alleles (negative = XQZB and positive = ZH9308).

2.3. Photoperiodic Response of *qHD7b*

The CSSL52 is one of the chromosome segment substitution lines, which contains two segments from XQZB in the ZH9308 background and harbors the early heading allele at the *qHD7b* region flanked by RM3859 and RM5875 markers (Figure S4). ZH9308 and CSSL52 were evaluated for five consecutive seasons under NLD and NSD conditions with different day lengths and temperatures (Figure 2A–D). There were significant HD differences ($p < 0.01$) between the parents in all the studied environments (Figure 2E–G) ZH9308 headed 24.6–26.4 days later than CSSL52 at Hangzhou NLD conditions (Figure 2E,G). At Hainan NSD conditions, ZH9308 headed 6.2–10.4 days later than CSSL52 (Figure 2F,G). Similarly, ZH9308 had a higher plant height than CSSL52 under all five environments (Figure 2H). Taken together, ZH9308 headed later than CSSL52 under both SD and LD conditions, and the phenotypic differences of days to heading were more significant under LD conditions. Due to longer days to heading, ZH9308 also exhibited a taller PH, longer panicle length, more numbers of internodes, longer internode length, a greater number of panicles per plant, and a more significant number of grains in the main panicle than CSSL52 (Figures S5 and S6).

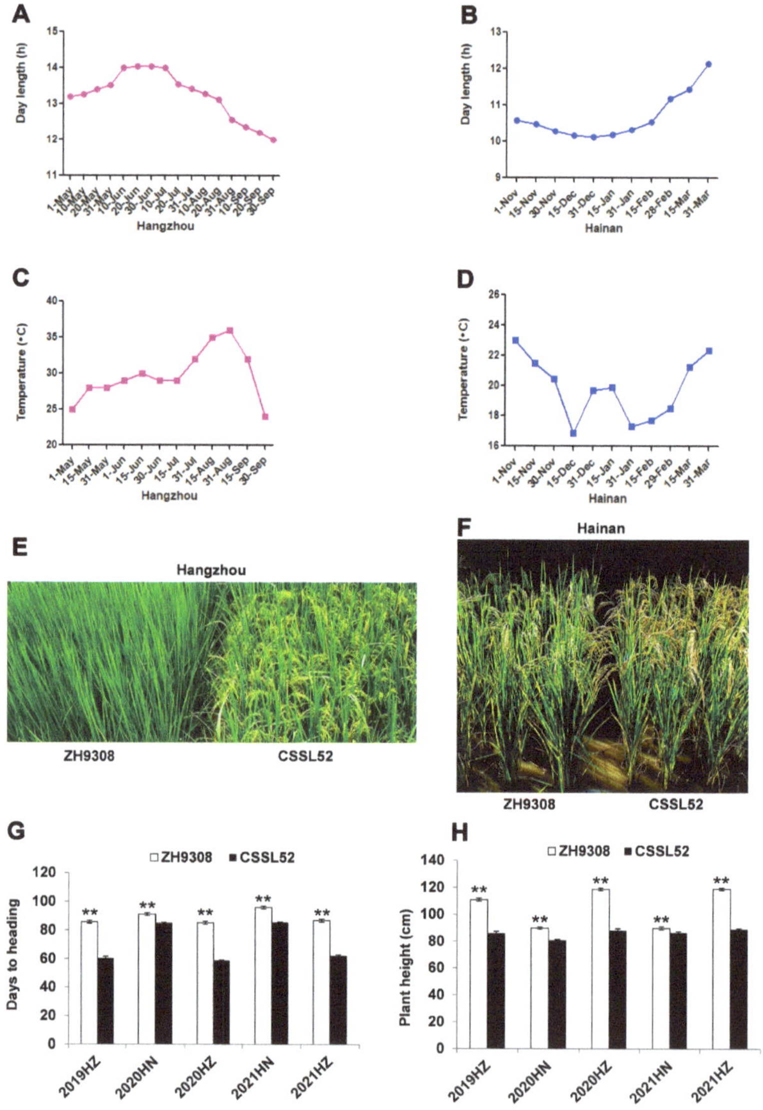

Figure 2. Comparison of heading date and plant height of ZH9308 and CSSL52 under five environmental conditions. (**A–D**) Daily photoperiod (**A**,**B**) and mean temperature (**C**,**D**) under Hangzhou and Hainan conditions in 2020 during the rice-growing season. (**E**,**F**) The phenotype of ZH9308 and CSSL52 under Hangzhou (**E**) and Hainan (**F**). The photo was taken at the CSSL52 heading stage in (**E**) and the CSSL52 maturation stage in (**F**). (**G**,**H**) Days to heading (**G**) and plant height (**H**) comparison between parents from 2019–2021 under NLD conditions in Hangzhou and NSD conditions in Hainan. The asterisks ** indicate significance between the parents at the $p < 0.01$, according to Student's *t*-test.

Heading date showed a highly-significant positive Pearson correlation with plant height (r = 0.83, $p < 0.01$), panicle length (r = 0.86, $p < 0.01$), and number of grains in the main panicle (r = 0.79, $p < 0.01$) (Table S2). Plant height showed high positive correlations with the panicle length (r = 0.83, $p < 0.01$) and number of grains in the main panicle (r = 0.80,

$p < 0.01$). Similarly, there was also a significant positive correlation between panicle length and number of grains in the main panicle ($r = 0.76$, $p < 0.01$) (Table S2).

2.4. QTL Mapping in $BC_5F_{2:3}$ Population

The *qHD7b* was initially delimited to the 7.1 Mb region between the RM3859 and RM5875 markers (Table 1). We used a BC_5F_2 population to validate and fine map the *qHD7b* QTL, which was developed by crossing CSSL52 that has the early-heading allele with the ZH9308 parent and then backcrossing the progenies five times with the ZH9308 to develop a secondary F_2 (BC_5F_2) population for fine mapping *qHD7b* (Figure S7).

Using a linkage map of 9 markers near the *qHD7b* QTL and heading data of a subset of 501 BC_5F_2 plants (Figure 3) evaluated for two consecutive seasons under Hangzhou NLD and two seasons under Hainan NSD conditions (Figure S8), we mapped the *qHD7b* QTL between the InDel4373 and InDel3 markers. The QTL was detected in all four environments (seasons) and was between 10.8% and 41.1% in each environment (mean = 26.0%), and had a LOD score ranging from 4.9 to 62.7 (mean = 33.8). Under Hangzhou conditions, however, the *qHD7b* had a very large effect (31.5–41.1%) and more significant LOD scores (59.8–62.7) than in the Hainan conditions (PVE = 10.8–21.2%, LOD = 4.9–7.5), which was about two-fold greater in the mean phenotypic effect and nearly ten-fold larger in the mean LOD score (Table S3). We then classified the 501 BC_5F_2 as late and early heading and calculated the mean difference in heading date. The average difference between the early and late segregating progenies was 9.6 and 25.5 days under NSD and NLD, respectively (Figure S8). A Chi-square analysis performed on the BC_5F_2 population fit the expected 3 (late):1 (early) ratio (386:115; $\chi^2 = 0.52$, $p = 0.47$), indicating that *qHD7b* behaves as a single dominant gene that is more functional under LD conditions than SD conditions.

2.5. Fine Mapping of qHD7b

To narrow down the confidence interval of the *qHD7b* identified using the 501 BC_5F_2 plants, we used a total of 2997 $BC_5F_{2:3}$ individuals genotyped with 7 markers that were polymorphic between the ZH9308 and CSSSL52, as well as the two flanking InDels (InDel4373 and InDel3) identified during the initial mapping. The genotype data revealed 14 homozygous recombinants (Figure 3D) that belong to 4 groups (G1 = 3 plants, G2 = 5 plants, G3 = 4 plants, and G4 = 2 plants). The recombinant plants in G1 had the same heading date (86.0 d) as the ZH9308 parent (85.2 d), while the remaining three recombinant groups had nearly the same heading date (56.2–58.0 d) as the CSSL52 parent (58.8 d). In the fine mapping, InDel4477, located at 9,075,693 bp, was the closest marker to *qHD7b*, while RM5436 and RM5499 were the flanking markers located at 9,075,636 bp and 9,988,139 bp on chromosome 7. In contrast to the G1 recombinants that had the ZH9308 parent genome between RM5436 and RM5499 at the *qHD7b* interval, the remaining three groups of recombinants (G2 to G4) all inherited the CSSL52 genome. The physical interval between RM5436 and RM5499 markers spans 912.7 kb (Figure 3D), which was confirmed using 20 $BC_5F_{3:4}$ progeny from each recombinant group.

2.6. Candidate Gene Analysis of qHD7b and Validation Using CRISPR/Cas9

A candidate gene search using the physical position of the two flanking markers identified during the fine mapping (RM5436 and RM5499) in the Gramene database (https://www.gramene.org/ (accessed on 29 August 2022)) using the *Oryza sativa japonica* group reference genome identified 29 predicted genes, of which 18 had *Oryza indica* homologues that fell within the 912.7-kb region (7:9,075,636–9,988,139) of *qHD7b* (Figure 3E and Table S4). ORF4 (*Os07g0261200*) is physically located at 9,152,377 bp on chromosome 7, encodes the CCT motif family protein, and has been annotated as *Ghd7 (Os07g0261200)*. We then sequenced *Os07g0261200* in the two parents using seven markers (M1 to M7), which revealed a 5.984-kb deletion in the ORF4 region in the CSSL52 parent but not in ZH9308 (Figure 3F). The expression levels of *Os07g0261200* in CSSL52 were also nearly zero as compared with ZH9308, which also suggests that *Ghd7* is a probable putative candidate gene for *qHD7b* (Figure S9).

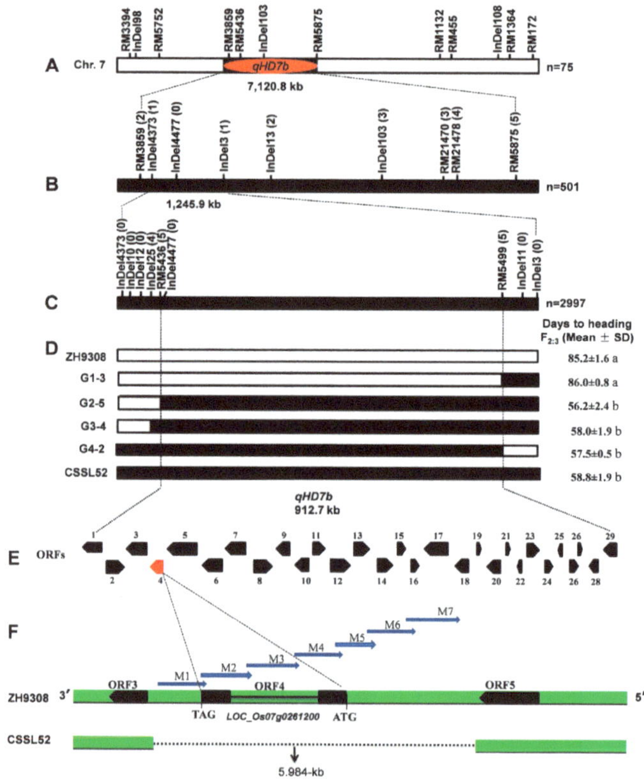

Figure 3. Coarse mapping and fine mapping of *qHD7b* on chromosome 7. (**A**) The position of *qHD7b* based on 76 CSSLs genotyped with 12 molecular markers on chromosome 7. (**B**) The position of *qHD7b* based on 501 BC_5F_2 plants genotyped with 7 markers located between RM3859 and RM5875 (the flanking markers identified using the CSSLs). (**C**) Fine mapping of *qHD7b* using $BC_5F_{2:3}$ plants genotyped with 7 markers that mapped between InDel4373 and InDel3 (the flanking markers identified using the 501 BC_5F_2 plants). (**D**) Genotypes and phenotypes of the two parents (ZH9308 and CSSL52) and 14 homozygous recombinant lines used for fine mapping of *qHD7b*. The ZH9308 and CSSL52 genotypic markers are represented by white and black bars, respectively. The 14 homozygous recombinants belong to four groups (G1 = 3 plants, G2 = 5 plants, G3 = 4 plants, and G4 = 2 plants). The superscripted letters (a and b) indicate statistically significant differences in the heading dates of recombinants relative to the parents. (**E**) Approximately 29 open reading frames (ORFs) were located between the two flanking markers identified during the fine mapping (RM5436 and RM5499), which are summarized in Table S4. (**F**) Sequence comparison between ZH9308 and CSSL52 with 5.984-kb deletion in CSSL52 at *Os07g0261200* using 7 markers (M1 to M7). The deleted region in CSSL52 is significant, and seven molecular markers (*Ghd7*-M1 to *Ghd7*-M7) linked with *qHD7b* are amplified in ZH9308.

To validate the mutant phenotype, we knocked out *Ghd7* in the Nipponbare genetic background utilizing the CRISPR/Cas9 system (Figure 4A). The HD of the wild type was 9.4 days later than the *ghd7* mutant under NLD conditions in Hangzhou (Figure 4B,C) and 2.0 days later under NSD conditions in Hainan (Figure 4C). The wild-type showed significant differences with the *ghd7* mutant for HD, plant height, and the number of grains per panicle (Figure 4D,E).

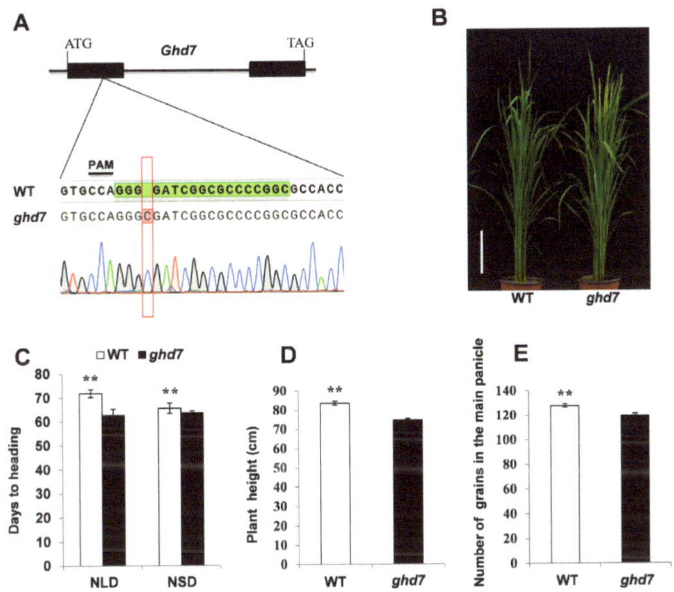

Figure 4. Heading date of wild-type (WT) and *ghd7* mutant in the Nipponbare genetic background. (**A**) Schematic of the *Ghd7* gene with the sgRNA:Cas9 targets (green) and corresponding protospacer-adjacent motif sequences (underlined). The insertion nucleotide is shown as a red letter. (**B**) The phenotype of WT and mutant *ghd7* at the heading stage under Hangzhou conditions. (**C**) Days to heading of WT and *ghd7* under natural long-day Hangzhou and natural short-day Hainan conditions. (**D**,**E**) Comparison of the WT and *ghd7* mutant for plant height (**D**) and the number of grains in the main panicle (**E**) under natural long-day conditions. The data are expressed as mean values ± SD. The asterisks ** indicate significance between WT and *ghd7* mutant at the $p < 0.01$, as determined by Student's *t*-test.

3. Discussion

Heading date is one of the most important agronomic traits and varies widely in rice depending on the genetic differences among genotypes, environmental conditions, day length, temperature, and their interactions [23]. A clear understanding of the genetic drivers of heading dates is essential for cultivating rice in different geographical regions and seasons [24]. Numerous genetic mapping studies were conducted to identify genes and QTLs associated with heading using bi-parental populations, such as F_2, backcross, doubled haploid lines, RILs, and CSSLs [25,26]. Hundreds of genomic regions related to HD have been reported in rice; however, few have been mapped and cloned [27]. *qHD7b* was identified using two types of populations that were 76 CSSLs and a BC_5F_2 population. Using 76 CSSL lines, the present study confirmed the *qHD7b* QTL that we previously reported between markers RM5436, located at 9,075,636 bp, and RM3670, located at 13,439,924 bp [22]. However, the physical interval of the QTL was about double in the CSSLs (7120.8 kb) as compared with the previous study (4364.3 kb), which is likely due to the small population size of the CSSLs. The use of 501 BC_5F_2 plants reduced the physical confidence interval of the QTL to 1245.9 kb, followed by 912.7 kb when it was fine-mapped using 2997 $BC_5F_{2:3}$ plants (Figure 3). RM5436 and RM5499 were the flanking markers of the *qHD7b* QTL after fine mapping, which were also previously reported as flanking markers for a major QTL (*qSSP7*), thereby controlling the number of spikelets per panicle [28], and another major QTL (*qSE7*), thereby influencing the stigma exertion rate [29] in rice. A candidate gene search conducted using the physical interval of the *qHD7b* (7:9,075,636–9,988,139) and the *Oryza sativa japonica* group reference genome in Gramene identified a total of 29 candidate genes. *Ghd7* (*Os07g0261200*) is one of the 29 candidate genes, which has been extensively studied for its effect on influencing heading date, grain yield, plant height, and other agronomic traits in rice [7,30]. However, other multiple genes should also be explored in the future.

One of the challenges in QTL mapping is identifying QTLs that are consistently detected across different environments, which was not the case for *qHD7b*. This QTL was consistently detected in all environments regardless of the type of mapping populations, but its effect and LOD scores tend to be very erratic. The LOD score for the *qHD7b* varied from 3.2 to 54.4 in the 76 CSSLs and from 4.9 to 62.7 in the 501 BC_5F_2 plants, while the phenotypic variance explained by the QTL varied from 8.1% to 32.8% in the CSSLs and from 10.8% to 41.1% in the BC_5F_2 plants (Table 2 and Table S3). A previous study showed that the QTL effect detected by the different mapping populations and environmental conditions is not necessarily the same [31]. In our case, the discrepancies may be due to differences in the population size and/or the type of mapping populations. CSSLs are powerful for QTL discovery studies due to the presence of multiple overlapping segments and several recombination events [32]. Still, the small size of the population in the current study may affect the QTL results. The BC_5F_2 population size, on the other hand, was ideal for QTL discovery studies in terms of population size but has limited recombination frequency, which makes it inferior in terms of mapping resolutions. The two phenotyping locations were also different in both temperature and day length, with Hangzhou's showing a higher temperature and longer day length than Hainan (Figure 2A–D), which resulted in a clear difference in phenotype between the parents in Hangzhou's than Hainan (Figure 2E–H) [33]. As a result, the proportion of phenotypic variance explained by *qHD7b* was very low in Hainan in both the CSSL and BC_5F_2 populations (Tables 2 and S3). Liu et al. [23] also reported relatively low effects for *qHD1b* in Hainan than in Hangzhou. The differences in heading dates between the wild-type and the *ghd7* mutant were also smaller in Hainan than in the Hangzhou growing conditions, and recent studies also support this result [34]. These results suggest *qHD7b* as a major HD QTL function, mainly under LD conditions.

Our sequencing result revealed that at the *Ghd7* locus, the $qHD7b^{XQZB}$ allele belongs to *Ghd7*-0, which is non-functional, and the $qHD7b^{ZH9308}$ allele belongs to *Ghd7*-4, which is functional together with *Ghd7*-1, *Ghd7*-2, and *Ghd7*-3 [35]. The non-functional Minghui 63 allele of *Ghd7* showed non-significant phenotype differences under SD conditions [7]. However, under SD conditions, the $qHD7b^{XQZB}$ allele flowered earlier than the $qHD7b^{ZH9308}$ allele (Figure 2G). These phenotypic differences may be caused by background differences, indicating *qHD7b* also functions in SD conditions, but the effect is smaller than that in LD conditions. The two flanking molecular markers for *qHD7b* (RM5436 and RM5499) and the seven molecular markers that mapped within the QTL confidence interval (*Ghd7*-M1 located at 9,150,263 to *Ghd7*-M7 located at 9,155,572 bp) can contribute towards the effort in the breeding of rice varieties using marker-assisted selection (MAS). For example, $qHD7b^{ZH9308}$ could be useful in breeding late-maturing cultivars. When the rice cultivars with *Ghd7*-0a, *Ghd7*-0, and *Ghd7*-2 alleles originating from northern China were introduced to southern China, their heading date would be significantly earlier. The $qHD7b^{ZH9308}$ allele could be introduced into these cultivars to prolong their heading date to make them have delayed heading and increase their yield. On the contrary, $qHD7b^{XQZB}$ could be useful in breeding early-maturing cultivars. When the rice cultivars with *Ghd7*-1, *Ghd7*-3, and *Ghd7*-4 alleles originating from the tropical and subtropical regions were introduced to northern China, their grains couldn't reach maturity due to later heading. The $qHD7b^{XQZB}$ allele could be introduced into these cultivars to make early heading to mature and harvest in time.

4. Materials and Methods

4.1. Population Development and Phenotyping

The present study was conducted using three populations. One of the populations was developed by crossing 134 RILs with Zhonghui9308 (ZH9308), which was then backcrossed three times, and selfed six times to form BC_4F_6 generation. The RILs were initially developed from a cross between an early-heading XieqingzaoB (XQZB) donor parent and a late-heading ZH9308 recipient parent and parental lines of Xieyou9308 (an *indica-japonica* subspecies super hybrid rice with 87.5% *indica* and 12.5% *japonica* genome) [22]. Seventy-six of the 134 BC_4F_6 lines were selected to represent CSSLs for QTL mapping, and the genotypes of the 76 CSSLs were also investigated [36]. The second population was developed by crossing one of the CSSLs (CSSL52) that exhibited early heading with the ZH9308 parent and then backcrossing the progenies four times with the ZH9308 to develop a secondary BC_5F_2 population for validation and a $BC_5F_{2:3}$ population for the fine-mapping of the *qHD7b* QTL (Figure S1).

A set of 76 CSSLs along with parental lines, ZH9308 and XQZB, were evaluated for heading date through a Randomized Complete Block (RCB) design consisting of 3 replications, with 4 rows × 8 plants for each line at Hangzhou and Hainan for 4 seasons. Each plot consisted of 4 rows of 1.32 square meters with 16.5 cm × 26.5 cm spacing between plants and rows. The BC_5F_2 population, along with ZH9308 and CSSL52 parents, was evaluated for two years under NLD (day length >14 h) conditions in Hangzhou, Zhejiang Province (120.0° E, 30.15° N), and NSD (day length <12 h) conditions in Lingshui, Hainan Province (110.0° E, 18.5° N). HD was recorded as the number of days from the sowing date to the emergence of the first heading. Plant height (PH) was measured from the ground level to the tip of the panicle at full physiological maturity. Internode length was recorded as the length between two nodes. Number of panicles per plant was recorded as the number of all panicles per plant. Panicle length (PL) was recorded from the neck to the apex of the panicle. The number of grains in the main panicle was recorded as the filled grain numbers of the main panicle. Paddy field management followed conventional practices [37].

4.2. Test of Day Length Response in Growth Chambers

To investigate the photoperiodic response, ZH9308 and CSSL52 plants were grown under CLD (14 h light, 30 °C/10 h darkness, 25 °C) and CSD (10 h light, 30 °C/14 h darkness, 25 °C). The humidity was 75%, and the light intensity was 300 μmol m^{-2} s^{-1} [38].

4.3. Molecular Markers Development and DNA Extraction

Polymorphisms between ZH9308 and CSSL52 were screened using Insertion/Deletion (InDel) and Simple Sequence Repeat (SSR) markers [39]. To determine the candidate genes of *qHD7b* QTL, the sequence between RM3859 and RM5875 was downloaded from the EnsemblePlants (http://plants.ensembl.org/index.html (accessed on 29 August 2022)) and Gramene website (https://www.gramene.org/ (accessed on 29 August 2022)) using the *Oryza sativa japonica* group reference genome by a blast search of the primer sequences. New InDel markers in the QTL region were designed based on differences in genomic sequence between Indica and Japonica using the Primer Premier 5.0 software (PREMIER Biosoft International, San Francisco, CA, USA). The markers are listed in Tables S5 and S6. The markers that amplified the mutated region between parents are listed in Table S7.

Genomic DNA was extracted from the fresh leaves of parents and the secondary F_2 (BC_5F_2) population using the cetyltrimethylammonium bromide (CTAB) method described by Luo et al. [40]. A polymerase chain reaction (PCR) was performed in a 12 μL volume that consisted of 5 μL of master mix, 3 μL of ddH_2O, 2 μL of 10 pmol μL^{-1} primers (1 μL forward primer and reverse primers), and 2 μL of template genomic DNA (200 ng). The PCR amplification protocol consisted of a pre-denaturation step (95 °C for 3 min), followed by 32 cycles (denaturation at 95 °C for 15 s, annealing at 55 °C for 15 s, and extension at 72 °C for 5 s), and final extension (72 °C for 10 min). The PCR products were separated using gel electrophoresis on 8% non-denaturing polyacrylamide gel and visualized with silver nitrate staining using a formaldehyde solution [41].

4.4. RNA Extraction and qRT-PCR

The total RNA was extracted from leaves of ZH9308 and CSSL52 using an RNAprep Pure Plant kit (Tiangen Biotech Co., Ltd., Beijing, China). A total of 50 µL of complementary DNA (cDNA) was synthesized using 5 µg of RNA with ReverTra Ace® qPCR RT Master Mix with gDNA Remover (Toyobo Co., Ltd., Osaka, Japan). Real-time quantitative RT-PCR (RT-qPCR, 20 µL reaction volume) was performed using 0.5 µL of cDNA, 0.2 µM of each gene-specific primer, and TB Green Premix ExTaq II (Takara Bio, Inc., Kusatsu, Shiga, Japan) in a LightCycler®480 II (Roche). The rice *Ubq* gene (*Os03g0234350*) was exploited as the endogenous control. The primers for qRT-PCR were designed using GeneScript (https://www.genscript.com/tools/real-time-pcr-taqman-primer-design-tool (accessed on 29 August 2022)) and are displayed in Table S7.

4.5. Generation of *ghd7* Mutant Using CRISPR/Cas9 System

The CRISPR/Cas9 system was used to knock out *Ghd7* according to the methods as described previously [42]. The 18-bp sgRNA:Cas9 target sequence of *Ghd7* was introduced into the pCas9-sgRNA vector at the *AarI* site. The final vector was transformed into Nipponbare using *Agrobacterium*-mediated transformation [43]. The primers used are displayed in Table S7.

4.6. Statistical Analysis

The experimental design in each environment was a randomized complete block design with three replications per environment/location. The best linear unbiased prediction (BLUP) values were obtained through META-R v6.03 [44], using the linear model:

$$Y_{ik} = \mu + Rep_i + Gen_k + \varepsilon_{ik} \text{ (within the environment)}$$

$$Y_{ijk} = \mu + Rep_i\ (Env_j) + Env_j \times Gen_k + Gen_k + Env_j + \varepsilon_{ijk} \text{ (across environments)}$$

where Y_{ik} is the trait of interest, μ is the mean effect, Rep_i is the effect of the $_i$th replicate, Gen_k is the effect of the $_k$th genotype, ε_{ik} is the error associated with the $_i$th replication, and the $_k$th genotype, which is assumed to be normally and independently distributed. For across environments, Y_{ijk} is the trait response, Env_j is the $_j$th environment, Rep_i (Env_j) is the effect of $_i$th replication in the $_j$th environment, and $Env_j \times Gen_k$ is the environment and genotype interaction. The resulting analysis produced the adjusted trait phenotypic values in the form of BLUP within and across environments. The BLUP model considers genotypes as random effects. Broad sense heritability (H^2) and repeatability (H) were calculated according to Alemu et al. [45] using META-R software.

$$H = \frac{\sigma^2 g}{\sigma^2 g + \sigma^2 e/\text{reps}} \text{ (within the environment)}$$

$$H = \frac{\sigma^2 g}{\sigma^2 g + \sigma^2 ge/\text{env} + \sigma^2 e/(\text{reps} \times \text{env})} \text{ (across environment)}$$

where $\sigma^2 g$ and $\sigma^2 e$ are the genotypic and error variance, $\sigma^2 ge$ is the genotype by environment interaction variance, rep is the number of replicates, and env is the number of environments.

QTL analysis was performed using the inclusive composite interval mapping (ICIM) function implemented in QTL IciMapping software [46]. A LOD threshold value $\geq$2.5 indicates the presence of QTL (selected by 1000 permutation tests to obtain a 0.05 genome-wide probability level of Type I error, with a search step of 1 cm). QTLs were named by placing a "q" at the beginning of the trait "HD", followed by the chromosome number. For more than one QTL on the same chromosome, a second identifier was placed after the chromosome number reported previously [47].

5. Conclusions

A total of fourteen significant (LOD $\geq$ 2.5) HD QTLs (*qHD2a*, *qHD4a*, *qHD4b*, *qHD5a*, *qHD6a*, *qHD6b*, *qHD7b*, *qHD7c*, *qHD8a*, *qHD10a*, *qHD10b*, *qHD11a*, *qHD12a*, and *qHD12b*) were detected in the 76 CSSL populations concerning the position and introgression segments under four different environments. We focused on *qHD7b* for further research due to its stability. A secondary F_2 (BC_5F_2) population was developed by backcrossing CSSL52 with recurrent parent ZH9308, and *qHD7b* was narrowed down to the 912.7-kb region, flanked by markers RM5436 and RM5499 using 2995 individuals from the secondary $F_{2:3}$ ($BC_5F_{2:3}$) population. The CSSL52 allele at the *qHD7b* locus negatively regulates HD under SD and LD conditions. Sequencing and expression analysis demonstrated that *Os07g0261200* encodes *Ghd7*, a suitable candidate gene for *qHD7b*. The *ghd7* mutant generated through CRISPR/Cas9 promoted the heading date and validated *Ghd7* as a putative candidate gene for HD. Further study on *qHD7b* will contribute to MAS and the developing late-maturing varieties that can be used in diverse geographical regions.

Supplementary Materials: The following supporting information can be downloaded at: https://www.mdpi.com/article/10.3390/plants11172288/s1, Figure S1: Breeding scheme for QTL identification and fine mapping, Figure S2: GGE Biplot of days to heading from 76 CSSLs tested in four environments, Figure S3: Linkage map of 76 CSSLs derived from ZH9308 × XQZB using 120 polymorphic markers, Figure S4: A schematic representation of ZH9308, XQZB, and CSSL52 plants to show differences on chromosome 7 near *qHD7b*, Figure S5: Agronomic trait phenotypes of ZH9308 and CSSL52, Figure S6: Measurement of agronomic traits of ZH9308 and CSSL52, Figure S7: Phenotypes and genotypes of ZH9308, CSSL52, and secondary F_2 (BC_5F_2) population, Figure S8: Frequency distribution of heading date in the secondary F_2 (BC_5F_2) population under Hainan and Hangzhou conditions, Figure S9: The expression levels of *Os07g0261200* in ZH9308 and CSSL52, Table S1: Chromosome-wise SNP markers and genetic map length of rice CSSL population, Table S2: Pearson correlation coefficients among heading date and yield-related traits, Table S3: Evaluation of heading date QTL *qHD7b* under Hainan and Hangzhou conditions, Table S4: Candidate genes within 912.7-kb physical regions of *qHD7b* on chromosome 7, Table S5: Polymorphic DNA markers used in heading-date QTL analysis in 76 CSSLs and the secondary F_2 (BC_5F_2) population, Table S6: DNA markers used in QTL analysis and fine mapping of *qHD7b*, Table S7: Markers used for sequencing, qRT-PCR, and CRISPR.

Author Contributions: Conceptualization, W.W., Y.Z., S.C. and L.C.; methodology, W.W.; software, A.S.; investigation, A.S., L.S., L.L., A.I., Z.Y., Q.Y., G.B.A., P.X., R.M.K., J.L. and X.S.; resources, Y.Z.; writing original draft preparation, A.S.; writing review and editing, A.S. and W.W.; supervision, W.W.; project administration, W.W.; funding acquisition, W.W. All authors have read and agreed to the published version of the manuscript.

Funding: This research was supported by grants from the National Key R&D Program of China (2020YFE0202300), the National Natural Science Foundation of China (31871604, 32071996, and 31961143016), the Fundamental Research Funds of Central Public Welfare Research Institutions (CPSIBRF-CNRRI-202102), Hainan Yazhou Bay Seed Lab (B21HJ0219), and the Agricultural Science and Technology Innovation Program of the Chinese Academy of Agricultural Sciences (CAAS-ASTIP2013-CNRRI).

Institutional Review Board Statement: Not applicable.

Informed Consent Statement: Not applicable.

Data Availability Statement: Not applicable.

Conflicts of Interest: The authors declare no conflict of interest.

References

1. Saito, H.; Okumoto, Y.; Tsukiyama, T.; Xu, C.; Teraishi, M.; Tanisaka, T. Allelic differentiation at the *E1/Ghd7* locus has allowed expansion of rice cultivation area. *Plants* **2019**, *8*, 550. [CrossRef] [PubMed]
2. Li, X.; Sun, Y.; Tian, X.; Ren, Y.; Tang, J.; Wang, Z.; Cheng, Y.; Bu, Q. Comprehensive identification of major flowering time genes and their combinations, which determined rice distribution in Northeast China. *Plant Growth Regul.* **2018**, *84*, 593–602. [CrossRef]

3. Wu, W.; Zheng, X.M.; Lu, G.; Zhong, Z.; Gao, H.; Chen, L.; Wu, C.; Wang, H.J.; Wang, Q.; Zhou, K.; et al. Association of functional nucleotide polymorphisms at *DTH2* with the northward expansion of rice cultivation in Asia. *Proc. Natl. Acad. Sci. USA* **2013**, *110*, 2275–2280. [CrossRef]
4. Syakudo, K.; Kawase, T. Studies on the quantitative inheritance (11): A. Rice (*Oryza sativa* L.) (d) Inheritance of the heading duration and the quantitative function of the causal genes in its determination. (1) On the quantitative function of the genes E1, E2 and D1. *Jpn. J. Breed.* **1953**, *3*, 6–12. [CrossRef]
5. Zhang, J.; Zhou, X.; Yan, W.; Zhang, Z.; Lu, L.; Han, Z.; Zhao, H.; Liu, H.; Song, P.; Hu, Y.; et al. Combinations of the *Ghd7*, *Ghd8* and *Hd1* genes largely define the ecogeographical adaptation and yield potential of cultivated rice. *N. Phytol.* **2015**, *208*, 1056–1066. [CrossRef] [PubMed]
6. Wei, H.; Wang, X.; Xu, H.; Wang, L. Molecular basis of heading date control in rice. *Abiotech* **2020**, *1*, 219–232. [CrossRef]
7. Xue, W.; Xing, Y.; Weng, X.; Zhao, Y.; Tang, W.; Wang, L.; Zhou, H.; Yu, S.; Xu, C.; Li, X.; et al. Natural variation in *Ghd7* is an important regulator of heading date and yield potential in rice. *Nat. Genet.* **2008**, *40*, 761–767. [CrossRef]
8. Yano, M.; Katayose, Y.; Ashikari, M.; Yamanouchi, U.; Monna, L.; Fuse, T.; Baba, T.; Yamamoto, K.; Umehara, Y.; Nagamura, Y.; et al. *Hd1*, a major photoperiod sensitivity quantitative trait locus in rice, is closely related to the *Arabidopsis* flowering time gene *CONSTANS*. *Plant Cell.* **2000**, *12*, 2473–2483. [CrossRef]
9. Takahashi, Y.; Ayahiko, S.; Takuji, S.; Masahiro, Y. *Hd6*, a rice quantitative trait locus involved in photoperiod sensitivity, encodes the α subunit of protein kinase CK2. *Proc. Natl. Acad. Sci. USA* **2001**, *98*, 7922–7927. [CrossRef]
10. Kojima, S.; Takahashi, Y.; Kobayashi, Y.; Monna, L.; Sasaki, T.; Araki, T.; Yano, M. *Hd3a*, a rice ortholog of the *Arabidopsis FT* gene, promotes transition to flowering downstream of *Hd1* under short-say conditions. *Plant Cell Physiol.* **2002**, *43*, 1096–1105. [CrossRef]
11. Doi, K.; Izawa, T.; Fuse, T.; Yamanouchi, U.; Kubo, T.; Shimatani, Z.; Yano, M.; Yoshimura, A. *Ehd1*, a B-type response regulator in rice, confers short-day promotion of flowering and controls *FT*-like gene expression independently of *Hd1*. *Genes Dev.* **2004**, *18*, 926–936. [CrossRef]
12. Wei, X.; Xu, J.; Guo, H.; Jiang, L.; Chen, S.; Yu, C.; Zhou, Z.; Hu, P.; Zhai, H.; Wan, J. *DTH8* suppresses flowering in rice, influencing plant height and yield potential simultaneously. *Plant Physiol.* **2010**, *153*, 1747–1758. [CrossRef]
13. Matsubara, K.; Tanaka, E.O.; Hori, K.; Ebana, K.; Ando, T.; Yano, M. Natural variation in *Hd17*, a homolog of Arabidopsis *ELF3* that is involved in rice photoperiodic flowering. *Plant Cell Physiol.* **2012**, *53*, 709–716. [CrossRef]
14. Tanaka, E.O.; Matsubara, K.; Yamamoto, S.I.; Nonoue, Y.; Wu, J.; Fujisawa, H.; Ishikubo, H.; Tanaka, T.; Ando, T.; Matsumoto, T.; et al. Natural variation of the *RICE FLOWERING LOCUS T 1* contributes to flowering time divergence in rice. *PLoS ONE* **2013**, *8*, e75959.
15. Ye, J.; Niu, X.; Yang, Y.; Wang, S.; Xu, Q.; Yuan, X.; Yu, H.; Wang, Y.; Wang, S.; Feng, Y.; et al. Divergent *Hd1*, *Ghd7*, and *DTH7* alleles control heading date and yield potential of Japonica rice in northeast China. *Front. Plant Sci.* **2018**, *9*, 35. [CrossRef]
16. Yano, M.; Harushima, Y.; Nagamura, Y.; Kurata, N.; Minobe, Y.; Sasaki, T. Identification of quantitative trait loci controlling heading date in rice using a high-density linkage map. *Theor. Appl. Genet.* **1997**, *95*, 1025–1032. [CrossRef]
17. Yuan, R.; Zhao, N.; Usman, B.; Luo, L.; Liao, S.; Qin, Y.; Nawaz, G.; Li, R. Development of chromosome segment substitution lines (CSSLs) derived from Guangxi wild rice (*Oryza rufipogon* Griff.) under rice (*Oryza sativa* L.) background and the identification of QTLs for plant architecture, agronomic traits and cold tolerance. *Genes* **2020**, *11*, 980. [CrossRef]
18. Fan, J.; Hua, H.; Luo, Z.; Zhang, Q.; Chen, M.; Gong, J.; Wei, X.; Huang, Z.; Huang, X.; Wang, Q. Whole-Genome sequencing of 117 chromosome segment substitution lines for genetic analyses of complex traits in rice. *Rice* **2022**, *15*, 5. [CrossRef]
19. Uehara, K.B.; Furuta, T.; Masuda, K.; Yamada, S.; Angeles-Shim, R.B.; Ashikari, M.; Takashi, T. Construction of rice chromosome segment substitution lines harboring *Oryza barthii* genome and evaluation of yield-related traits. *Breed. Sci.* **2017**, *67*, 408–415. [CrossRef]
20. Balakrishnan, D.; Surapaneni, M.; Mesapogu, S.; Neelamraju, S. Development and use of chromosome segment substitution lines as a genetic resource for crop improvement. *Theor. Appl. Genet.* **2019**, *132*, 1–25. [CrossRef]
21. Shen, W.Q.; Zhao, B.B.; Yu, G.L.; Li, F.F.; Zhu, X.Y.; Ma, F.Y.; Li, Y.F.; He, G.H.; Zhao, F.M. Identification of an excellent rice chromosome segment substitution line Z746 and QTL mapping and verification of important agronomic traits. *Acta Agron. Sin.* **2021**, *47*, 451–461. [CrossRef]
22. Liang, Y.; Zhan, X.; Gao, Z.; Lin, Z.; Yang, Z.; Zhang, Y.; Shen, X.; Cao, L.; Cheng, S. Mapping of QTLs associated with important agronomic traits using three populations derived from a super hybrid rice Xieyou9308. *Euphytica* **2012**, *184*, 1–13. [CrossRef]
23. Liu, L.; Zhang, Y.; Yang, Z.; Yang, Q.; Zhang, Y.; Xu, P.; Li, J.; Islam, A.; Shah, L.; Zhan, X.; et al. Fine mapping and candidate gene analysis of *qHD1b*, a QTL that promotes flowering in common wild rice (*Oryza rufipogon*) by up-regulating *Ehd1*. *Crop J.* **2022**, *10*, 1083–1093. [CrossRef]
24. Hori, K.; Yasunori, N.; Ono, N.; Shibaya, T.; Ebana, K.; Matsubara, K.; Ogiso-Tanaka, E.; Tanabata, T.; Sugimoto, K.; Taguchi-Shiobara, F.; et al. Genetic architecture of variation in heading date among Asian rice accessions. *BMC Plant Biol.* **2015**, *15*, 115. [CrossRef] [PubMed]
25. Ma, L.; Yang, C.; Zeng, D.; Cai, J.; Li, X.; Ji, Z.; Xia, Y.; Qian, Q.; Bao, J. Mapping QTLs for heading synchrony in a doubled haploid population of rice in two environments. *J. Genet. Genom.* **2009**, *36*, 297–304. [CrossRef]

26. Surapaneni, M.; Balakrishnan, D.; Mesapogu, S.; Addanki, K.R.; Yadavalli, V.R.; Tripura Venkata, V.G.N.; Neelamraju, S. Identification of major effect QTLs for agronomic traits and CSSLs in rice from Swarna/*Oryza nivara* derived backcross inbred lines. *Front. Plant Sci.* **2017**, *8*, 1027. [CrossRef]
27. Zhu, H.; Li, Y.; Liang, J.; Luan, X.; Xu, P.; Wang, S.; Zhang, G.; Liu, G. Analysis of QTLs on heading date based on single segment substitution lines in rice (*Oryza Sativa* L.). *Sci. Rep.* **2018**, *8*, 13232. [CrossRef]
28. Xing, Y.Z.; Tang, W.J.; Xue, W.Y.; Xu, C.G.; Zhang, Q. Fine mapping of a major quantitative trait loci, *qSSP7*, controlling the number of spikelets per panicle as a single Mendelian factor in rice. *Theor. Appl. Genet.* **2008**, *116*, 789–796. [CrossRef]
29. Zhang, K.; Zhang, Y.; Wu, W.; Zhan, X.; Anis, G.B.; Rahman, M.H.; Hong, Y.; Riaz, A.; Zhu, A.; Cao, Y. *qSE7* is a major quantitative trait locus (QTL) influencing stigma exsertion rate in rice (*Oryza sativa* L.). *Sci. Rep.* **2018**, *8*, 14523. [CrossRef]
30. Wang, X.; Zhou, T.; Li, G.; Yao, W.; Hu, W.; Wei, X.; Che, J.; Yang, H.; Shao, L.; Hua, J.; et al. A *Ghd7*-centered regulatory network provides a mechanistic approximation to the optimal heterosis in an elite rice hybrid. *Plant J.* **2022**, *Epub ahead of print*. [CrossRef]
31. Kang, Y.; Zhang, M.; Zhang, Y.; Wu, W.; Xue, P.; Zhan, X.; Cao, L.; Cheng, S.; Zhang, Y. Genetic mapping of grain shape associated QTL utilizing recombinant inbred sister lines in high yielding rice (*Oryza sativa* L.). *Agronomy* **2021**, *11*, 705. [CrossRef]
32. Zhu, J.; Niu, Y.; Tao, Y.; Wang, J.; Jian, J.; Tai, S.; Li, J.; Yang, J.; Zhong, W.; Zhou, Y. Construction of high-throughput genotyped chromosome segment substitution lines in rice (*Oryza sativa* L.) and QTL mapping for heading date. *Plant Breed.* **2015**, *134*, 156–163. [CrossRef]
33. Sun, Z.; Zhu, Y.; Chen, J.; Zhang, H.; Zhang, Z.; Niu, X.; Fan, Y.; Zhuang, J. Minor-effect QTL for heading date detected in crosses between indica rice cultivar Teqing and near isogenic lines of IR24. *Crop J.* **2018**, *6*, 291–298. [CrossRef]
34. Wang, G.; Wang, C.; Lu, G.; Wang, W.; Mao, G.; Habben, J.E.; Greene, T.W. Knockouts of a late flowering gene via CRISPR–Cas9 confer early maturity in rice at multiple field locations. *Plant Mol. Biol.* **2020**, *104*, 137–150. [CrossRef]
35. Lu, L.; Yan, W.; Xue, W.; Shao, D.; Xing, Y. Evolution and association analysis of *Ghd7* in rice. *PLoS ONE* **2012**, *7*, e34021.
36. Riaz, A.; Huimin, W.; Zhenhua, Z.; Zegeye, A.W.; Yanhui, L.; Hong, W.; Pao, X.; Zequn, P.; Xihong, S.; Shihua, C.; et al. Development of chromosome segment substitution lines and genetic dissection of grain size related locus in rice. *Rice Sci.* **2021**, *28*, 322–324. [CrossRef]
37. Yang, H.; Yang, Q.; Kang, Y.; Zhang, M.; Zhan, X.; Cao, L.; Cheng, S.; Wu, W.; Zhang, Y. Finding Stable QTL for plant height in super hybrid rice. *Agriculture* **2022**, *12*, 165. [CrossRef]
38. Sun, B.; Zhan, X.D.; Lin, Z.C.; Wu, W.X.; Yu, P.; Zhang, Y.X.; Sun, L.P.; Cao, L.Y.; Cheng, S.H. Fine mapping and candidate gene analysis of *qHD5*, a novel major QTL with pleiotropism for yield-related traits in rice (*Oryza sativa* L.). *Theor. Appl. Genet.* **2017**, *130*, 247–258. [CrossRef]
39. Murray, M.G.; Thompson, W.F. Rapid isolation of high molecular weight plant DNA. *Nucleic Acids Res.* **1980**, *8*, 4321–4325. [CrossRef]
40. Luo, Z.Y.; Zhou, G.; Chen, X.H.; Lu, Q.H.; Hu, W.X. Isolation of high-quality genomic DNA from plants. *Bull. Hunan Med. Univ.* **2001**, *26*, 178–180.
41. Creste, S.; Tulmann, N.A.; Figueira, A. Detection of single sequence repeat polymorphisms in denaturing polyacrylamide sequencing gels by silver staining. *Plant Mol. Biol. Rep.* **2001**, *19*, 299–306. [CrossRef]
42. Miao, J.; Guo, D.; Zhang, J.; Huang, Q.; Qin, G.; Zhang, X.; Wan, J.; Gu, H.; Qu, L.J. Targeted mutagenesis in rice using CRISPR-Cas system. *Cell Res.* **2013**, *23*, 1233–1236. [CrossRef] [PubMed]
43. Hiei, Y.; Komari, T.; Kubo, T. Transformation of rice mediated by *Agrobacterium tumefaciens*. *Plant Mol. Biol.* **1997**, *35*, 205–218. [CrossRef] [PubMed]
44. Alvarado, G.; Rodríguez, F.M.; Pacheco, A.; Burgueño, J.; Crossa, J.; Vargas, M.; Pérez-Rodríguez, P.; Lopez-Cruz, M.A. META-R: A software to analyze data from multi-environment plant breeding trials. *Crop J.* **2020**, *8*, 745–756. [CrossRef]
45. Alemu, A.; Brazauskas, G.; Gaikpa, D.S.; Henriksson, T.; Islamov, B.; Jorgensen, L.N.; Koppel, M.; Koppel, R.; Liatukas, Z.; Svensson, J.T.; et al. Genome-wide association analysis and genomic prediction for adult-plant resistance to septoria tritici blotch and powdery mildew in winter wheat. *Front. Genet.* **2021**, *12*, 661742. [CrossRef]
46. Meng, L.; Li, H.; Zhang, L.; Wang, J. QTL IciMapping: Integrated software for genetic linkage map construction and quantitative trait locus mapping in biparental populations. *Crop J.* **2015**, *3*, 269–283. [CrossRef]
47. McCouch, S.R. Gene nomenclature system for Rice. *Rice* **2008**, *1*, 72–84. [CrossRef]

plants

MDPI

Article

YTH Domain Proteins Play an Essential Role in Rice Growth and Stress Response

Weiwei Ma, Song Cui, Zhenfei Lu, Xiaofeng Yan, Long Cai and Yongfa Lu et al.

[1] Institute of Crop Sciences, Ningbo Academy of Agricultural Sciences, Ningbo 315000, China
[2] State Key Laboratory for Crop Genetics and Germplasm Enhancement, Jiangsu Plant Gene Engineering Research Center, Nanjing Agricultural University, Nanjing 210095, China
* Correspondence: srzhou@njau.edu.cn (S.Z.); linda2086415@163.com (X.W.)

Abstract: As the most prevalent epi-transcriptional modification, m^{6}A modifications play essential roles in regulating RNA fate. The molecular functions of YTH521-B homology (YTH) domain proteins, the most known READER proteins of m^{6}A modifications, have been well-studied in animals. Although plants contain more YTH domain proteins than other eukaryotes, little is known about their biological importance. In dicot species Arabidopsis thaliana, the YTHDFA clade members ECT2/3/4 and CPSF30-L are well-studied and important for cell proliferation, plant organogenesis, and nitrate transport. More emphasis is needed on the biological functions of plant YTH proteins, especially monocot YTHs. Here we presented a detailed phylogenetic relationship of eukaryotic YTH proteins and clustered plant YTHDFC clade into three subclades. To determine the importance of monocot YTH proteins, YTH knockout mutants and RNAi-induced knockdown plants were constructed and used for phenotyping, transcriptomic analysis, and stress treatments. Knocking out or knocking down *OsYTHs* led to the downregulation of multicellular organismal regulation genes and resulted in growth defects. In addition, loss-of-function *ythdfa* mutants led to better salinity tolerance whereas *ythdfc* mutants were more sensitive to abiotic stress. Overall, our study establishes the functional relevance of rice *YTH* genes in plant growth regulation and stress response.

Keywords: YTH domain protein; CRISPR; rice; growth; stress response

Citation: Weiwei Ma, Song Cui, Zhenfei Lu, Xiaofeng Yan, Long Cai and Yongfa Lu et al. YTH Domain Proteins Play an Essential Role in Rice Growth and Stress Response. *Plants* **2022**, *11*, 2206. https://doi.org/10.3390/plants11172206

Academic Editors: Xiangjin Wei, Yingxin Zhang, Weixun Wu and Guiai Jiao

Received: 19 July 2022
Accepted: 22 August 2022
Published: 25 August 2022

1. Introduction

In the past 30 years, epigenomic modifications (DNA modifications, RNA modifications, and histone modifications) have been widely studied and shown to play essential roles in regulating gene expression, stress resistance, development, and other key biological processes [1–4]. Current studies demonstrate that the epi-transcriptional modifications, especially the most prevalent m^{6}A modification, exist widely in eukaryotic RNAs and function in regulating RNA fate, such as alternative splicing, RNA export, 3′ untranslated region (UTR) processing, and RNA stability [5]. In plants, m^{6}A modification is an essential and extensive molecular mechanism in regulating organ development [6–14], circadian rhythm [15], fruit ripening [16], and stress tolerance [11,12,17–20]. A recent study also revealed that FTO-mediated m^{6}A demethylation in rice caused a more than threefold increase in grain yield [21]. As a critical post-transcriptional regulatory pathway, dynamic reversible m^{6}A modification after RNA transcription is an exciting and efficient new way to regulate gene expression, which provides a new approach to understanding epigenetic regulation in plants and further improves agronomic traits in plant breeding.

As the most known READER proteins of m^{6}A modifications, YT521-B homology (YTH) domain proteins are also found in most eukaryotic species and are well-studied, especially in vertebrates [5]. Three categories can be classified in animal YTH proteins, namely the YTHDF (YTH domain-containing family protein) family, YTHDC1 (YTH domain-containing protein 1, also called DC1), and YTHDC2 (YTH domain-containing protein 2, also called DC2). Most vertebrates have three YTHDF members, whereas only one YTHDF protein was found in invertebrate species. By influencing RNA stability, alternative splicing, or translation, YTHDFs mainly take part in stem cell differentiation/cancer progression [22–24] and neuronal function [25–31]. In contrast, YTHDC1 and YTHDC2 function in sex determination by affecting the alternative splicing and 3′ untranslated region (UTR) processing [28–30,32–38].

Compared to the restricted members of YTH proteins in animals, plants contain much more YTH proteins, whereas little is known about their biological importance. In Arabidopsis, the ECT2/3/4 are well-studied and important for cell proliferation and plant organogenesis. ECT2 forms a complex in vivo with m^6A RNA via its YTH domain, whereas plants depleted of ECT2 show a trichome branching defect [12]. Weaker trichome branching defects can also be observed in ect3 single mutants, whereas ect2/ect3 double mutants exhibit more severe trichome defects and delayed post-embryonic leaf formation [13]. By additional mutation of ECT4, ect2/ect3/ect4 triple mutants have exacerbated the slow growth of leaf primordia and serrated-edged leaves, which resembles that of mta knockdown plants [39]. Recently, Arribas-Hernández, Rennie et al., reveal the mode of RNA recognition by YTH domain-containing proteins (ECT2/3) using HyperTRIBE which coheres the genetic and molecular studies of m^6A-YTH function [20,40]. Other than YTHDF proteins, AtCPSF30L, as one of the YTHDC members, modulates nitrate content by regulating nitrate transport and assimilation in plants [41]. This raises the question, what are the functions of the rest of plant YTH proteins? Why does the plant need so many YTH proteins?

To answer these questions, we present our research on YTH proteins in a monocot model plant—rice. In this research, a detailed phylogenetic relationship of eukaryotic YTH proteins was solved, and a total of five subclades could be found within the plant YTHDF clade, and no YTHDC2 was found in plant species. To determine the importance of rice YTH proteins, YTH knockout mutants and RNAi-induced knockdown plants were constructed. Agricultural trait changes were observed within these transgenic plants, and transcriptomic data supported that the expression level of multicellular organismal regulation genes has changed. In addition, transcriptomic data showed that most mutants enriched stimulus response genes, indicating that *OsYTH* genes may also contribute to a stress response which was further supported by a stress tolerance assay. In conclusion, we hypothesize that the large number of YTH proteins in vascular plant species participated in balancing self-growth and adaptation to variant growth environmental changes that plants cannot avoid.

2. Results

2.1. Phylogenetic Analysis of YTH Proteins

Previous studies have conducted much work on the phylogenetic relationship of YTH family proteins, but either these works focused only on a specific group of species or YTH proteins in which the researchers are interested [42,43]. In this research, 561 YTH domain-containing proteins (Supplementary Tables S1–S3) from 37 plant species, 47 animal species, and two yeasts were collected, covering most eukaryotic categories. YTH domains were used for phylogenetic tree construction.

In coincidence with previous studies on animal YTH proteins, the phylogenetic tree of YTH proteins can be classified into three categories (Figure 1), which are YTHDC1, YTHDC2, and YTHDF family, representing the three categories described in animals [6]. All the plant YTHDC proteins are more closely related to YTHDC1 proteins, and no YTHDC2 homolog was found in plant species (Figure 1). Besides, the plant YTHDF proteins are paralogs to animal YTHDF proteins. The YTHDF genes of non-vascular plant species such as algae, liverworts, or mosses formed the outgroup of all plant YTHDF proteins whereas vascular plant YTHDF genes formed three well-supported clades YTHDFA/B/C (Support value = 95, 97, 98, respectively) as the previous study suggested [12]. Furthermore, by adding the basal angiosperm species *Amborella trichopoda* into the construction of the phylogenetic tree, three subclades can be recognized in the YTHDFC clade and named DFCI (including

OsYTHDF2C, OsYTHDF3C, ECT6, and ECT7), DFCII (including OsYTHDF1C, and ECT8), DFCIII (including OsYTHDF4C, OsYTHDF5C, and ECT11) with high support value (98, 92, 99, respectively) (Figure 1). Each YTHDFC subclade contains one AtrYTHDFC protein and 0–4 YTHDFC proteins from other angiosperm species (Supplementary Table S1).

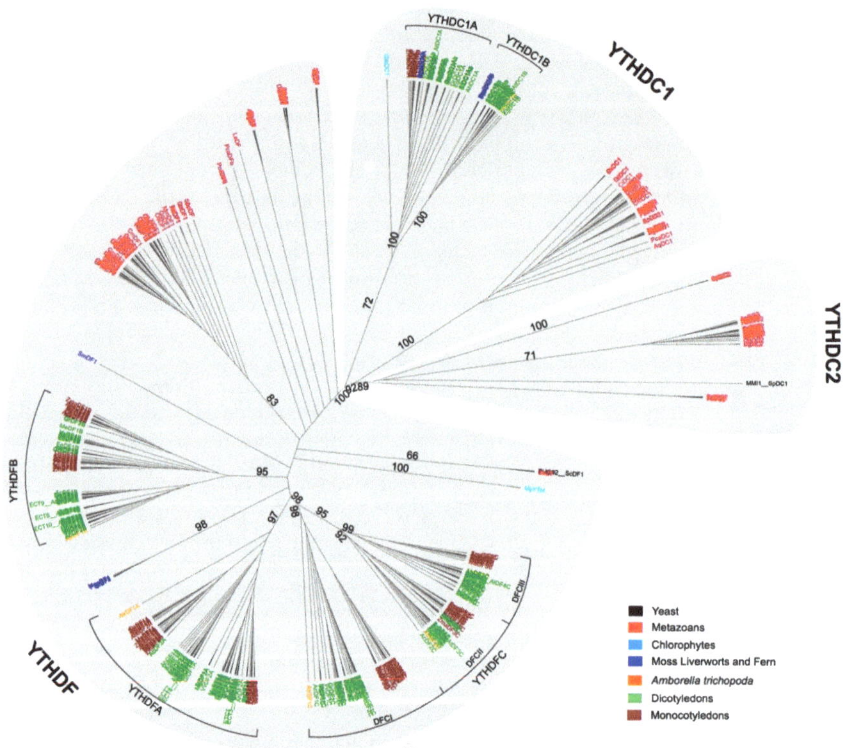

Figure 1. Phylogenetic tree of YTH domains from animal and plant YTH proteins using maximum likelihood method. Phylogenetic analysis of the eukaryotic YTH family. An unrooted maximum likelihood tree was constructed with iqtree focusing on major model species representing a wide taxonomic range. Numbers on branches indicate the percentage of trees in which associated taxa clustered together during bootstrap analysis (1000 bootstraps). Only the values on the main clades are shown.

Unlike the conserved YTHDC1 genes in all species collected in our research (most animal species have only one copy of YTHDC1), plant YTHDC1 can be divided into two subclades, DC1A and DC1B (Figure 1). Moss, liverworts, fern, and most dicots have both DC1A and DC1B proteins, whereas monocot plants have only DC1A protein, which indicates that a loss event has occurred in the common ancestor of modern monocot species. Other than YTHDC1 proteins, the number of YTHDFs also varied from species to species (Supplementary Tables S1 and S2). Most non-vascular or non-vertebrate species have only limited YTHDF genes (usually 1–2), whereas vascular or vertebrate species have more. Within Chondrichthyes and vertebrate animals, exactly three YTHDF genes can be recognized. The only exception is in species from Salmoniformes and Neoteleostei families, which have an additional copy of the YTHDF1 gene (Figure 1 and Supplementary Table S2). In contrast, vascular plants have more YTHDF genes, which usually vary from 4 to 20 (Supplementary Tables S1 and S2).

Numerous works have been conducted to reveal the molecular function of animal YTHs, whereas a few focused on Arabidopsis YTH proteins (ECT2/3/4 and AtCPSF30L). To obtain an initial understanding of the functions of YTH domain proteins in monocot plants, rice YTH proteins were used for further study.

2.2. Expression Analysis of the OsYTH Genes in Different Tissues/Organs

Seven tissues/organs (young seedling, root, stem, leaf, sheath, young panicle, and matured panicle) at different development stages were used for tissue expression pattern analysis. The result showed that most *OsYTHs* are ubiquitously expressed across plant development, except for *OsYTHDF5C*, which is merely expressed in tissues other than young seedlings (Figure 2). Among the tissues tested, the *OsYTHs* showed relatively higher expression levels in sheath and panicles than other tissues. *OsYTHDF1A* showed the highest expression level of all the *YTH* genes (Figure 2A). The other two *YTHDFA* members also showed higher expression levels than *YTHDFB* and *YTHDFC* genes (Figure 2). These results are consistent with the previous study that *ECT2* and *ECT3* are the most highly and widely expressed members of the YTH family in *Arabidopsis* [13]. The rest of the *YTH* members showed relatively lower expression levels, with *OsYTHDF2B* and *OsYTHDF1C* performing the highest expression level in each clade (Figure 2B–D).

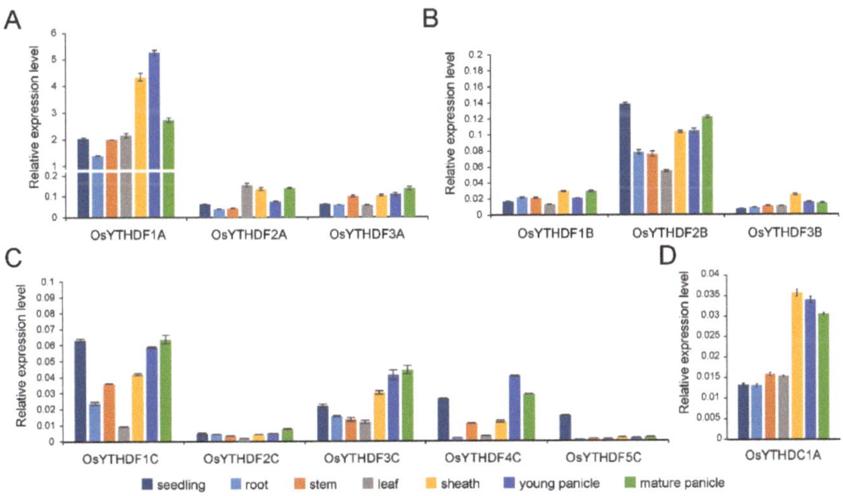

Figure 2. The tissue expression patterns of the *YTHDFA* (**A**), *YTHDFB* (**B**), and *YTHDFC* (**C**) members and *YTHDC1A* (**D**) in different tissues of rice. Different colors denoted different tissues. Values are means ± SD of three independent biological replicates.

2.3. Characterization of Loss-of-Function Mutants and Knockdown Plants

To learn the function of YTH proteins in rice, the CRISPR-Cas9 system was used to generate loss-of-function mutations. Except for *OsYTHDF2C* and *OsYTHDC1A*, ten loss-of-function mutants with either an insertion or deletion of one or few bases at the target sequences which caused early termination were successfully gained and confirmed by sequencing (Supplementary Figures S1–S4). For *OsYTHDF2C*, although several lines carrying a loss-of-function mutation were detected in T0 generation and seeds were collected (Supplementary Figures S1C and S4B), all the T1 seeds failed to germinate. On the other hand, osythdc1a mutations were detected in multiple transgenic lines, but all the mutations did not lead to frameshift changes (Supplementary Figures S1D and S4F). These results indicate that OsYTHDF2C and OsYTHDC1A might be very important for plant development, whose loss-of-function mutation will be lethal for plants. For these two genes, three lines of siRNA-induced knockdown plants were carried out for each gene whose expression level has been confirmed by RT–PCR results (Supplementary Figure S5) and used for further studies.

In all the mutants/knockdown plants, agricultural trait changes have been observed (Figure 3 and Supplementary Figure S6). Most mutants/knockdown plants showed decreases in plant height, the number of spikelets per panicle, setting rate, and grain weight (Figure 3A,D–F and Supplementary Figures S6A,D–F), whereas the heading date did not change too much in either mutation or knockdown plants (Figure 3C and Supplementary Figure S6C). Loss-of-function of *YTHDFA* clade protein led to a significantly decreased tiller number whereas other mutants/knockdown plants performed conversely (Figure 3B and Supplementary Figure S6B). These results indicated that the rice YTH proteins are essential for shaping the plant architecture.

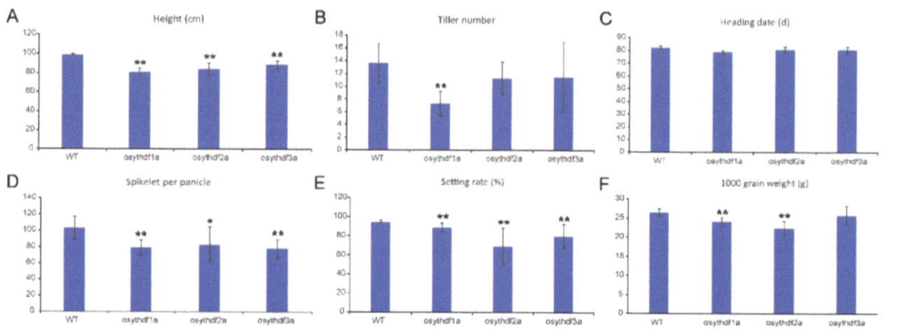

Figure 3. Quantification of several agricultural traits of *DFA* clade mutants including plant height (**A**), tiller number (**B**), heading date (**C**), spikelets number per panicle (**D**), setting rate (**E**), and 1000 grain weight (**F**). Values are shown as mean ± SD. The significance of difference was determined by Student's *t*-test (*, $p < 0.05$; **, $p < 0.01$).

2.4. OsYTHs Are Essential for Plant Growth

Transcriptomic analyses were carried out for a deeper view of "how does lack of YTH proteins interference with the plant". Thousands of significantly differentially expressed genes (DEGs) were found within *yth* mutants or knockdown plants compared to the wildtype transcriptomic data (Figure 4A and Supplementary Figure S7).

Within the *YTHDFA* clade, which showed the most significant agricultural trait changes, 1139 different expression genes (DEGs) were shared by all three *ythdfa* mutants (Figure 4A), with the multicellular organismal development process enriched by GO analysis (Figure 4B). A total of 77 organismal development-related genes exhibited expression changes in *YTHDFA* clade mutants, out of which 65 genes were significantly decreased, including *OsYABBY5* [44], *OsCD1* [45,46], *OsAPO2* [47,48], *OsARF5* [49], *OsAGO1c* [50], *SD1* [51], *OsAS2* [52], whose mutants/knockdown plants were previously reported and showed developmental defects (Figure 4C and Supplementary Table S4). Other than *YTHDFA* members, the multicellular organismal development process was also enriched in *osythdf4c* and *RNAi-OsYTHDC1A* plants (Supplementary Figure S8). In combination with the mutation phenotype of these genes, these results indicate that rice YTH domain-containing proteins, mainly *YTHDFA* clade members and *OsYTHDC1A*, affect plant architecture by regulating the expression of organismal development-related genes.

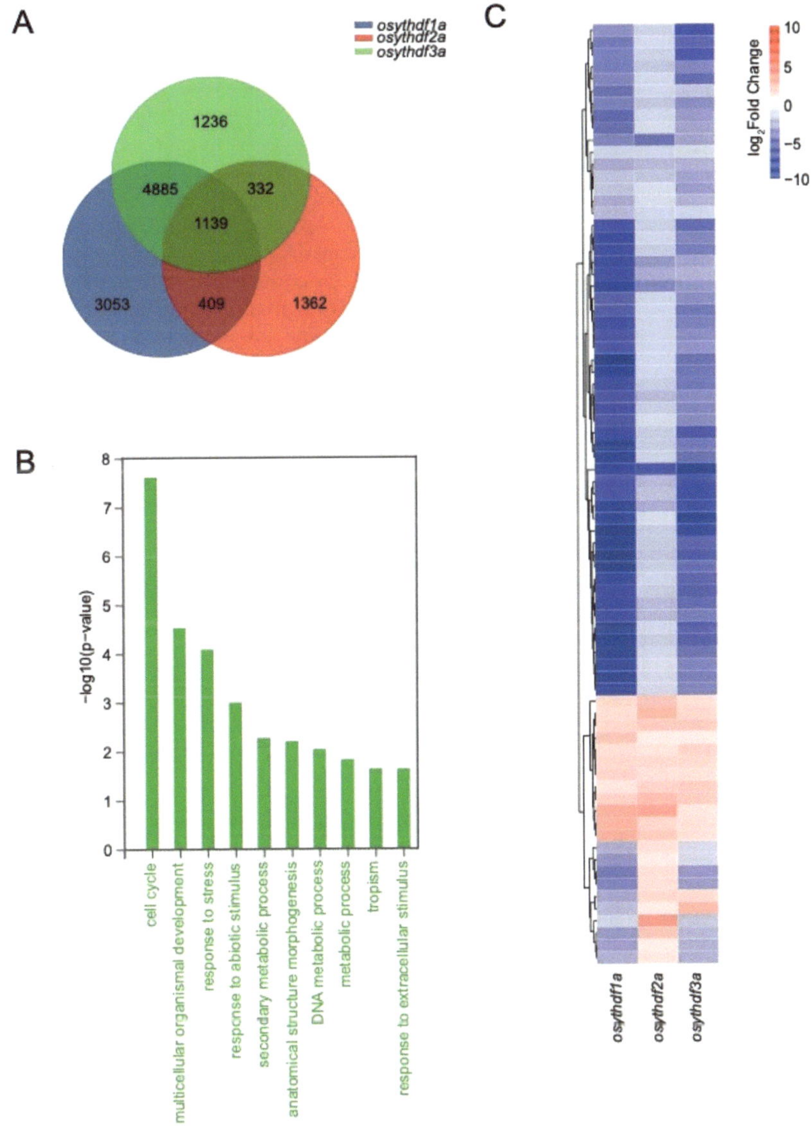

Figure 4. The Venn diagram, GO enrichment analysis, and heatmap of DEGs in YTHDFA clade mutants. (**A**) The Venn diagram showing unique and shared DEGs in YTHDFA mutants. (**B**) The GO enrichment analysis of 1139 DEGs shared within YTHDFA clade. (**C**) Expression heatmap of 77 organismal development-related genes. Heatmap was drawn according to log_2FC values, which were calculated pairwise based on the expression level of each mutant and wild-type plant.

2.5. YTH Proteins Are Required for Stress Response

Previous studies have proven that m^6A modifications and ECT2 take part in stress response [11,12,17–19]. Interestingly, "response to biotic stimulus" and "response to abiotic stimulus" were also commonly enriched by GO analysis in most mutants/knockdown plants, except for *osythdf3a* (Supplementary Figure S8), indicating that the rice YTH proteins also take part in stress response progresses. Among the DEGs related to "response to biotic stimulus" and "response to abiotic stimulus", 38 abiotic response genes (including response to drought, salt, cold, heat, ABA, Ion deficiency, and UV damage), 17 biotic response genes (including response to rice blast, bacterial blight, sheath blight, rice dwarf, planthopper, etc.) and 38 LRR disease resistance genes were recognized. Heat map clustering showed that the mutants that carried YTHDFA clade mutations have a similar pattern in both biotic and abiotic stress response genes. In contrast, mutants that carried YTHDFB or YTHDFC clade mutations gathered into another clade (Figure 5, Supplementary Tables S6 and S7) (only mutants were used for heat map analysis due to relatively low expression changes in knockdown plants). Around half of the abiotic stress response genes (17 out of 38) and a third of the biotic stress response genes (20 out of 55) showed an opposite expression change from ythdfa clade mutants to others (Figure 5). A total of 21 LRR genes were significantly upregulated with three downregulated in ythdfb and ythdfc mutants whereas the ythdfa mutants did not show significant changes (Figure 5B).

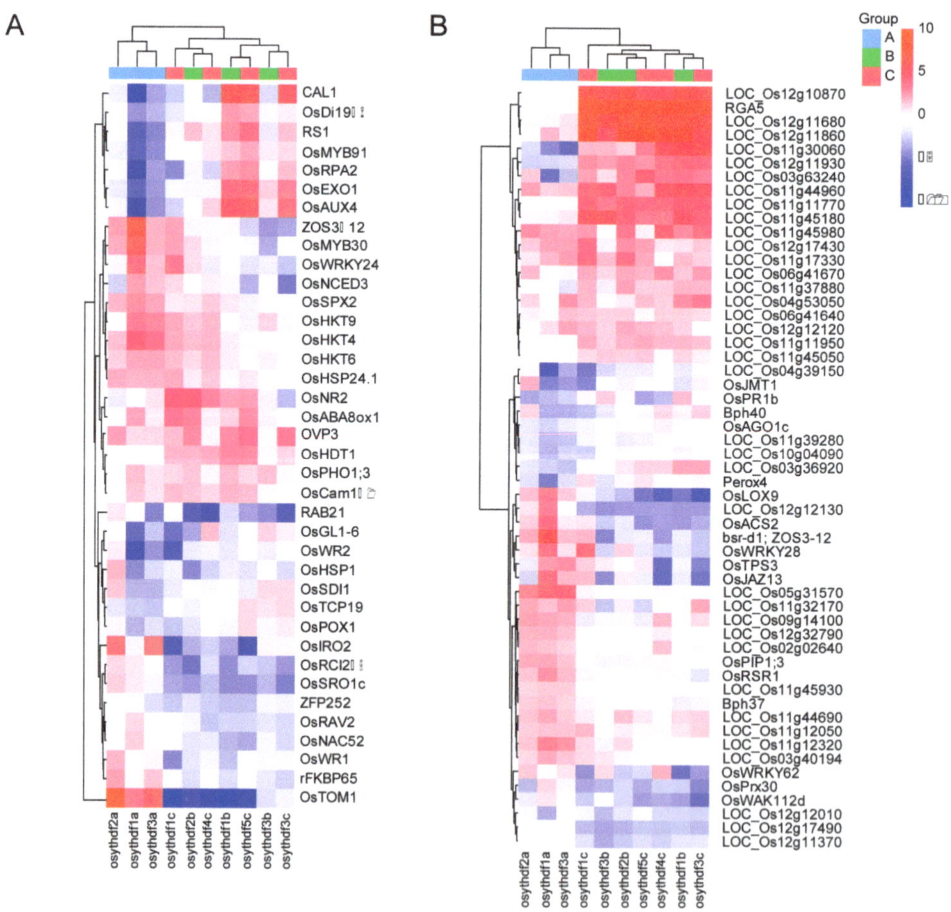

Figure 5. Expression heatmap of 38 abiotic response genes (**A**) and 55 biotic response genes (**B**) which have been previously reported. Heatmap was drawn according to log2FC values, which were calculated pairwise based on the expression level of each mutant and wild-type plant.

As the transcriptomic result reveals that the OsYTHs take part in abiotic stress responses, four abiotic stress response patterns were tested in a cultivated variety of Nipponbare, including salt, drought/PEG-induced osmotic stress, and cold. As ABA plays a vital role in the stress response and tolerance of plants, the ABA response was also determined. Most *OsYTH* genes significantly upregulated after abiotic stress treatment for 3–6 h, whereas some *OsYTHs* responded later (Supplementary Figure S9). In contrast, most YTH genes didn't show significant expression changes under ABA treatment. These results indicated that the *OsYTH* genes take part in stress response and might not through the ABA signal pathway.

To further investigate the connection between YTH proteins and phenotypes under abiotic stress, loss-of-function mutants and knockdown plants were treated with 100 mM/150 mM NaCl as described in the methods section. At the end of the 100 mM NaCl treatment, all plants showed complete growth retardation with shorter shoots/roots, lower water content, and lower biomass (Figure 6). In comparison to the WT (DJ for *YTHDFA* clade mutants and NIP for *YTHDFB/YTHDFC* clade mutants and *RNAi-OsYTHDC1A*), all the *ythdfa* mutants showed significantly lower growth retardation on shoot/root length and biomass, whereas the *ythdfb* and *ythdfc* mutants showed marked reduction (Figure 6A,B,D). The percentage of water content was not substantially different between WT and most mutants (Figure 6C). In order to determine the survival rate under salt stress, 150 mM NaCl was applied to all the wild-type plants and mutants with different treatment times (7 days for assays 1 and 2, 8 days for assay 3). Most *ythdfc* mutants and *RNAi-OsYTHDC1A* plants showed lower survival rates than the wild-type for all the three assays, whereas *ythdfb* mutants remained similar to NIP (Supplementary Figure S10A). The *ythdfa* mutants did not show significant changes under the 7-day treatment, but the survival rates of *osythdf2a* and *osythdf3a* were significantly higher when the salt treatment time was extended to 8 days (Supplementary Figure S10A,B). In conclusion, the loss of *ythdfa* genes results in better tolerance to salinity treatment, with *ythdfc* and *ythdc1a* mutants being more sensitive to salt stress. These results indicated that apart from regulating plant growth, which is mainly related to the function of YTHDFA clade members and YTHDC1A, OsYTHs also contribute to stress response progresses.

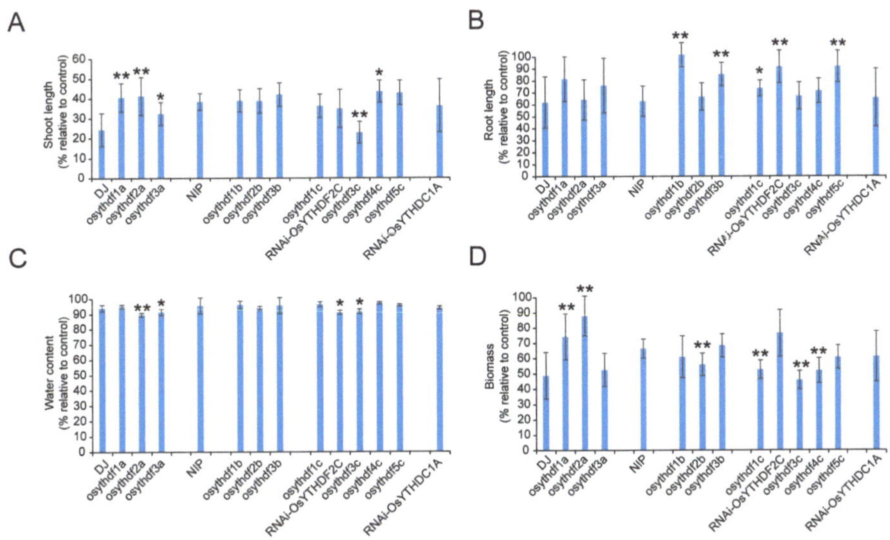

Figure 6. Phenotypic evaluation of YTH mutants/knockouts under salt stress. The shoot length (**A**), root length (**B**), water content (**C**), and biomass (**D**) values correspond to a % of the change in salt relative to control. Values are shown as mean ± SD. The significance of the difference was determined by Student's *t*-test (*, $p < 0.05$; **, $p < 0.01$).

3. Discussion

In this research, we constructed a fine phylogenetic tree of YTH proteins covering most of the eukaryotic categories, which helped us to solve the relationship between animal YTH proteins and plant YTH proteins, clustered the plant YTH proteins, and further raised our interest in the importance of numerous YTH proteins in plants, especially monocot plants. Expression analysis showed that most *YTH* genes have ubiquitous expression patterns with the exception of *OsYTHDF5C*. The transcriptomic analysis of knockout mutants and knockdown plants revealed that *OsYTH* genes might take part in growth regulation and stress responses progress. The in-field phenotype proved that the loss of *YTHDFA* clade members and *OsYTHDC1A* would lead to defects in plant architecture development. Other than growth regulation, the *OsYTHs* also contribute to stress response progress which was proved by stress response analysis and salinity tolerance assay.

One of the main functions of animal YTHs is to regulate stem cell differentiation/cancer progression [22–24]. *Arabidopsis* ECT2/3/4 are also reported to be important for organogenesis and cell proliferation through binding to the m^{6}A site and affecting mRNA stability [6,12,13,53,54]. Single mutants in *AtYTH* genes do not exhibit obvious developmental phenotypes other than abnormal trichome branching, and double or triple mutations in *ECT2/ECT3/ECT4* lead to delayed growth and aberrant morphology [6,12,13,53]. In this study, most single mutants of *OsYTHs* showed multiple growth defects, including decreased plant height, changed tiller number, and smaller panicles, showing that rice YTH proteins play similar roles in regulating plant architecture. The transcriptomic results also supported that the knockout of rice *YTHDFA* clade members, *OsYTHDF4C*, and knockdown of *OsYTHDC1A*, will lead to expression changes of multicellular organismal development-related genes. Notably, the heading date did not change in both mutation and knockdown plants, indicating that the rice YTH proteins do not take part in regulating the timing of flowering. In addition, we also found that *OsYTHDF2C* and *OsYTHDC1A* loss-of-function mutations are lethal for plants. Previous studies have also proved that *ect2/ect3/ect4* triple mutations slow down the growth of leaf primordia, but do not affect the initiation timing of leaf primordia [53]. These results indicate that the m^{6}A -YTH regulatory module plays a vital role in plant organogenesis by regulating the expression level of organogenesis-related genes, especially in monocot species.

Qian and his team member have reported that the nuclear YTHDF2 preserves 5′UTR methylation of stress-induced transcripts under heat stress, further promoting cap-independent translation initiation [55]. Aside from the response to abiotic stress, m^{6}A modification and YTHDF2 also act as critical regulators of immune cell homeostasis by mediating the gene expression of immune-related signaling pathways [56,57]. In plants, stress response genes were upregulated in ect2/3/4 mutants, although possibly through indirect effects [20], and the expression pattern of *YTH* genes can also be induced by biotic and abiotic stress [58,59]. In this study, we found that not only do the expression levels of *OsYTHs* respond to stress stimulus but also that the *OsYTHs* take part in a stress response by regulating the expression of stress response genes. Transcriptomic results showed that the expression levels of numbers of the stress response genes were changed in most *osyth* mutants. NaCl treatment also demonstrated that *OsYTHs* are important for abiotic stress resistance, especially for YTHDFA clade and YTHDFC clade members. Aside from abiotic stress, biotic stimulus response genes were also enriched in transcriptomic results. The function of OsYTHs in biotic stress remains further studied. In addition, the extra YTHDF1-like gene found in five species of Salmoniformes and Neoteleostei (Supplementary Table S2) which usually lived in harsh environments (such as highly variable salinity, temperature fluctuations, low oxygen levels, and heavily polluted ecosystems) also reminds us that this additional YTHDF1-like gene might contribute to their adaptation to different environments. After all, these findings indicate that YTH domain-containing proteins and m^{6}A modifications might be related to the adaptability to the environment not only in animals but also in plants.

With the help of the phylogenetic tree we constructed in this study, we can discover that the expansion event mainly occurred after the initiation of land plants, which have more complicated organs (leaf, stem, roots, flowers, and fruits) and live in more complicated environments (Figure 1). Transcriptomic analysis and phenotype evidence, together with the previous study on ECT2/3/4 [6,13,14,54], indicated that the land plant YTHDFA clade members contributed to the organogenesis progress. In addition, knocking out YTHDFA clade members results in growth defects and increased tolerance to abiotic stress, whereas knocking out YTHDFC clade members (especially YTHDFCI and YTHDFCII subclade members) results in more tiller number and more sensitivity to abiotic treatments. Taken together, we hypothesized that the numerous numbers of plant YTH proteins balanced the development of multicellular organisms and their responses to environmental stimuli, which allowed the land plants to generate various tissue types and adapt to different living environments.

Numerous studies have shown that m^6A and YTH domain-containing proteins play important roles in animals, including cell differentiation [22–24], learning, memory, and fitness to environment [25–31]. It attracts attention to the study of m^6A and its readers in plants. In this study, we mainly focused on the phenotype and transcriptomic changes (especially organogenesis and stress response) of loss-of-function mutants and knockdown plants and showed that the plant YTH proteins are not only crucial for plant architecture but also take part in stress response, which, given an initial work on the function of YTH domain-containing proteins in rice, provides an advancement in the further studies of the biological functions of plant YTH proteins. Substantial work is left to be conducted on screening out the target genes of YTH proteins and exploring the molecular mechanisms of YTH-mediated post-transcriptional regulating systems.

4. Materials and Methods

4.1. Sequence Selection, Multiple Sequence Alignments, and Phylogenetic Reconstruction

BLAST searches (blastp) were performed starting from known Human and Arabidopsis YTH domains on 86 species representing the diversity of the Eukaryote lineage at the Uniprot (https://www.uniprot.org/blast/ (accessed on 18 August 2022)), NCBI (https://www.ncbi.nlm.nih.gov/ (accessed on 18 August 2022)), and Phytozome 13 (https://phytozome-next.jgi.doe.gov/ (accessed on 1 November to 15 December 2018)). Each time a new YTH was found in a given species, it was used as a query in a new BLAST search until no new YTH gene was found. Sequences annotated as "partial sequence" or "low-quality sequence" were discarded, and other YTH sequences were aligned using the MUSCLE v3.8.425 [60] on Geneious v2021 software using standard parameters. Maximum likelihood (ML) trees were reconstructed using the iqtree web server [61] (http://iqtree.cibiv.univie.ac.at/ (accessed on 18 August 2022)) with the JTT + I + G4 amino acids replacement matrix. Main branch support values were estimated from bootstrap analyses of 1000 replicates. All manipulations on phylogenetic trees were performed with Figtree (v1.3.1).

4.2. Growth Conditions and Abiotic Stress Treatment

Oryza sativa sub. Japonica var. Nipponbera (NIP) and Dongjin (DJ) were used as the wild-type ecotypes. Transgenic and wild-type seeds were soaked in water for germination at 37 °C for three days and then grown in a paddy field in Nanjing. For tissue expression pattern analysis and transcriptomic analysis, wild-type and transgenic plants were grown in climate chambers (BES1500QH-LED, BOERSI) after germination with 70% humidity and 14 h light/10 h dark photoperiod for two weeks before sampling. For determination of YTH gene expression response pattern to abiotic stress, 14-day-old NIP seedlings were grown in nutrient solution supplemented with 150 mM NaCl/100 mM ABA/20% PEG4000 for salt/ABA/PEG treatment for previously determined time, respectively. Cold treatment was conducted under 5 °C. For determining the salinity tolerance, germinated seeds were grown in the nutrient solution with or without (mock) 100 mM NaCl under the same growth condition at 25 °C under 14 h light/10 h dark in a growth room. The plant height, root length, water content, and biomass were measured after 14 days. For determining the survival rates under salt stress, 14-day-old seedlings were subjected to the nutrient solution containing 150 mM NaCl for seven days (for assays 1 and 2) or eight days (for assay 3) at 25 °C under 14 h light/10 h dark in a growth room and subsequently grown in the nutrient solution for recovery. The survival rates were calculated after recovery for seven days.

4.3. Vector Construction and Rice Transformation

To knock out the OsYTH genes, 18-bp gene-specific spacer sequences of each target gene were cloned into the entry vector pCRAC harboring both Cas9 and sgRNA segments. Mutation type was determined by PCR and sequencing.

To knock down OsYTHDF2C and OsYTHDC1A, both anti-sense and sense versions of specific 416 bp or 418 bp fragments from the coding region of either gene were amplified and successively inserted into the SacI and SnaBI restriction site of the LH-FAD1390RNAi binary vector to form the RNAi construct.

All the resulting plasmids were transformed into rice variety Nipponbera (NIP, as transgenic receptor for ythdfb/ythdfc/ythdc1a knockout/knockdown plants) and Dongjin (DJ, as transgenic receptor for ythdfa knockout mutants) by Agrobacterium-mediated transformation and subsequently confirmed by PCR sequencing. Primer sequences for vector constructions and sequencing are listed in Supplementary Table S5.

4.4. Measurement of Morphological Traits

For measuring the plant height, tiller number, and heading date of each loss-of-function mutant/knockdown, 6 individuals were selected and measured manually in the field. For panicle traits, at least 9 panicles from 3 individuals were selected and dried at 50 °C for 5 days. The number of empty grains was counted manually and the number of filled grains and 1000-grain weight were measured by SC-G automatic seeds test and thousand kernels weightier (WANSHEN). The setting rate was calculated by numbers of filled grains/numbers of total grains.

For the salinity-tolerant assay, the shoot length (cm) was measured from the base of the stem to the tip of the topmost leaf of the plants. Root length (cm) was measured for each plant. Fresh weight (FW) was determined immediately after collecting the samples, followed by tissue drying at 50 °C for 7 days for the determination of dry weight (DW). Water content was estimated as follows $\%WC = (FW-DW)/DW) \times 100$ and biomass refers to the dry weight. All parameters are presented as percentages relative to control plants.

4.5. Total RNA Isolation and RT–PCR Analysis

Young seedlings were collected 14 days after germination and leaf blades, sheath, stems, and panicles were collected 60 days after germination. Total RNA were isolated from frozen tissues using a ZR Plant RNA MiniPrep Kit (ZYMO Research, Beijing, China) and reverse transcribed using a QuantiTect reverse transcription kit (Qiagen, Shanghai, China) according to the manufacturer's protocol. RT–PCR was performed with an SYBR premix Ex Taq Kit (TaKaRa, Kusatsu, Japan) according to the operation manual and amplified in an ABI 7500 using primers listed in Supplementary Table S5. Data from three biological replicates were analyzed following the $\Delta\Delta$CT method.

4.6. Transcriptomic Analysis

For transcriptomic analysis, three 2-week-old seedlings from two or three different transgenic lines were sampled and total RNA were extracted. Experiments were conducted following standard procedures of Shanghai Applied Protein Technology, Co., Ltd. (APT, Shanghai, China). Briefly, total RNAs were extracted with RNAprep Pure Plant Kit (TIANGEN, Beijing, China) and quantified by NanoDrop 2000C spectrophotometer (Thermo Fisher Scientific, Waltham, MA, USA). RNA integrity was checked by RNA Nano 6000 Assay Kit of the Agilent Bioanalyzer 2100 system (Agilent Technologies, Santa Clara, CA, USA). For cDNA synthesis, a total 1 μg RNA for each sample was treated with DNase

I to eradicate the genomic DNA and then used as a template for reverse transcription (QuantiTect Reverse Transcription Kit, Qiagen, China).

We added fragment buffer to break into short segments using short segment RNA as template. Sequencing libraries were created using NEB Next Ultra RNA Library Prep Kit following manufacturer's instructions. The index codes were added to each sample. Paired-end cDNA libraries with an insert size of 300 bp were constructed for transcriptome sequencing and sequenced on Illumina HiSeq 4000 platform (Illumina Inc., San Diego, CA, USA) at Shanghai Applied Protein Technology, Co., Ltd. (APT, Shanghai, China).

Supplementary Materials: The following supporting information can be downloaded at: https://www.mdpi.com/article/10.3390/plants11172206/s1, Figure S1: CRISPR/Cas9-induced mutation types of YTHDFA clade (A), YTHDFB clade (B), YTHDFC clade (C), and YTHDC clade (D) mutants in rice T1 transgenic plants (18bp target sequences were showed in blue with PAM in red). Figure S2: The multiple-sequence alignment of wild-type and mutation-type YTHDFA clade proteins used in this study. CRISPR-Cas9 system induced mutations lead to a serial of amino acid changes and early termination. Red box above the sequences shows the YTH domain. Figure S3: The multiple-sequence alignment of wild-type and mutation-type YTHDFB clade proteins used in this study. CRISPR-Cas9 system induced mutations lead to a serial of amino acid changes and early termination. Red box above the sequences shows the YTH domain. Figure S4: The multiple-sequence alignment of wild-type and mutation-type YTHDFC clade proteins (A–E) and OsYTHDC1A (F) used in this study. CRISPR-Cas9 system induced mutations lead to a serial of amino acid changes and early termination. Red box above the sequences shows the YTH domain. Figure S5: (A,B) Expression fold change of OsYTHDF2C (A) and OsYTHDC1A (B) mRNA in knockdown plants. The values obtained with wild type plants were referred to as 1 and shown as mean ± SD. Figure S6: Quantification of several agricultural traits of mutants/RNAi plants except DFA clade mutants, including plant height (A), tiller number (B), heading date (C), spikelets number per panicle (D), setting rate (E), and 1000 grain weight (F). Values are shown as mean ± SD. The significance of difference was determined by Student's-test (*, $p < 0.05$; **, $p < 0.01$). Figure S7: The Venn diagrams show unique and shared DEGs in YTHDFB (A) and YTHDFC (B) mutants. Figure S8: GO enrichment analysis of YTH mutants/knockdown plants. A–L represent for the GO enrichment results for genes differentially expressed in osythdf1a (A), osythdf2a (B), osythdf3a (C), osythdf1b (D), osythdf2b (E), osythdf3b (F), osythdf1c (G), RNAi-OsYTHDF2C (H), osythdf3c (I), osythdf4c (J), osythdf5c (K), and RNAi-OsYTHDC1A (L) compared with wild type. "Multicellular organismal development", "Response to stress", "Response to abiotic stimulus", and "Response to biotic stimulus" were marked in black. Figure S9: Expression patterns of OsYTH genes under different abiotic stresses (salt, drought, PEG, cold) and exogenous ABA application. Expression data are means of three independent biological replicate and normalized with the relative expression levels at the same treatment time under mock treatment. Figure S10: OsYTH genes contribute to salt stress response progress (A) percentage survival rates (Percentage is the ratio of number of plants with green new leaves to number of total plants) of WT/mutant seedlings after 150 mM salt stress treatment in three assays. About 20 seedlings were used in each assay. (B) phenotypes of WT/mutant seedlings under salt stress for 8 days and recovery for 7 days (assay 3), Table S1: Gene numbers of plant YTH genes. Table S2: Gene numbers of animal and yeast YTH genes. Table S3: YTH proteins used in phylogenetic analysis. Table S4: the relative expression level of 77 organismal-development-related genes used for heatmap analysis. Table S5: Primers used in this study. Table S6: Log2 expression changes of abiotic response genes in OsYTH mutants. Table S7: Log2 expression changes of biotic-response genes in OsYTH mutants.

Author Contributions: Conceptualization, methodology, software, formal analysis: W.M., X.W., S.C. and Z.L.; investigation, X.Y. and L.C.; validation, X.W.; resources, Y.L., K.C., H.Z. and R.M.; data curation, S.Z.; writing—original draft preparation, W.M.; writing—review and editing, X.W.; supervision, S.Z.; project administration, X.W.; funding acquisition, W.M. All authors have read and agreed to the published version of the manuscript.

Funding: This research was funded by the NAAS on NAAS-CAAS cooperation innovation project (2019CXGC001), Ningbo Science & Technology bureau on Ningbo modern seed industry technology innovation project (2019B10001), and Science technology department of Zhejiang Province on Special Project for breeding of new agricultural varieties in Zhejiang province (2021C02063-1-4).

Data Availability Statement: The raw sequence data can be downloaded at https://www.ncbi.nlm.nih.gov/geo/query/acc.cgi?acc=GSE207213 (accessed on 18 August 2022).

Acknowledgments: W.M. is thankful to Dan Ohtan Wang, Laboratory for Neuroepitranscriptomics, RIKEN Center, Kobe, Japan for inspiring and encouraging him on this research.

Conflicts of Interest: The authors declare no conflict of interest.

References

1. Bewick, A.J.; Schmitz, R.J. Gene body DNA methylation in plants. *Curr. Opin. Plant Biol.* **2017**, *36*, 103–110. [CrossRef] [PubMed]
2. Zhang, H.; Lang, Z.; Zhu, J. Dynamics and function of DNA methylation in plants. *Nat. Rev. Mol. Cell Biol.* **2018**, *19*, 489–506. [CrossRef] [PubMed]
3. Zheng, H.; Sun, X.; Zhang, X.; Sui, N. m6A Editing: New Tool to Improve Crop Quality? *Trends Plant Sci.* **2020**, *25*, 859–867. [CrossRef] [PubMed]
4. Whitaker, J.W.; Chen, Z.; Wang, W. Predicting the human epigenome from DNA motifs. *Nat. Methods* **2015**, *12*, 265–272. [CrossRef] [PubMed]
5. Patil, D.P.; Pickering, B.F.; Jaffrey, S.R. Reading m6A in the Transcriptome: m6A-Binding Proteins. *Trends Cell Biol.* **2018**, *28*, 113–127. [CrossRef]
6. Wei, L.; Song, P.; Wang, Y.; Lu, Z.; Tang, Q.; Yu, Q.; Xiao, Y.; Zhang, X.; Duan, H.; Jia, G. The m6A Reader ECT2 Controls Trichome Morphology by Affecting mRNA Stability in Arabidopsis. *Plant Cell* **2018**, *30*, 968–985. [CrossRef] [PubMed]
7. Zhong, S.; Li, H.; Bodi, Z.; Button, J.; Vespa, L.; Herzog, M.; Fray, R.G. MTA is an Arabidopsis messenger RNA adenosine methylase and interacts with a homolog of a sex-specific splicing factor. *Plant Cell* **2008**, *20*, 1278–1288. [CrossRef]
8. Bodi, Z.; Zhong, S.; Mehra, S.; Song, J.; Li, H.; Graham, N.; May, S.; Fray, R.G. Adenosine methylation in Arabidopsis mRNA is associated with the 3′ end and reduced levels cause developmental defects. *Front. Plant Sci.* **2012**, *3*, 48. [CrossRef]
9. Shen, L.; Liang, Z.; Gu, X.; Chen, Y.; Teo, Z.W.N.; Hou, X.; Cai, W.M.; Dedon, P.C.; Liu, L.; Yu, H. N6-methyladenosine RNA modification regulates shoot stem cell fate in Arabidopsis. *Dev. Cell* **2016**, *38*, 186–200. [CrossRef]
10. Růžička, K.; Zhang, M.; Campilho, A.; Bodi, Z.; Kashif, M.; Saleh, M.; Eeckhout, D.; El-Showk, S.; Li, H.; Zhong, S.; et al. Identification of factors required for m6A mRNA methylation in Arabidopsis reveals a role for the conserved E3 ubiquitin ligase HAKAI. *New Phytol.* **2017**, *215*, 157–172. [CrossRef]
11. Anderson, S.J.; Kramer, M.C.; Gosai, S.J.; Yu, X.; Vandivier, L.E.; Nelson, A.D.L.; Anderson, Z.D.; Beilstein, M.A.; Fray, R.G.; Lyons, E.; et al. N6-Methyladenosine Inhibits Local Ribonucleolytic Cleavage to Stabilize mRNAs in Arabidopsis. *Cell Rep.* **2018**, *25*, 1146–1157. [CrossRef]
12. Scutenaire, J.; Deragon, J.; Jean, V.; Benhamed, M.; Raynaud, C.; Favory, J.; Merret, R.; Bousquet-Antonelli, C. The YTH domain protein ECT2 is an m6A reader required for normal trichome branching in Arabidopsis. *Plant Cell* **2018**, *30*, 986–1005. [CrossRef]
13. Arribas-Hernández, L.; Bressendorff, S.; Hansen, M.H.; Poulsen, C.; Erdmann, S.; Brodersen, P. An m6A-YTH module controls developmental timing and morphogenesis in Arabidopsis. *Plant Cell* **2018**, *30*, 952–967. [CrossRef]
14. Chen, M.; Urs, M.J.; Sánchez-González, I.; Olayioye, M.A.; Herde, M.; Witte, C. m6A RNA Degradation Products Are Catabolized by an Evolutionarily Conserved N6-Methyl-AMP Deaminase in Plant and Mammalian Cells. *Plant Cell* **2018**, *30*, 1511–1522. [CrossRef]
15. Hofmann, N.R. Epitranscriptomics and Flowering: mRNA Methylation/Demethylation Regulates Flowering Time. *Plant Cell* **2018**, *29*, 2949–2950. [CrossRef]
16. Zhou, L.; Tian, S.; Qin, G. RNA methylomes reveal the m6A-mediated regulation of DNA demethylase gene SlDML2 in tomato fruit ripening. *Genome Biol.* **2019**, *20*, 156. [CrossRef]
17. Martínez-Pérez, M.; Aparicio, F.; López-Gresa, M.P.; Bellés, J.M.; Sánchez-Navarro, J.A.; Pallás, V. Arabidopsis m6A demethylase activity modulates viral infection of a plant virus and the m6A abundance in its genomic RNAs. *Proc. Natl. Acad. Sci. USA* **2017**, *114*, 10755–10760. [CrossRef]
18. Li, Z.; Shi, J.; Yu, L.; Zhao, X.; Ran, L.; Hu, D.; Song, B. N6-methyl-adenosine level in Nicotiana tabacum is associated with tobacco mosaic virus. *Virol. J.* **2018**, *15*, 87. [CrossRef]
19. Miao, Z.; Zhang, T.; Qi, Y.; Song, J.; Han, Z.; Ma, C. Evolution of the RNA N6-Methyladenosine Methylome Mediated by Genomic Duplication. *Plant Physiol.* **2020**, *182*, 345–360. [CrossRef]
20. Arribas-Hernández, L.; Rennie, S.; Schon, M.; Porcelli, C.; Enugutti, B.; Andersson, R.; Nodine, M.D.; Brodersen, P. The YTHDF proteins ECT2 and ECT3 bind largely overlapping target sets and influence target mRNA abundance, not alternative polyadenylation. *eLife* **2021**, *10*, e72377. [CrossRef]
21. Yu, Q.; Liu, S.; Yu, L.; Xiao, Y.; Zhang, S.; Wang, X.; Xu, Y.; Yu, H.; Li, Y.; Yang, J.; et al. RNA demethylation increases the yield and biomass of rice and potato plants in field trials. *Nat. Biotechnol.* **2021**, *39*, 1581–1588. [CrossRef]
22. Geula, S.; Moshitch-Moshkovitz, S.; Dominissini, D.; Mansour, A.A.; Kol, N.; Salmon-Divon, M.; Hershkovitz, V.; Peer, E.; Mor, N.; Manor, Y.S. m6A mRNA methylation facilitates resolution of naïve pluripotency toward differentiation. *Science* **2015**, *347*, 1002–1006. [CrossRef]
23. Zhang, Z.; Theler, D.; Kaminska, K.H.; Hiller, M.; de la Grange, P.; Pudimat, R.; Rafalska, I.; Heinrich, B.; Bujnicki, J.M.; Allain, F.H. The YTH domain is a novel RNA binding domain. *J. Biol. Chem.* **2010**, *285*, 14701–14710. [CrossRef]

24. Hou, J.; Zhang, H.; Liu, J.; Zhao, Z.; Wang, J.; Lu, Z.; Hu, B.; Zhou, J.; Zhao, Z.; Feng, M.; et al. YTHDF2 reduction fuels inflammation and vascular abnormalization in hepatocellular carcinoma. *Mol. Cancer* **2019**, *18*, 163. [CrossRef]
25. Merkurjev, D.; Hong, W.; Iida, K.; Oomoto, I.; Goldie, B.J.; Yamaguti, H.; Ohara, T.; Kawaguchi, S.; Hirano, T.; Martin, K.C.; et al. Synaptic N6-methyladenosine (m6A) epitranscriptome reveals functional partitioning of localized transcripts. *Nat. Neurosci.* **2018**, *21*, 1004–1014. [CrossRef]
26. Shi, H.; Zhang, X.; Weng, Y.; Lu, Z.; Liu, Y.; Lu, Z.; Li, J.; Hao, P.; Zhang, Y.; Zhang, F.; et al. m6A facilitates hippocampus-dependent learning and memory through YTHDF1. *Nature* **2018**, *563*, 249–253. [CrossRef]
27. Weng, Y.; Wang, X.; An, R.; Cassin, J.; Vissers, C.; Liu, Y.; Liu, Y.; Xu, T.; Wang, X.; Wong, S.Z.H.; et al. Epitranscriptomic m6A Regulation of Axon Regeneration in the Adult Mammalian Nervous System. *Neuron* **2018**, *97*, 313–325. [CrossRef]
28. Kan, L.; Grozhik, A.V.; Vedanayagam, J.; Patil, D.P.; Pang, N.; Lim, K.; Huang, Y.; Joseph, B.; Lin, C.; Despic, V. The m 6 A pathway facilitates sex determination in Drosophila. *Nat. Commun.* **2017**, *8*, 15737. [CrossRef]
29. Haussmann, I.U.; Bodi, Z.; Sanchez-Moran, E.; Mongan, N.P.; Archer, N.; Fray, R.G.; Soller, M. m 6 A potentiates Sxl alternative pre-mRNA splicing for robust Drosophila sex determination. *Nature* **2016**, *540*, 301. [CrossRef]
30. Lence, T.; Akhtar, J.; Bayer, M.; Schmid, K.; Spindler, L.; Ho, C.H.; Kreim, N.; Andrade-Navarro, M.A.; Poeck, B.; Helm, M. m 6 A modulates neuronal functions and sex determination in Drosophila. *Nature* **2016**, *540*, 242. [CrossRef]
31. Liu, X.; Shimada, T.; Otowa, T.; Wu, Y.Y.; Kawamura, Y.; Tochigi, M.; Iwata, Y.; Umekage, T.; Toyota, T.; Maekawa, M. Genome-wide association study of autism spectrum disorder in the east Asian populations. *Autism Res.* **2016**, *9*, 340–349. [CrossRef] [PubMed]
32. Berlivet, S.; Scutenaire, J.; Deragon, J.; Bousquet-Antonelli, C. Readers of the m6A epitranscriptomic code. *Biochim. Biophys. Acta Gene Regul. Mech.* **2019**, *1862*, 329–342. [CrossRef] [PubMed]
33. Bailey, A.S.; Batista, P.J.; Gold, R.S.; Chen, Y.G.; de Rooij, D.G.; Chang, H.Y.; Fuller, M.T. The conserved RNA helicase YTHDC2 regulates the transition from proliferation to differentiation in the germline. *eLife* **2017**, *6*, e26116. [CrossRef] [PubMed]
34. Wojtas, M.N.; Pandey, R.R.; Mendel, M.; Homolka, D.; Sachidanandam, R.; Pillai, R.S. Regulation of m6A Transcripts by the $3' \rightarrow 5'$ RNA Helicase YTHDC2 Is Essential for a Successful Meiotic Program in the Mammalian Germline. *Mol. Cell* **2017**, *68*, 374–387. [CrossRef]
35. Soh, Y.Q.S.; Mikedis, M.M.; Kojima, M.; Godfrey, A.K.; de Rooij, D.G.; Page, D.C. Meioc maintains an extended meiotic prophase I in mice. *PLoS Genet.* **2017**, *13*, e1006704. [CrossRef]
36. Abby, E.; Tourpin, S.; Ribeiro, J.; Daniel, K.; Messiaen, S.; Moison, D.; Guerquin, J.; Gaillard, J.; Armengaud, J.; Langa, F.; et al. Implementation of meiosis prophase I programme requires a conserved retinoid-independent stabilizer of meiotic transcripts. *Nat. Commun.* **2016**, *7*, 10324. [CrossRef]
37. Xiao, W.; Adhikari, S.; Dahal, U.; Chen, Y.; Hao, Y.; Sun, B.; Sun, H.; Li, A.; Ping, X.; Lai, W. Nuclear m 6 A reader YTHDC1 regulates mRNA splicing. *Mol. Cell* **2016**, *61*, 507–519. [CrossRef]
38. Wang, C.; Zhu, Y.; Bao, H.; Jiang, Y.; Xu, C.; Wu, J.; Shi, Y. A novel RNA-binding mode of the YTH domain reveals the mechanism for recognition of determinant of selective removal by Mmi1. *Nucleic Acids Res.* **2015**, *44*, 969–982. [CrossRef]
39. Arribas-Hernandez, L.; Brodersen, P. Occurrence and Functions of m(6)A and Other Covalent Modifications in Plant mRNA. *Plant Physiol.* **2020**, *182*, 79–96. [CrossRef]
40. Arribas-Hernández, L.; Rennie, S.; Köster, T.; Porcelli, C.; Lewinski, M.; Staiger, D.; Andersson, R.; Brodersen, P. Principles of mRNA targeting via the Arabidopsis m6A-binding protein ECT2. *eLife* **2021**, *10*, e72. [CrossRef]
41. Li, Z.; Wang, R.; Gao, Y.; Wang, C.; Zhao, L.; Xu, N.; Chen, K.E.; Qi, S.; Zhang, M.; Tsay, Y.F. The Arabidopsis CPSF30-L gene plays an essential role in nitrate signaling and regulates the nitrate transceptor gene NRT 1.1. *New Phytol.* **2017**, *216*, 1205–1222. [CrossRef]
42. Yue, H.; Nie, X.; Yan, Z.; Weining, S. N6-methyladenosine regulatory machinery in plants: Composition, function and evolution. *Plant Biotechnol. J.* **2019**, *17*, 1194–1208. [CrossRef]
43. Liang, Z.; Riaz, A.; Chachar, S.; Ding, Y.; Du, H.; Gu, X. Epigenetic Modifications of mRNA and DNA in Plants. *Mol. Plant* **2020**, *13*, 14–30. [CrossRef]
44. Tanaka, W.; Toriba, T.; Ohmori, Y.; Yoshida, A.; Kawai, A.; Mayama-Tsuchida, T.; Ichikawa, H.; Mitsuda, N.; Ohme-Takagi, M.; Hirano, H. The YABBY gene TONGARI-BOUSHI1 is involved in lateral organ development and maintenance of meristem organization in the rice spikelet. *Plant Cell* **2012**, *24*, 80–95. [CrossRef]
45. Luan, W.; Liu, Y.; Zhang, F.; Song, Y.; Wang, Z.; Peng, Y.; Sun, Z. OsCD1 encodes a putative member of the cellulose synthase-like D sub-family and is essential for rice plant architecture and growth. *Plant Biotechnol. J.* **2011**, *9*, 513–524. [CrossRef]
46. Ding, Z.; Lin, Z.; Li, Q.; Wu, H.; Xiang, C.; Wang, J. DNL1, encodes cellulose synthase-like D4, is a major QTL for plant height and leaf width in rice (*Oryza sativa* L.). *Biochem. Biophys. Res. Commun.* **2015**, *457*, 133–140. [CrossRef]
47. Huang, L.; Hua, K.; Xu, R.; Zeng, D.; Wang, R.; Dong, G.; Zhang, G.; Lu, X.; Fang, N.; Wang, D. The LARGE2-APO1/APO2 regulatory module controls panicle size and grain number in rice. *Plant Cell* **2021**, *33*, 1212–1228. [CrossRef]
48. Kyozuka, J.; Konishi, S.; Nemoto, K.; Izawa, T.; Shimamoto, K. Down-regulation of RFL, the FLO/LFY homolog of rice, accompanied with panicle branch initiation. *Proc. Natl. Acad. Sci. USA* **1998**, *95*, 1979–1982. [CrossRef]
49. Sakamoto, T.; Morinaka, Y.; Inukai, Y.; Kitano, H.; Fujioka, S. Auxin signal transcription factor regulates expression of the brassinosteroid receptor gene in rice. *Plant J. Cell Mol. Biol.* **2013**, *73*, 676–688. [CrossRef]

50. Wu, L.; Zhang, Q.; Zhou, H.; Ni, F.; Wu, X.; Qi, Y. Rice MicroRNA effector complexes and targets. *Plant Cell* **2009**, *21*, 3421–3435. [CrossRef]
51. Spielmeyer, W.; Ellis, M.H.; Chandler, P.M. Semidwarf (sd-1), "green revolution" rice, contains a defective gibberellin 20-oxidase gene. *Proc. Natl. Acad. Sci. USA* **2002**, *99*, 9043–9048. [CrossRef]
52. Ma, Y.; Wang, F.; Guo, J.; Zhang, X.S. Rice OsAS2 gene, a member of LOB domain family, functions in the regulation of shoot differentiation and leaf development. *J. Plant Biol.* **2009**, *52*, 374–381. [CrossRef]
53. Arribas-Hernández, L.; Simonini, S.; Hansen, M.H.; Paredes, E.B.; Bressendorff, S.; Dong, Y.; Østergaard, L.; Brodersen, P. Recurrent requirement for the m6A-ECT2/ECT3/ECT4 axis in the control of cell proliferation during plant organogenesis. *Development* **2020**, *147*, v189134. [CrossRef]
54. Zhao, Y.; Liu, Y.; Wu, R.; Bi, Z.; Yao, Y.; Liu, Q.; Wang, Y.; Wang, X. Understanding m 6 A function through uncovering the diversity roles of YTH domain-containing proteins. *Mol. Biotechnol.* **2019**, *61*, 355–364. [CrossRef] [PubMed]
55. Zhou, J.; Wan, J.; Gao, X.; Zhang, X.; Jaffrey, S.R.; Qian, S. Dynamic m 6 A mRNA methylation directs translational control of heat shock response. *Nature* **2015**, *526*, 591. [CrossRef]
56. Ma, S.; Yan, J.; Barr, T.; Zhang, J.; Chen, Z.; Wang, L.; Sun, J.C.; Chen, J.; Caligiuri, M.A.; Yu, J. The RNA m6A reader YTHDF2 controls NK cell antitumor and antiviral immunity. *J. Exp. Med.* **2021**, *218*, e20210279. [CrossRef]
57. Zhang, C.; Fu, J.; Zhou, Y. A review in research progress concerning m6A methylation and immunoregulation. *Front. Immunol.* **2019**, *10*, 922. [CrossRef]
58. Li, D.; Zhang, H.; Hong, Y.; Huang, L.; Li, X.; Zhang, Y.; Ouyang, Z.; Song, F. Genome-wide identification, biochemical characterization, and expression analyses of the YTH domain-containing RNA-binding protein family in Arabidopsis and rice. *Plant Mol. Biol. Rep.* **2014**, *32*, 1169–1186. [CrossRef]
59. Meyer, K.D.; Patil, D.P.; Zhou, J.; Zinoviev, A.; Skabkin, M.A.; Elemento, O.; Pestova, T.V.; Qian, S.; Jaffrey, S.R. 5′ UTR m 6 A promotes cap-independent translation. *Cell* **2015**, *163*, 999–1010. [CrossRef]
60. Edgar, R.C. MUSCLE: A multiple sequence alignment method with reduced time and space complexity. *BMC Bioinform.* **2004**, *5*, 113. [CrossRef] [PubMed]
61. Trifinopoulos, J.; Nguyen, L.; von Haeseler, A.; Minh, B.Q. W-IQ-TREE: A fast online phylogenetic tool for maximum likelihood analysis. *Nucleic Acids Res.* **2016**, *44*, W232. [CrossRef] [PubMed]

MDPI AG
Grosspeteranlage 5
4052 Basel
Switzerland
Tel.: +41 61 683 77 34

Plants Editorial Office
E-mail: plants@mdpi.com
www.mdpi.com/journal/plants

www.ingramcontent.com/pod-product-compliance
Lightning Source LLC
Chambersburg PA
CBHW041350210926
51850CB00013B/3035